Synthesis Lectures on Mathematics & Statistics

Series Editor

Steven G. Krantz, Department of Mathematics, Washington University, Saint Louis, USA

This series includes titles in applied mathematics and statistics for cross-disciplinary STEM professionals, educators, researchers, and students. The series focuses on new and traditional techniques to develop mathematical knowledge and skills, an understanding of core mathematical reasoning, and the ability to utilize data in specific applications.

Liangliang Li · Yu Huang · Goong Chen

Chaotic Maps, Fractals, and Rapid Fluctuations

With Applications to Chaotic Vibration of the Wave Equation

Second Edition

 Springer

Liangliang Li
IFCEN
Sun Yat-Sen University
Zhuhai, China

Yu Huang
School of Science
Sun Yat-Sen University
Shenzhen, China

Goong Chen
Texas A&M University
College Station, USA

ISSN 1938-1743   ISSN 1938-1751 (electronic)
Synthesis Lectures on Mathematics & Statistics
ISBN 978-3-031-84827-8   ISBN 978-3-031-84828-5 (eBook)
https://doi.org/10.1007/978-3-031-84828-5

1st edition: © Springer Nature Switzerland AG 2022
2nd edition: © The Editor(s) (if applicable) and The Author(s), under exclusive license to Springer Nature Switzerland AG 2025

This work is subject to copyright. All rights are solely and exclusively licensed by the Publisher, whether the whole or part of the material is concerned, specifically the rights of translation, reprinting, reuse of illustrations, recitation, broadcasting, reproduction on microfilms or in any other physical way, and transmission or information storage and retrieval, electronic adaptation, computer software, or by similar or dissimilar methodology now known or hereafter developed.

The use of general descriptive names, registered names, trademarks, service marks, etc. in this publication does not imply, even in the absence of a specific statement, that such names are exempt from the relevant protective laws and regulations and therefore free for general use.

The publisher, the authors and the editors are safe to assume that the advice and information in this book are believed to be true and accurate at the date of publication. Neither the publisher nor the authors or the editors give a warranty, expressed or implied, with respect to the material contained herein or for any errors or omissions that may have been made. The publisher remains neutral with regard to jurisdictional claims in published maps and institutional affiliations.

This Springer imprint is published by the registered company Springer Nature Switzerland AG
The registered company address is: Gewerbestrasse 11, 6330 Cham, Switzerland

If disposing of this product, please recycle the paper.

Preface to the Second Edition

This is a revised edition of the book: *Chaotic Maps: Dynamics, Fractals and Rapid Fluctuations*, first published in 2011. We have slightly adjusted the title to be: *Chaos, Fractals and Rapid Fluctuations, with Applications to the Chaotic Vibration of the Wave Equation*. We are pleased that Springer Nature is interested to publish this enlarged version of the book.

In the first edition, there were nine chapters and two appendices. Now those two appendices have been adapted into the new Chaps. 10 and 11. Furthermore, three additional chapters have been added, expanding the book of the second edition to a total of 14 chapters. The newly added material all deals with the study of chaotic vibration of the wave equation subject to various types of nonlinear boundary conditions, mostly based on the authors' work on this subject during the most recent few years. What has motivated us to do so, on the one hand, is that there have been very few books studying chaos in partial differential equations (PDEs) and, on the other hand, is because of our special interest on this topic of chaotic vibration in PDEs. But obviously, our work has only touched a tiny portion of this immense field of chaos in PDEs. We hope this book is a useful resource for beginners.

Elsewhere, a large number of previous typographical errors have been corrected. Some additional homework problems have also been added to the exercises.

We thank all of our past students and collaborators for their indirect contributions to our research contained in this book. We thank the publisher Springer Nature again for their kind assistance regarding the publication matters.

We gratefully acknowledge the funding supports from the Chinese National Natural Science Foundation Grants 12171492 (for Yu Huang), 11671410 (for Liangliang Li) and the Natural Science Foundation of Guangdong Province, China (No. 2022A1515012153) (for Liangliang Li), that have been extremely beneficial to the preparation of the new edition of the book.

Zhuhai, Guangdong Province, China   Liangliang Li
Shenzhen, Guangdong Province, China   Yu Huang
College Station, TX, USA   Goong Chen

Preface to the First Edition

The understanding and analysis of chaotic systems are considered as one of the most important advances of the 20th Century. Such systems behave contrary to the ordinary belief that the universe is always orderly and predicable as a grand ensemble modelizable by differential equations. The great mathematician and astronomer Pierre-Simon Laplace (1749–1827) once said:

> "We may regard the present state of the universe as the effect of its past and the cause of its future. An intellect which at a certain moment would know all forces that set nature in motion, and all positions of all items of which nature is composed, if this intellect were also vast enough to submit these data to analysis, it would embrace in a single formula the movements of the greatest bodies of the universe and those of the tiniest atom; for such an intellect nothing would be uncertain and the future just like the past would be present before its eyes." (Laplace, A Philosophical Essay on Probabilities [1].)

Laplace had a conviction that, knowing all the governing differential equations and the initial conditions, we can predict everything in the universe in a deterministic way. But we now know that Laplace has underestimated the complexities of the equations of motion. The truth is that rather simple systems of ordinary differential equations can have behaviors that are extremely sensitive to initial conditions as well as manifesting randomness. In addition, in quantum mechanics, even though the governing equation, the Schrodinger equation, is deterministic, the outcomes from measurements are probabilistic.

The term "chaos," literally, means confusion, disarray, disorder, disorganization. turbulence, turmoil, etc. It appears to be the antithesis of beauty, elegance, harmony, order, organization, purity, and symmetry that most of us are all indoctrinated to believe that things should rightfully be. And, as such, chaos seems inherently to defy an organized description and a systematic study for a long time. Henri Poincaré is most often credited as the founder of modern dynamical systems and the discoverer of chaotic phenomena. In his study of the three-body problem during the 1880s, he found that celestial bodies can have orbits which are nonperiodic, and yet for any choices of period of motion, that period will not be steadily increasing nor approaching a fixed value. Poincare's interests have stimulated the development of ergodic theory, studied and developed by prominent mathematicians G. D. Birkhoff, A. N. Kolmogorov, M. L. Cartwright, J. E. Littlewood, S. Smale, etc., mainly from the nonlinear differential equations point of view.

With the increasing availability of electronic computers during the 1960s, scientists and engineers could begin to play with them and in so doing have discovered phenomena never known before. Two major discoveries were made during the early 1960s: the Lorenz Attractor, by Edward Lorenz in his study of weather prediction, and fractals, by Benoit Mandelbrot in the study of fluctuating cotton prices. These discoveries have deeply revolutionized the thinking of engineers, mathematicians, and scientists in problem solving and the understanding of nature and shaped the future directions in the research and development of nonlinear science.

The actual coinage of "chaos" for the field is due to a 1975 paper by T. Y. Li and J. A. Yorke [2] entitled "Period Three Implies Chaos". It is a perfect, captivating phrase for a study ready to take off, with enthusiastic participants from all walks of engineering and natural and social sciences. More rigorously speaking, a system is said to be chaotic if

(1) it has sensitive dependence on initial conditions;
(2) it must be topologically mixing; and
(3) the periodic orbits are dense.

Several other similar, but non-equivalent, definitions are possible and are used by different groups.

Today, nonlinear science is a highly active established discipline (and interdiscipline), where bifurcations, chaos, pattern formations, self-organizations, self-regulations, stability and instability, fractal structures, universality, synchronization, and peculiar nonlinear dynamical phenomena are some of the most intensively studied topics.

The topics of dynamical systems and chaos have now become a standard course in both the undergraduate and graduate mathematics curriculum of most major universities in the world. This book is developed from the lecture notes on dynamical systems and chaos the two authors taught at the Mathematics Departments of Texas A&M University and Zhongshan (Sun Yat-Sen) University in Guangzhou, China during 1995–2011.

The materials in the notes are intended for a semester-long introductory course. The main objective is to familiarize the students with the theory and techniques for (discrete-time) maps from mainly an analysis viewpoint, aiming eventually to also provide a stepping stone for nonlinear systems governed by ODEs and PDEs. The book is divided into ten chapters and two appendices. They cover the following major themes:

(I) Interval maps: Their basic properties (Chapter 1), Sharkovski's Theorem on periodicities (Chapter 3), bifurcations (Chapter 4), and homoclinicity (Chapter 5).

(II) General dynamical systems and Smale Horseshoe: The 2- and k-symbol dynamics, topological conjugacy and shift invariant sets (Chapter 6), and the Smale Horseshoe (Chapter 7).
(III) Rapid fluctuations and fractals: Total variations and heuristics (Chapter 2), fractals (Chapter 8), and rapid fluctuations of multi-dimensional maps and infinite-dimensional maps (Chapters 9 and 10).

Chapter 11 is provided in order to show some basic qualitative behaviors of higher-dimensional differential equation systems, and how to study continuous-time dynamical systems, which are often described by nonlinear ordinary differential equations, using its Poincaré section, which is a map. This would, hopefully, give the interested reader some head start toward the study of continuous-time dynamical systems.

Chapter 12 offers an example of a concrete case of an infinite-dimensional system described by the one-dimensional wave equation with a van der Pol type nonlinear boundary condition, and shows how to use interval maps and rapid fluctuations to understand and prove chaos.

For these three major themes (I)–(III) above, much of the contents in (I) and (II) are rather standard. But the majority of the materials in Theme (III) is taken from the research done by the two authors and our collaborators during the recent years. This viewpoint of regarding chaos as exponential growth of total variations on a strange attractor of some fractional Hausdorff dimensions is actually mostly stimulated by our research on the chaotic vibration of the wave equation introduced in Chapter 12.

There are already a good number of books and monographs on dynamical systems and chaos on the market. In developing our own instructional materials, we have referenced and utilized them extensively and benefited immensely. We mention, in particular, the excellent books by Afraimovich and Hsu [3], Devaney [4], Guckenheimer and Holmes [5], Meyer and Hall [6], Robinson [7], and Wiggins [8]. In addition, we have also been blessed tremendously from two Chinese sources: Wen [9] and Zhou [10]. To these book authors, and in addition, our past collaborators and students who helped us either directly or indirectly in many ways, we express our sincerest thanks.

Professor Steven G. Krantz, editor of the book series, and Mr. Joel Claypool, book publisher, constantly pushed us, tolerated the repeated long delays, but kindly expedited the publication process. We are truly indebted.

The writing of this book was supported in part by the Texas Norman Hackman Advanced Research Program Grant #010366-0149-2009 from the Texas Higher Education Coordinating Board, Qatar National Research Fund (QNRF) National Priority Research Program Grants #NPRP09-462-1-074 and #NPRP4-1162-1-181, and the Chinese National Natural Science Foundation Grants #10771222 and 11071263.

College Station, TX, USA Goong Chen
Shenzhen, Guangdong Province, China Yu Huang
July 2011

References

1. P.-S. Laplace, *A Philosophical Essay on Probabilities*, translated from the 6th French edition by Frederick Wilson Truscott and Frederick Lincoln Emory, Dover Publications, New York, 1951.
2. T.Y. Li and J.A. Yorke, Period three implies chaos, *Amer. Math. Monthly* **82** (1975), 985–992. https://doi.org/10.2307/2318254
3. V.S. Afraimovich and S.-B. Hsu, *Lectures on chaotic dynamical systems* (AMS/IP Studies in Advanced Mathematics), *American Mathematical Society, Providence, R.I.*, 2002.
4. R.L. Devaney, *An Introduction to Chaotic Dynamical Systems*, 2nd ed., Addison-Wesley, New York, 1989. Cited on page(s) 19, 36, 75, 86, 209, 211.
5. J. Guckenheimer and P. Holmes, *Nonlinear Oscillations, Dynamical Systems and Bifurcations of Vector Fields*, Springer, New York, 1983.
6. K.R. Meyer and G.R. Hall, *Introduction to Hamiltonian Dynamical Systems and the N-Body Problem*, Springer, New York, 1992.
7. C. Robinson, *Dynamical Systems, Stability, Symbolic Dynamics and Chaos*, CRC Press, Boca Raton, FL, 1995, pp. 67–69.
8. S. Wiggins, *Introduction to Applied Nonlinear Dynamical Systems and Chaos*, 2nd ed., Springer, New York, 2003.
9. Zhi-Ying Wen, *Mathematical Foundations of Fractal Geometry*, Advanced Series in Nonlinear Science, Shanghai Scientific and Technological Education Publishing House, Shanghai, 2000 (in Chinese).
10. Z.L. Zhou, *Symbolic Dynamics*, Shanghai Scientific and Technological Education Publishing House, Shanghai, China, 1997 (in Chinese).

Contents

1	**Simple Interval Maps and Their Iterations**	1
	1.1 Introduction	1
	1.2 The Inverse and Implicit Function Theorems	8
	1.3 Visualizing from the Graphics of Iterations of the Quadratic Map	11
	References	21
2	**Total Variations of Iterates of Maps**	23
	2.1 The Use of Total Variations as a Measure of Chaos	23
	Reference	30
3	**Ordering Among Periods: The Sharkovski Theorem**	31
	References	35
4	**Bifurcation Theorems for Maps**	37
	4.1 The Period-Doubling Bifurcation Theorem	37
	4.2 Saddle-Node Bifurcation	43
	4.3 The Pitchfork Bifurcation	46
	4.4 Hopf Bifurcation	50
	References	61
5	**Homoclinicity. Lyapunov Exponents**	63
	5.1 Homoclinic Orbits	63
	5.2 Lyapunov Exponents	67
	References	75
6	**Symbolic Dynamics, Conjugacy, and Shift Invariant Sets**	77
	6.1 The Itinerary of an Orbit	77
	6.2 Properties of the Shift Map σ	79
	6.3 Symbolic Dynamic Systems Σ_k and Σ_k^+	87
	6.4 The Dynamics of (Σ_k^+, σ^+) and Chaos	89
	6.5 Topological Conjugacy and Semi-conjugacy	101

	6.6	Shift Invariant Sets	106
	6.7	Construction Of Shift Invariant Sets	106
	6.8	Snap-Back Repeller as a Shift Invariant Set	113
	References		117
7	**The Smale Horseshoe**		**119**
	7.1	The Standard Smale Horseshoe	119
	7.2	The General Horseshoe	124
	References		133
8	**Fractal**		**135**
	8.1	Examples of Fractals	135
	8.2	Hausdorff Dimension and the Hausdorff Measure	136
	8.3	Iterated Function Systems (IFS)	139
	References		148
9	**Rapid Fluctuations of Chaotic Maps on $\mathbb{R}^N$**		**149**
	9.1	Total Variation for Vector-Value Maps	149
	9.2	Rapid Fluctuations of Maps on $\mathbb{R}^N$	153
	9.3	Rapid Fluctuations of Systems with Quasi-Shift Invariant Sets	155
	9.4	Rapid Fluctuations of Systems Containing Topological Horseshoes	158
	9.5	Examples of Applications of Rapid Fluctuations	160
	References		169
10	**Infinite-Dimensional Systems Induced by Continuous-Time Difference Equations**		**171**
	10.1	Infinite-Dimensional Discrete Dynamical System (I3DS)	171
	10.2	Rates of Growth of Total Variations of Iterates	172
	10.3	Properties of the Set $B(f)$	174
	10.4	Properties of the Set $U(f)$	176
	10.5	Properties of the Set $E(f)$	184
	References		187
11	**Introduction to Continuous-Time Dynamical Systems**		**189**
	11.1	The Local Behavior of Two-Dimensional Nonlinear Systems	190
	11.2	Index for Two-Dimensional Systems	203
	11.3	The Poincaré Map for a Periodic Orbit in $\mathbb{R}^N$	206
	References		215
12	**Chaotic Vibration of the Wave Equation Due to Energy Pumping and van der Pol Boundary Conditions**		**217**
	12.1	The Mathematical Model and Motivations	217
	12.2	Chaotic Vibration of the Wave Equation	221
	References		226

13	**Necessary Conditions for Chaotic Vibrations**	227
	13.1 Necessary Conditions for the Onset of Chaos	231
	13.2 Sufficient Conditions for the Onset of Chaos	233
	13.3 Applications	242
	13.4 Numerical Simulations	243
	References	244
14	**Chaotic Vibrations of a Multi-Dimensional Wave Equation**	247
	14.1 Preliminary Analysis	255
	14.2 Chaotic Dynamics of the Composite Maps	262
	14.3 Chaotic Vibration Phenomenon of the PDE System	263
	14.4 Numerical Simulations	265
	References	266

Index 269

Simple Interval Maps and Their Iterations

The discovery of many nonlinear phenomena and their study by systematic methods are a major breakthrough in science and mathematics of the 20th Century, leading to the research and development of *nonlinear science*, which is at the forefront of science and technology of the 21st Century. Chaos is an extreme form of nonlinear dynamical phenomena. But what exactly is *chaos*? This is the main focus of this book.

Mathematical definitions of chaos can be given in many different ways. Though we will give the first of such definitions at the end of Chap. 2 (in Definition 2.7), during much of the first few chapters the term "chaos" (or its adjective "chaotic") should be interpreted in a rather liberal and intuitive sense that it stands for some irregular behaviors or phenomena. This vagueness should automatically take care of itself once more rigorous definitions are given.

1.1 Introduction

We begin by considering some population models. A simple one is the Malthusian law of linear population growth:

$$\begin{cases} x_0 > 0 \text{ is given;} \\ x_{n+1} = \mu x_n; \quad n = 0, 1, 2, \ldots \end{cases} \quad (1.1)$$

where

$$x_n = \text{the population size of certain biological species at time } n,$$

and $\mu > 0$ is a constant. For example, $\mu = 1.03$ if

$$0.03 = 3\% = \text{net birth rate} = \text{birth rate} - \text{death rate}.$$

The solution of (1.1) is

$$x_n = \mu^n x_0, \quad n = 1, 2, \ldots.$$

Therefore

$$\begin{cases} x_n \to \infty, & \text{if } \mu > 1, \text{ as } n \to \infty, \\ x_n = x_0, & \forall n = 1, 2, \ldots, \text{ if } \mu = 1, \\ x_n \to 0, & \text{if } \mu < 1, \text{ as } n \to \infty. \end{cases} \quad (1.2)$$

Thus, the long-term, or *asymptotic behavior*, of the system (1.1) is completely answered by (1.2). The model (1.1) from the population dynamics point of view is quite naive. An improved model of (1.1) is the following:

$$\begin{cases} x_{n+1} = \mu x_n - a x_n^2, \\ x_0 > 0 \text{ is given,} \end{cases} \quad a > 0, \quad n = 0, 1, 2, \ldots \quad (1.3)$$

where the term $-ax_n^2$ models conflicts (such as competition for the same resources) between members of the species. It has a negative effect on population growth. Equation (1.3) is called the *modified Malthusian law* for population growth.

For a nonlinear system with a single power law non-linearity, we can always "scale out" the coefficient associated with the nonlinear term. Let $x_n = k y_n$ and substitute it into (1.3). We obtain:

$$k y_{n+1} = \mu (k y_n) - a(k y_n)^2$$
$$y_{n+1} = \mu y_n - a k y_n^2.$$

Set $k = \mu/a$. We have:

$$y_{n+1} = \mu y_n - \mu y_n^2 = \mu y_n (1 - y_n).$$

Rename $y_n = x_n$. We obtain

$$x_{n+1} = \mu x_n (1 - x_n) \equiv f(\mu, x_n), \quad (1.4)$$

where

$$f(\mu, x) = f_\mu(x) = \mu x (1 - x). \quad (1.5)$$

The map f is called the *quadratic map* or *logistic map*. It played a very important role in the development of chaos theory due to the study of the British biologist, Robert May, who noted (1975) that as μ changes, the system does not attain simple steady states as those in (1.2). One of our main interests here is to study the *asymptotic behavior* of the iterates of (1.4) as $n \to \infty$. Iterations of the type $x_{n+1} = f(x_n)$ happen very often elsewhere in applications, too. We look at another example below.

1.1 Introduction

Fig. 1.1 Newton's algorithm

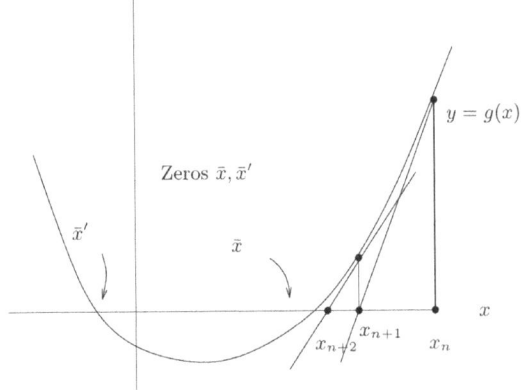

Example 1.1 Newton's algorithm for finding the zero of a given function $g(x)$ (Fig. 1.1).

Newton's algorithm provides a useful way for approximating solutions of an equation $g(x) = 0$ iteratively. Start from an initial point x_0, we compute $x_1, x_2, \ldots$, as follows. At each point x_n, draw a tangent line to the curve $y = g(x)$ passing through $(x_n, g(x_n))$:

$$y - g(x_n) = g'(x_n)(x - x_n).$$

This line intersects the x-axis at $x = x_{n+1}$:

$$0 - g(x_n) = g'(x_n)(x_{n+1} - x_n).$$

So

$$x_{n+1} = x_n - \frac{g(x_n)}{g'(x_n)} \equiv f(x_n),$$

where

$$f(x) \equiv x - \frac{g(x)}{g'(x)}.$$

The above iterations can encounter difficulty, for example, when:

(1) At $\bar{x}$, where $g(\bar{x}) = 0$, we also have $g'(\bar{x}) = 0$;
(2) The iterates x_n converge to a different (undesirable solution) $\bar{x}'$ instead of $\bar{x}$;
(3) The iterates x_j jump between two values x_n and x_{n+1}, such as what Fig. 1.2 shows in the following.

If any of the above happens, we have:

$$\lim_{n \to \infty} x_n \neq \bar{x}, \text{ for the desired solution } \bar{x}.$$

□

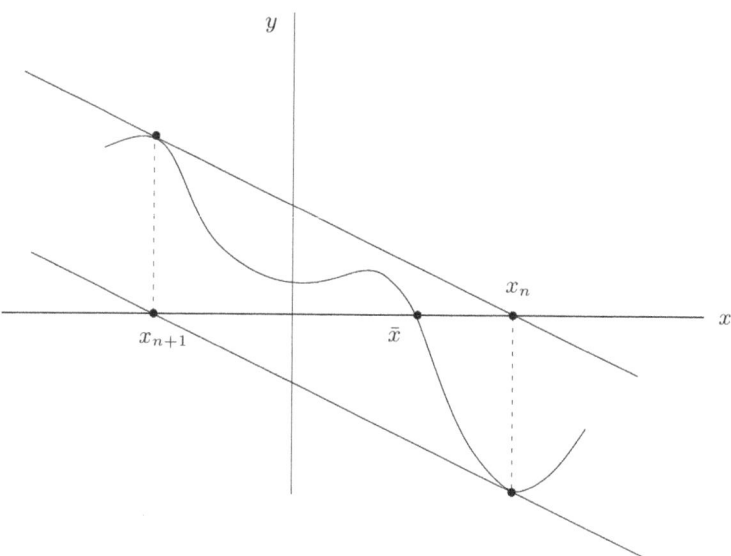

Fig. 1.2 Newton's algorithm becomes stagnant at x_n and x_{n+1}

From now on, for any real-valued function f, we will use f^n to denote the n-th iterate of f defined by

$$f^n(x) = \underbrace{f(f(f\cdots(f(x)))\cdots)}_{n\text{-times}} = \underbrace{f \circ f \circ f \cdots \circ f}_{n\text{-times}}(x),$$

if each $f^j(x)$ lies in the domain of definition of f for $j = 1, 2, \ldots, n-1$.

Exercise 1.2 Consider the iteration of the quadratic map:

$$\begin{cases} x_{n+1} = f_\mu(x_n); & f_\mu(x) = \mu x(1-x), \\ x_0 \in I \equiv [0, 1]. \end{cases}$$

(1) Choose $\mu = 3.2, 3.5, 3.55, 3.58, 3.65, 3.84$, and 3.94. For each given μ, plot the graphs:

$$y = f_\mu(x),\ y = f_\mu^2(x),\ y = f_\mu^3(x),\ y = f_\mu^4(x),\ y = f_\mu^5(x),\ y = f_\mu^{400}(x),\ x \in I,$$

where

$$f_\mu^2(x) = f_\mu(f_\mu(x)); \quad f_\mu^3(x) = f_\mu(f_\mu(f_\mu(x))), \text{ etc.}$$

(2) Let μ begin from $\mu = 2.9$ and increase μ to $\mu = 4$ with increment $\Delta\mu = 0.01$, with μ as the horizontal axis. For each μ, choose:

1.1 Introduction

$$x_0 = \frac{k}{100}; \quad k = 1, 2, 3, \ldots, 99.$$

Plot $f_\mu^{400}(x_0)$ (i.e., a dot) for these values of x_0 on the vertical axis. □

Example 1.3 The quadratic map f_μ as defined in Exercise 1.2 and shown in Fig. 1.3 is an example of a *unimodal map*. A map $f: I \equiv [a, b] \to I$ is said to be unimodal if it satisfies

$$f(a) = f(b) = a,$$

and f has a unique critical point $\bar{c}$: $a < \bar{c} < b$. The quadratic map f_μ is very representative of the dynamical behavior of unimodal maps. □

Let f be a continuous function such that $f: I \to I$ on a closed interval I. A point $\bar{x}$ is said to be a *fixed point* of a map $y = f(x)$ if

$$\bar{x} = f(\bar{x}). \tag{1.6}$$

The set of all fixed points of f is denoted as Fix(f). A point $\bar{x}$ is said to be a *periodic point* with *prime period* k, if

$$\bar{x} = f^k(\bar{x}), \tag{1.7}$$

and k is the smallest positive integer to satisfy (1.7). The set of all periodic points of prime period k of f is denoted as Per$_k(f)$, and that of all periodic points of f is denoted as Per(f). In the analysis of the iterations $x_{n+1} = f(x_n)$, *fixed points and periodic points play a critical role.*

Look at the quadratic map $f_\mu(x)$ in Fig. 1.3.
The fixed point $\bar{x} = \frac{\mu-1}{\mu}$ can be *attracting* or *repelling*, as Fig. 1.4a and b have shown.

Definition 1.4 Let $\bar{x}$ be a periodic point of prime period p of a differentiable real-valued map f: $f^p(\bar{x}) = \bar{x}$. We say that $\bar{x}$ is attracting (resp., repelling) if

$$|(f^p)'(\bar{x})| < 1 \quad (\text{resp.}, |(f^p)'(\bar{x})| > 1).$$

A periodic point $\bar{x}$ of prime period n is said to be *hyperbolic* if $|(f^p)'(\bar{x})| \neq 1$, i.e., $\bar{x}$ must be either attracting or repelling. □

We present a few fundamental theorems.

Theorem 1.5 (Brouwer's Fixed Point Theorem) *Let $I = [a, b]$, and let f be a continuous function on I*
such that either (i) $f(I) \subseteq I$ *or* (ii) $f(I) \supseteq I$.
Then f has at least one fixed point on I.

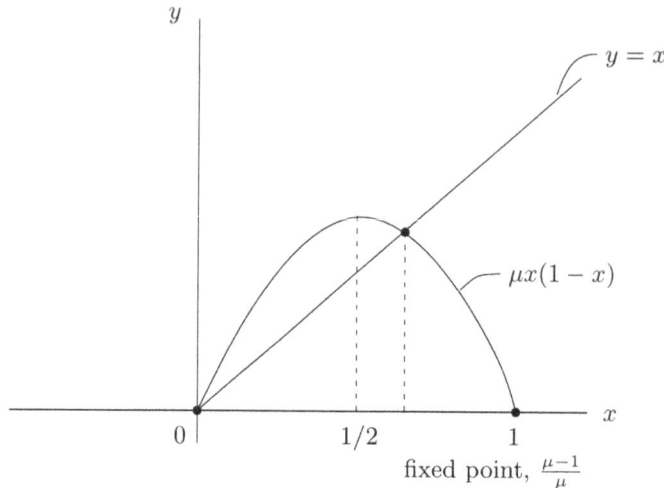

Fig. 1.3 Graph of the Quadratic Map $y = f_\mu(x)$, $\mu = 2.7$. Its maximum always occurs at $x = 1/2$. It has a (trivial) fixed point at $x = 0$, and another fixed point at $x = \frac{\mu-1}{\mu}$

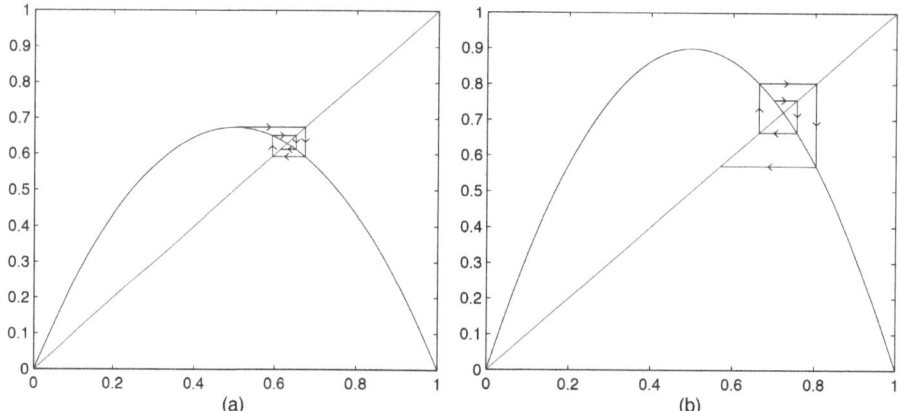

Fig. 1.4 a The fixed point has slope $f'(\bar{x})$ such that $|f'(\bar{x})| < 1$. The iterates are attracted to the fixed point. **b** The fixed point has slope $f'(\bar{x})$ such that $|f'(\bar{x})| > 1$. The iterates are moving away from $\bar{x}$

Proof Consider (i) first:
Define: $g(x) = x - f(x)$. Because $f(I) \subseteq I$, i.e., $f([a, b]) \subseteq [a, b]$, so $f(a) \in [a, b]$, and $f(b) \in [a, b]$, thus $a \leq f(a) \leq b$, $a \leq f(b) \leq b$. Then

$$g(a) = a - f(a) \leq 0,$$
$$g(b) = b - f(b) \geq 0.$$

1.1 Introduction

If equality does not hold in either of the two relations above, then

$$g(a) < 0; \quad g(b) > 0.$$

By the Intermediate Value Theorem, we get

$$g(c) = 0 \quad \text{for some} \quad c: a < c < b.$$

Thus, c is a fixed point of f. Next consider (ii):

$$f(I) \supseteq I, \quad [a, b] \subseteq f([a, b]).$$

Therefore, there exist $x_1, x_2 \in [a, b]$ such that

$$f(x_1) \leq a < b \leq f(x_2).$$

Again, define $g(x) = x - f(x)$. Then

$$g(x_1) = x_1 - f(x_1) \geq x_1 - a \geq 0,$$
$$g(x_2) = x_2 - f(x_2) \leq x_2 - b \leq 0.$$

Therefore, there exists a point $\tilde{x}$ in either $[x_1, x_2]$ or $[x_2, x_1]$, such that $g(\tilde{x}) = 0$. Thus, $\tilde{x}$ is a fixed point of f. □

Theorem 1.6 *Let $f: I \to I$ be continuous where $I = [a, b]$, such that f' is also continuous, satisfying:*

$$|f'(x)| < 1 \quad \text{on} \quad I.$$

Then f has a unique fixed point on I.

Proof The existence of a fixed point has been proved in Theorem 1.5, so we need only prove uniqueness. Suppose both x_0 and y_0 are fixed points of f:

$$x_0 = f(x_0),$$
$$y_0 = f(y_0).$$

Then

$$\frac{f(y_0) - f(x_0)}{y_0 - x_0} = \frac{y_0 - x_0}{y_0 - x_0} = 1 = f'(c)$$

by the Mean Value Theorem, for some $c: x_0 < c < y_0$. But

$$|f'(c)| < 1.$$

This is a contradiction. □

1.2 The Inverse and Implicit Function Theorems

From now on, we denote vectors and vector-valued functions by bold letters.

We state without proof two theorems which will be useful in future discussions.

Theorem 1.7 (The Inverse Function Theorem) *Let U and V be two open sets in $\mathbb{R}^N$ and $\boldsymbol{f}: U \to V$ is C^r for some $r \geq 1$. Assume that*

(i) $\boldsymbol{x}^0 \in U$, $\boldsymbol{y}^0 \in V$, and $\boldsymbol{f}(\boldsymbol{x}^0) = \boldsymbol{y}^0$;
(ii) $\nabla \boldsymbol{f}(\boldsymbol{x})|_{\boldsymbol{x}=\boldsymbol{x}^0}$ *is nonsingular, where*

$$\nabla \boldsymbol{f}(\boldsymbol{x}) = \begin{bmatrix} \dfrac{\partial f_1(\boldsymbol{x})}{\partial x_1} & \cdots & \dfrac{\partial f_1(\boldsymbol{x})}{\partial x_N} \\ \dfrac{\partial f_2(\boldsymbol{x})}{\partial x_1} & \cdots & \dfrac{\partial f_2(\boldsymbol{x})}{\partial x_N} \\ \vdots & & \vdots \\ \dfrac{\partial f_N(\boldsymbol{x})}{\partial x_1} & \cdots & \dfrac{\partial f_N(\boldsymbol{x})}{\partial x_N} \end{bmatrix}.$$

Then there exists an open neighborhood $N(\boldsymbol{x}^0) \subseteq U$ of $\boldsymbol{x}^0$ and an open neighborhood $N(\boldsymbol{y}^0) \subseteq V$ of $\boldsymbol{y}^0$ and a C^r-map $\boldsymbol{g}$:

$$\boldsymbol{g}: N(\boldsymbol{y}^0) \longrightarrow N(\boldsymbol{x}^0),$$

such that

$$\boldsymbol{f}(\boldsymbol{g}(\boldsymbol{y})) = \boldsymbol{y},$$

i.e., $\boldsymbol{g}$ is a local inverse of $\boldsymbol{f}$. □

Theorem 1.8 (The Implicit Function Theorem) *Let*

$$\left. \begin{array}{l} f_1(x_1, \ldots, x_m, y_1, \ldots, y_n) = 0, \\ f_2(x_1, \ldots, x_m, y_1, \ldots, y_n) = 0, \\ \quad \vdots \qquad\qquad\qquad \vdots \\ f_n(x_1, \ldots, x_m, y_1, \ldots, y_n) = 0 \end{array} \right\} \quad (1.8)$$

be satisfied for all $\boldsymbol{x} = (x_1, \ldots, x_m) \in U$ and $\boldsymbol{y} = (y_1, \ldots, y_n) \in V$, where U and V are open sets in, respectively, $\mathbb{R}^m$ and $\mathbb{R}^N$, and

$$f_i : U \times V \to \mathbb{R} \text{ is } C^r, \text{ for some } r \geq 1, \text{ for all } i = 1, 2, \ldots, n.$$

Assume that for $\boldsymbol{x}^0 = (x_1^0, x_2^0, \ldots, x_m^0) \in U$ and $\boldsymbol{y}^0 = (y_1^0, y_2^0, \ldots, y_n^0) \in V$,

1.2 The Inverse and Implicit Function Theorems

$$f_i(x^0, y^0) = 0 \quad \text{for} \quad i = 1, 2, \ldots, n,$$

$$[\nabla_y f_i(x^0, y^0)] = \begin{bmatrix} \dfrac{\partial f_1}{\partial y_1} & \cdots & \dfrac{\partial f_1}{\partial y_n} \\ \vdots & & \vdots \\ \dfrac{\partial f_n}{\partial y_1} & \cdots & \dfrac{\partial f_n}{\partial y_n} \end{bmatrix}_{\substack{at\ x=x^0 \\ y=y^0}} \quad \text{is nonsingular.}$$

Then there exist an open neighborhood $N(x^0) \subseteq U$ of x^0 and an open neighborhood $N(y^0) \subseteq V$ of y^0, and a C^r-map $g: N(x^0) \longrightarrow N(y^0)$, such that $y^0 = g(x^0)$ and

$$f_i(x, g(y)) = 0, \quad \forall x \in N(x^0), \forall y \in N(y^0), \quad i = 1, 2, \ldots, n,$$

i.e., locally, y is solvable in terms of x by $y = g(x)$. □

If we write equations in (1.8) as

$$F(x, y) = (f_1(x, y), f_2(x, y), \ldots, f_n(x, y)) = 0,$$

then taking the differential around $x = x^0$ and $y = y^0$, we have

$$\nabla_x F \cdot dx + \nabla_y F \cdot dy = 0,$$

where $dx \approx x - x^0$ and $dy \approx y - y^0$. Thus

$$\nabla_x F \cdot (x - x^0) + \nabla_y F \cdot (y - y^0) = 0.$$

An approximate solution of y in terms of x near $y = y_0$ is

$$y \approx y^0 + [\nabla_y F(x^0, y^0)]^{-1} \nabla_x F \cdot (x - x_0).$$

This explains intuitively why the invertibility of $\nabla_y F(x^0, y^0)$ is useful.

The Implicit Function Theorem can be proved using the Inverse Function Theorem, but the proofs of both theorems can be found in most advanced calculus books so we omit them here.

Example 1.9 Consider the relation

$$f(x, y) = ax^2 + bx + c + y = 0. \tag{1.9}$$

We have

$$\frac{\partial f}{\partial x} = 2ax + b \neq 0, \quad \text{if } x \neq -\frac{b}{2a};$$

$$\frac{\partial f}{\partial y} = 1 \neq 0.$$

Thus, y is always solvable in terms of x:

$$y = -(ax^2 + bx + c).$$

Here, we actually see that if $x = x^0 = -\frac{b}{2a}$, then x is *not uniquely solvable* in terms of y in a neighborhood of $x^0 = -\frac{b}{2a}$ because by the quadratic formula applied to (1.9), we have

$$x = \frac{-b \pm \sqrt{b^2 - 4a(c+y)}}{2a} = -\frac{b}{2a} \pm \frac{\sqrt{b^2 - 4a(c+y)}}{2a},$$

i.e., x is not unique.

On the other hand, if $x^0 \neq -\frac{b}{2a}$, then in a neighborhood of x^0, x is uniquely solvable in terms of y. For example, for $a = b = c = 1$, with $x^0 = 2$ and $y^0 = -7$,

$$x^0 = 2 \neq -\frac{b}{2a} = -\frac{1}{2}.$$

The (unique) solution of x in terms of y in a neighborhood of x is thus

$$x = \frac{-b + \sqrt{b^2 - 4a(c+y)}}{2a} = \frac{-1 + \sqrt{1 - 4(1+y)}}{2}.$$

We discard the branch

$$x = \frac{-b - \sqrt{b^2 - 4a(c+y)}}{2a} = \frac{-1 - \sqrt{1 - 4(1+y)}}{2}$$

because it doesn't satisfy

$$x^0 = 2 = \frac{-1 + \sqrt{1 - 4(1-7)}}{2} = \frac{-1 + 5}{2}. \quad \square$$

Exercise 1.10 Assume that $a, b, c, d \in \mathbb{R}$, and $a \neq 0$. Consider the relation

$$f(x, y) = ax^3 + bx^2 + cx + d - y = 0, \quad x, y \in \mathbb{R}$$

(i) Discuss the local solvability of real solutions x for given y by using the implicit function theorem.

(ii) Under what conditions does the function
$$y = g(x) = ax^3 + bx^2 + cx + d$$
have a local inverse? A global inverse? □

1.3 Visualizing from the Graphics of Iterations of the Quadratic Map

In the next few pages, we discuss the computer graphics from the previous Exercise 1.2. This type of graphics in Fig. 1.5 is called an *orbit diagram*. Note that the first period doubling happens at $\mu_0 = 3$. Then the second and third happen, respectively, at $\mu_1 \approx 3.45$ and $\mu_2 \approx 3.542$, and more period doublings happen in a cascade. We have

$$\frac{\mu_1 - \mu_0}{\mu_2 - \mu_1} \approx \frac{3.45 - 3}{3.542 - 3.45} = \frac{0.45}{0.092} \approx 4.8913\ldots.$$

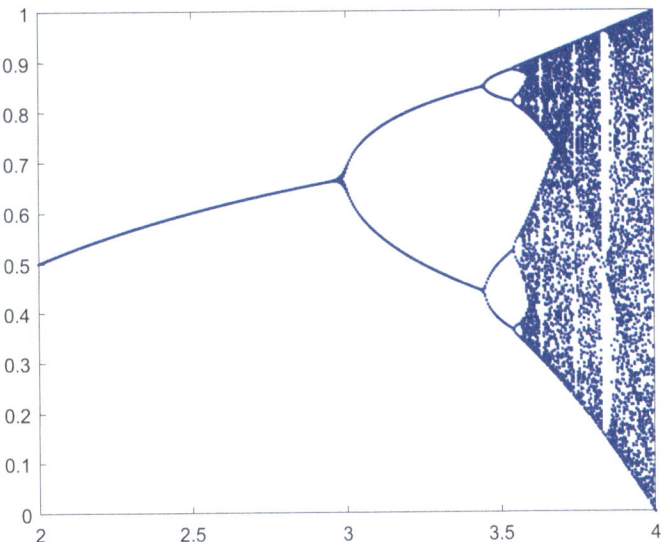

Fig. 1.5 The orbit diagram of $f(\mu, x) = f_\mu(x) = \mu x(1-x)$

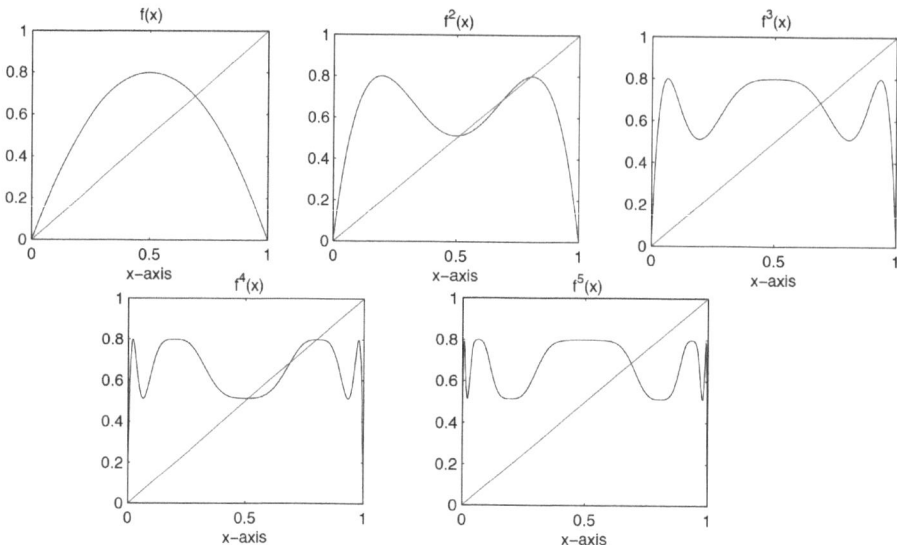

Fig. 1.6 The graphics of $f_\mu(x)$, $f_\mu^2(x)$, $f_\mu^3(x)$, $f_\mu^4(x)$ and $f_\mu^5(x)$, where $\mu = 3.2$. Note that the intersections of the curves with the diagonal line $y = x$ represent either a fixed point or a periodic point

It has been found that for *any period-doubling cascade*,

$$\lim_{n \to \infty} \frac{\mu_n - \mu_{n-1}}{\mu_{n+1} - \mu_n} = 4.669202\ldots.$$

This number, a *universal constant* due to M. Feigenbaum, is called the Feigenbaum constant. Also, note that there is a "window" area near $\mu = 3.84$ in Fig. 1.5.

Visualization from these graphics, Figs. 1.5, 1.6, 1.7, 1.8, 1.9, 1.10, 1.11, 1.12, 1.13 and 1.14, will provide inspirations for the study of the oscillatory behaviors related to chaos in this book.

Exercise 1.11 Define

$$y = f_\mu(x) = \mu \sin \pi x, \qquad x \in I = [0, 1].$$

(1) Vary $\mu \in [0, 1]$, plot the graphics of $f_\mu, f_\mu^2, f_\mu^3, \ldots, f_\mu^{10}$.
(2) Plot the orbit diagrams of $f_\mu(x)$.
(3) Describe what happens if $\mu > 1$. □

Exercise 1.12 Pick your arbitrary favorite continuous function of the form

$$y = f(\mu, x), \qquad x \in [0, 1],$$

1.3 Visualizing from the Graphics of Iterations of the Quadratic Map

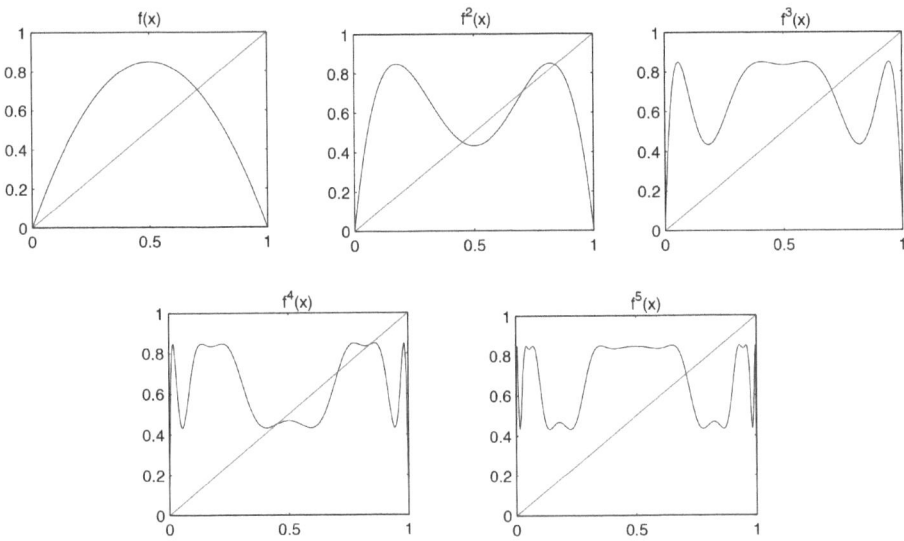

Fig. 1.7 The graphics of $f_\mu(x)$, $f_\mu^2(x)$, $f_\mu^3(x)$, $f_\mu^4(x)$ and $f_\mu^5(x)$, where $\mu = 3.40$

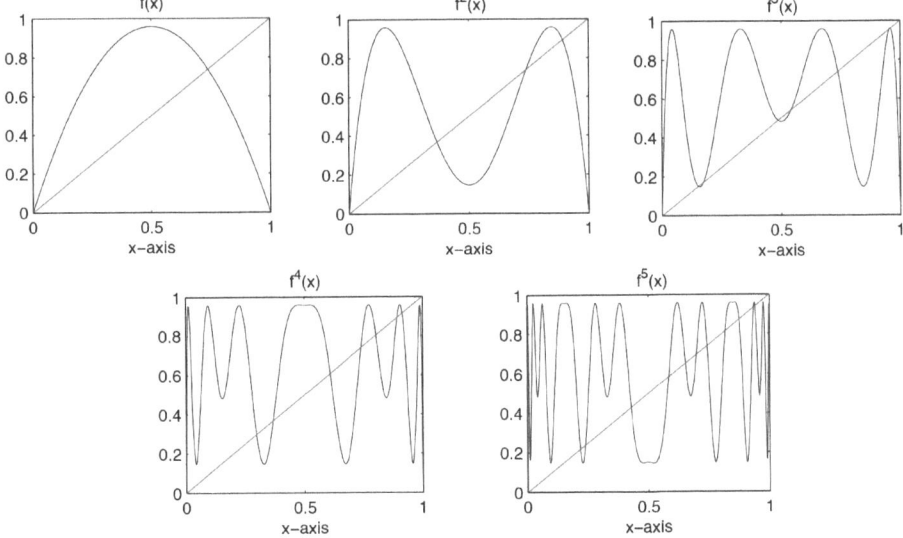

Fig. 1.8 The graphics of $f_\mu(x)$, $f_\mu^2(x)$, $f_\mu^3(x)$, $f_\mu^4(x)$ and $f_\mu^5(x)$ where $\mu = 3.84$. We see that the curves have become more oscillatory (in comparison with those in Fig. 1.7). They intersect with the diagonal line $y = x$ at more points, implying that there are more periodic points. Note that $f_\mu^3(x)$ intersects with $y = x$ at $P_1, P_2, \ldots, P_5$ and P_6 (in addition to the fixed point $x = 0$). Each point P_i, $i = 1, 2, \ldots, 6$, has period 3. *If a continuous map has period 3, then it has period n for any* $n = 1, 2, 3, 4, \ldots$

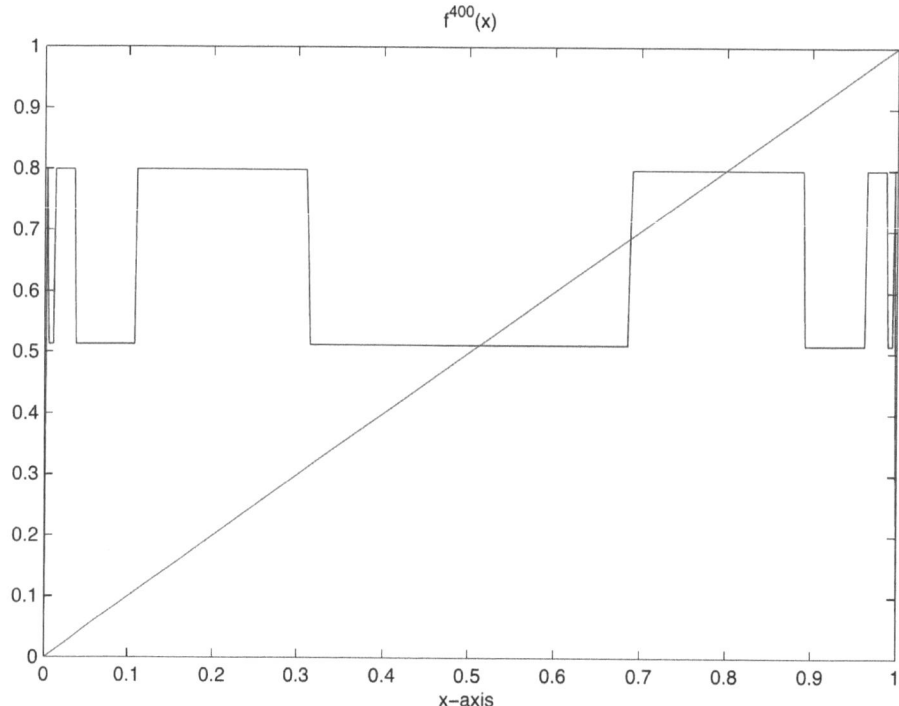

Fig. 1.9 The graph of $f_\mu^{400}(x)$, where $\mu = 3.2$. It looks like a step function. The two horizontal levels correspond to the period-2 bifurcation curves in Fig. 1.5. Question: In the x-ranges close to $x = 0$ and $x = 1$, how oscillatory is the curve?

such that $f(\mu, \cdot)$ maps I into I for the parameter μ lying within a certain range. Plot the graphics of the various iterates of $f(\mu, \cdot)$ as well as orbit diagrams of $f(\mu, \cdot)$. ☐

Exercise 1.13 Let $I = [a, b]$ be a nontrivial interval. Assume $f : I \to I$ is continuous.

(i) Prove that f has at least one fixed point.
(ii) Furthermore, assume that the f has only one fixed point, denoted by c, and has no periodic points whose periods are larger than 1.

 (a) If $c = a$, prove that $f([a, x]) \subsetneq [a, x]$ for every $x \in (a, b]$.
 (b) If $c = b$, prove that $f([x, b]) \subsetneq [x, b]$ for every $x \in [a, b)$.
 (c) If $a < c < b$, prove that $f(J) \subsetneq J$, where J is a subinterval of I and $c \in \mathring{J}$.

1.3 Visualizing from the Graphics of Iterations of the Quadratic Map

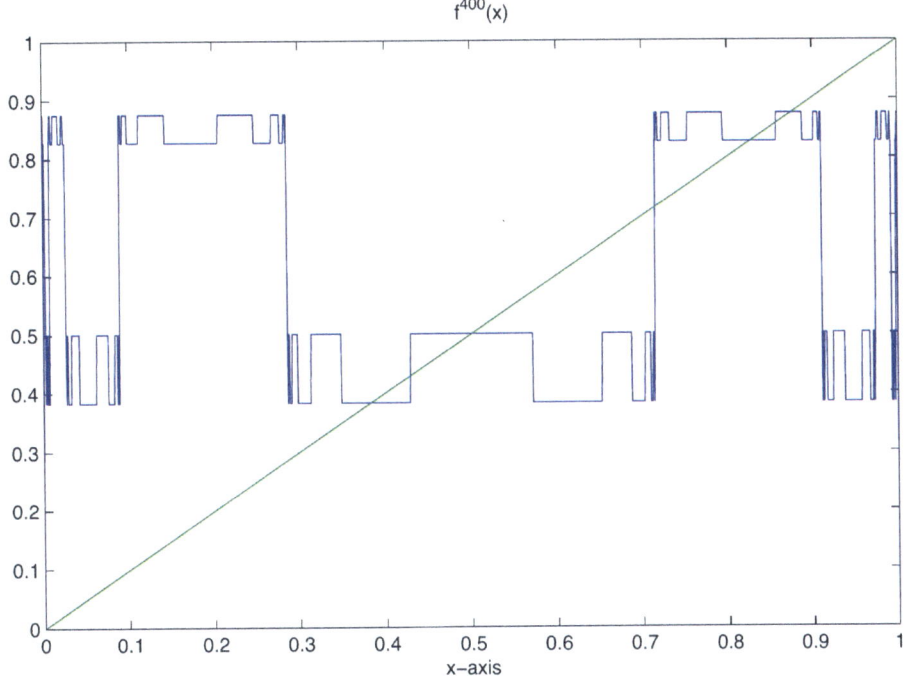

Fig. 1.10 The graph of $f_\mu^{400}(x)$, where $\mu = 3.5$. It again looks like a step function, but with four horizontal levels

(d) Deduce that for every $x \in I$,
$$\lim_{n \to +\infty} f^n(x) = c.$$
□

Exercise 1.14 Recall that we call x a *fixed point* of f if $f(x) = x$. We have known from Theorem 1.6 that a differentiable function $f : I \to I$ has a fixed point provided that $|f'| < 1$ and I is a bounded and closed interval. However, the result is no longer true if I is unbounded. Let f be a real-valued function on $\mathbb{R}$.

(i) If f is differentiable and $f'(x) \neq 1$ for every $x \in \mathbb{R}$, prove that f has at most one fixed point.
(ii) Show that the function g given by
$$g(t) = t - \arctan(t) + \frac{\pi}{2}$$
has no fixed point, even though $0 \leq g'(t) < 1$ for every $t \in \mathbb{R}$.

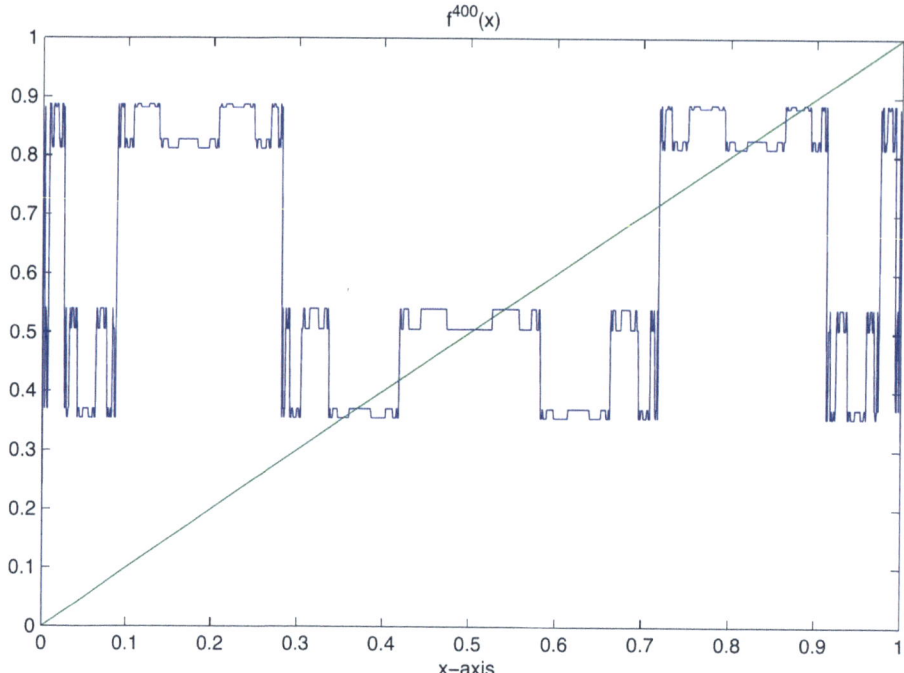

Fig. 1.11 The graph of $f_\mu^{400}(x)$, with $\mu = 3.55$. This curve actually has *eight* horizontal levels

(iii) However, if there is a constant $C_0 < 1$ such that $|f'| \leq C_0$, prove that a fixed point x of f exists, and that $x = \lim_{n \to +\infty} x_n$, where the real sequence (x_n) is defined by

$$x_0 \in \mathbb{R}, \quad \forall n \in \mathbb{N}, \quad x_{n+1} = f(x_n).$$

(iv) Find a function satisfying the condition in (iii), and plot the zig-zag path

$$(x_0, x_1) \to (x_1, x_1) \to (x_1, x_2) \to (x_2, x_2) \to (x_2, x_3) \cdots.$$

(v) Consider a function h defined by

$$h(x) = \frac{x^3 + 1}{3}.$$

For arbitrarily chosen $x_0 \in \mathbb{R}$, define a sequence (x_n) of iterates $x_{n+1} = h(x_n)$.

(a) Prove that h has three different fixed points, say P_1, P_2 and P_3, and assume that $P_1 < P_2 < P_3$.
(b) If $x_0 < P_1$, prove that $x_n \to -\infty$ as $n \to +\infty$.

1.3 Visualizing from the Graphics of Iterations of the Quadratic Map

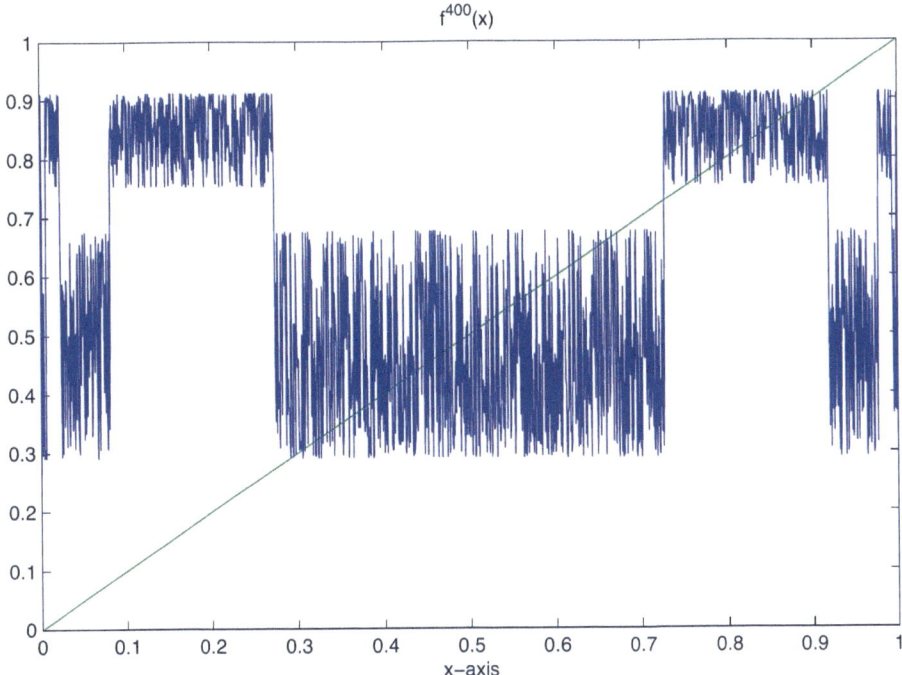

Fig. 1.12 The graph of $f_\mu^{400}(x)$, $\mu = 3.65$. This value of μ is already in the chaotic regime. The curve has exhibited highly oscillatory behavior

(c) If $P_1 < x_0 < P_2$, prove that $x_n \to P_2$ as $n \to +\infty$.
(d) If $x_0 > P_2$, prove that $x_n \to +\infty$ as $n \to +\infty$.

Hence the point P_2 can be located by this method of iteration, but P_1 and P_2 cannot. Why?

(iii) Define two functions

$$u(x) = \frac{1}{2}\left(x + \frac{2}{x}\right), \quad v(x) = \frac{2+x}{1+x}.$$

For chosen constants $x_0 > \sqrt{2}$ and $y_0 > \sqrt{2}$, define two sequences of iterates (x_n) and (y_n) by setting

$$x_{n+1} = u(x_n), \quad y_{n+1} = v(y_n).$$

(a) Prove that both u and v have $\sqrt{2}$ as their only fixed point in $(0, \infty)$.
(b) Prove that $x_n \to \sqrt{2}$, $y_n \to \sqrt{2}$ as $n \to +\infty$.

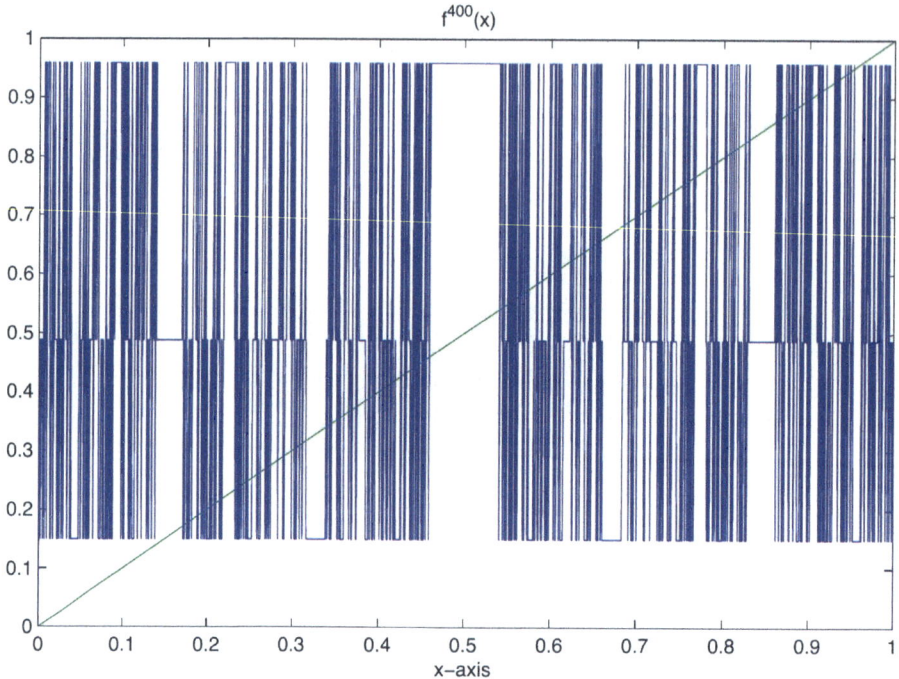

Fig. 1.13 The graph of $f_\mu^{400}(x)$, $\mu = 3.84$. Note that this value of μ corresponds to the "window" area in Fig. 1.5. The curve is highly oscillatory, but it appears to take only three horizontal values

(c) Compare the rates of convergence of (x_n) and (y_n).
(d) Compare u' and v', draw the zig-zags suggested in question (iv). Explain why the convergence of (x_n) is faster than (y_n)'s. □

Exercise 1.15 We revisit Newton's algorithm for finding a zero point of a given function. Let f be a real-valued function of class $\mathcal{C}^2$ on $[a, b]$. Assume that f also satisfies the following properties:

(a) $f(a) \cdot f(b) < 0$;
(b) $f' > 0$ on $[a, b]$;
(c) $f'' > 0$ on $[a, b]$.

Finally, we set $g(x) = x - \frac{f(x)}{f'(x)}$.

1. Show that Newton's method amounts to finding a fixed point of g.
2. Prove that there is a unique $p_0 \in (a, b)$ such that $f(p_0) = 0$.
3. Show that g is of class $\mathcal{C}^1$ on $[a, b]$, and that p_0 is the only fixed point of g.

1.3 Visualizing from the Graphics of Iterations of the Quadratic Map

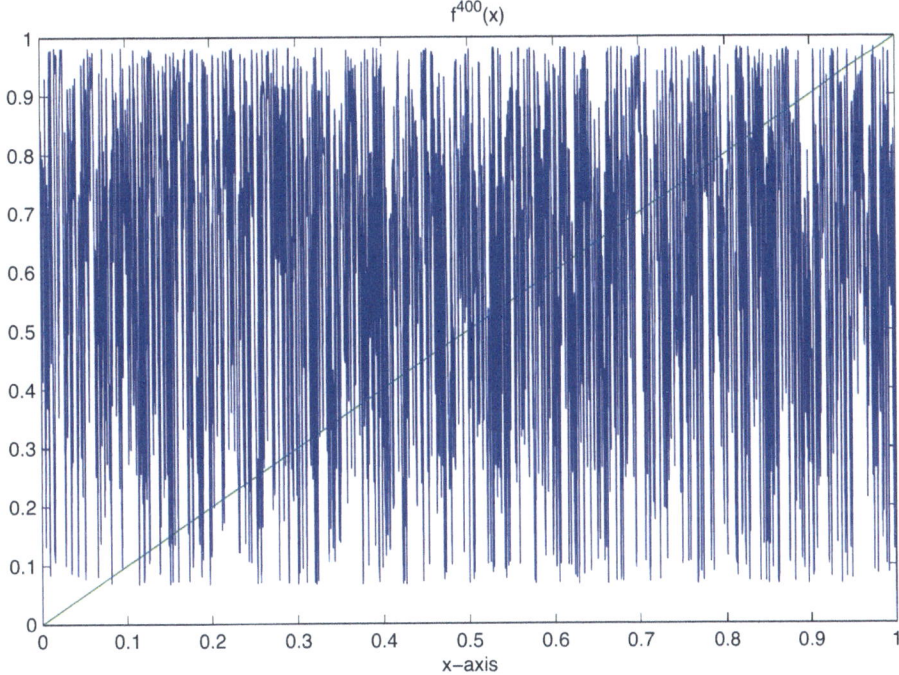

Fig. 1.14 The graph of $f_\mu^{400}(x)$, with $\mu = 3.93$. This value of μ is also in the "chaotic regime"

4. Study the monotonicity of g on $[p_0, b]$ and prove that $[p_0, b]$ is stabilized by g, i.e., $g([p_0, b]) \subset [p_0, b]$.
5. For arbitrarily chosen $x_0 \in [p_0, b]$, define a sequence (x_n) by setting $x_{n+1} = g(x_n)$.

    (a) Prove that (x_n) converges to p_0.
    (b) Draw a picture where you will sketch a graph of f and then explain how to construct x_{n+1} from x_n.
    (c) We now additionally assume that f is of class C^3.

     i. Prove that f is of class C^2.
     ii. Prove that there is a constant $M \geq 0$ that may depend on g, g' and g'', and that you will determine, such that
     $$\forall x \in [a, b], |g(x) - p_0| \leq \frac{M}{2}(x - c)^2.$$

iii. Can we have $M = 0$? If so, what happens in that case?
iv. Assume here after that $M > 0$. Prove that there is a constant $\delta > 0$ such that

$$x_0 \in [p_0, p_0 + \delta] \Rightarrow \forall n \in \mathbb{N}, 0 \le x_n - p_0 \le \frac{2}{M}\left(\frac{1}{2}\right)^{2^n}.$$

6. Application: set $f(x) = \tan(x) - 1$.

   (a) Apply the previous method to find an approximation of π to the nearest 10^{-12}.
   (b) Write a computer program with either Matlab or Python codes to perform the iteration x_n. □

Exercise 1.16 Suppose $f : \mathbb{R}^2 \to \mathbb{R}^2$ is given by

$$f(x, y) = (e^x \cos(y), e^x \sin(y)).$$

Let $X = (x, y) \in \mathbb{R}^2$.

(i) Compute $\nabla f(X)$, and determine whether $\nabla f(X)$ can be singular.
(ii) Prove that f is invertible near X, i.e., f is locally invertible.
(iii) Does f has a global inverse? □

Notes for Chapter 1

An *interval map* is formed by the 1-step scalar *equation of iteration* $x_{n+1} = f(x_n)$ for a continuous map f. Thus, it constitutes the simplest model for iterations. For example, Newton's method for finding roots of a nonlinear equation, and the time-marching of a 1-step explicit Euler finite-difference scheme for a first-order scalar ordinary differential equation, both can result in an interval map. Interestingly, even for partial differential equations such as the nonlinear initial-boundary value problem of the wave equation in Chap. 12, interval maps have found good applications.

Most of the textbooks on dynamical systems use the quadratic (or logistic) map (1.4) as a standard example to illustrate many peculiar, amazing behaviors of the iterates of the quadratic map. In fact, those iterates manifest strong chaotic phenomena which facilitates the understanding of what chaos is for pedagogical purposes. The focus of the first five chapters of this book is almost exclusively on interval maps.

The books by Devaney [1] and Robinson [2] contain excellent treatments of interval maps. The monograph by Block and Coppel [3] contains a more detailed account and further references about interval maps.

References

1. R.L. Devaney, *An Introduction to Chaotic Dynamical Systems*, 2nd ed., Addison-Wesley, New York, 1989. Cited on page(s) 19, 36, 75, 86, 209, 211.
2. C. Robinson, *Dynamical Systems, Stability, Symbolic Dynamics and Chaos*, CRC Press, Boca Raton, FL, 1995, pp. 67–69.
3. L. Block and W.A. Coppel, *Dynamics in One Dimension*, Lecture Notes in Mathematics, Vol. 1513, Springer Verlag, New York-Heidelberg-Berlin, 1992.

Total Variations of Iterates of Maps

2.1 The Use of Total Variations as a Measure of Chaos

Let $f: I = [a, b] \to \mathbb{R}$ be a given function; f is not necessarily continuous. A partition of I is defined as

$$P = \{x_0, x_1, \ldots, x_n \mid x_j \in I, \text{ for } j = 0, 1, \ldots, n; x_0 < x_1 < x_2 < \cdots < x_n\}$$

be an arbitrary finite collection of points on I. Define

$$V_I(f) = \text{the total variation of } f \text{ on } I$$

$$= \sup_{\text{all } P} \left\{ \sum_{i=1}^{n} |f(x_i) - f(x_{i-1})| \,\Big|\, x_i \in P \right\}. \tag{2.1}$$

If f is continuous on I and f has finitely many maxima and minima on I, such as indicated in Fig. 2.1. Then it is easy to see that

$$V_I(f) = |f(\tilde{x}_1) - f(\tilde{x}_0)| + |f(\tilde{x}_2) - f(\tilde{x}_1)| + \cdots + |f(\tilde{x}_n) - f(\tilde{x}_{n-1})|,$$

where each interval $[\tilde{x}_i, \tilde{x}_{i+1}]$ is a maximal interval where f is either increasing or decreasing.

Let I_1 and I_2 be two closed intervals and let f be continuous such that

$$f(I_1) \supseteq I_2.$$

We write the above as

$$I_1 \xrightarrow{f} I_2 \quad \text{or} \quad I_1 \longrightarrow I_2$$

and say that I_1 f-covers I_2.

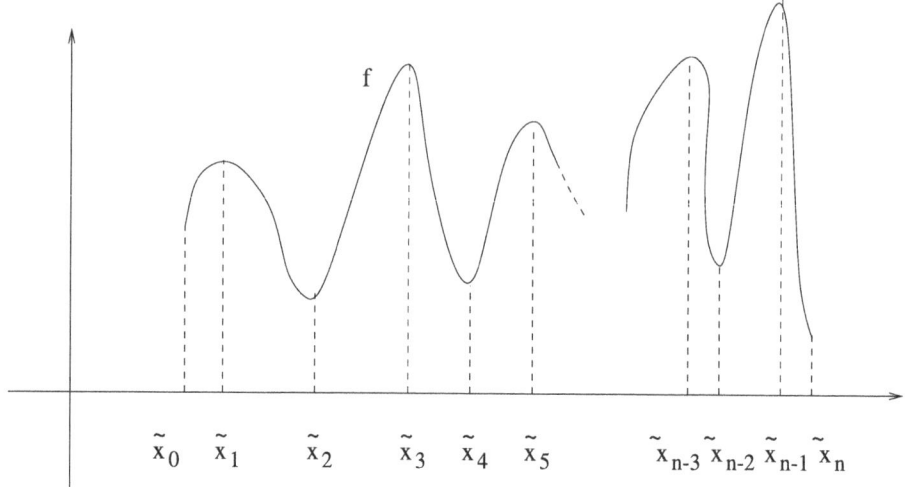

Fig. 2.1 A continuous function with finitely many maxima and minima at $\tilde{x}_0, \tilde{x}_1, \ldots, \tilde{x}_n$

Lemma 2.1 *If* $I_1 \xrightarrow{f} I_2$, *then* $V_{I_1}(f) \geq |I_2| \equiv \text{length of } I_2$.

Proof This follows easily from the observation of Fig. 2.2. □

Lemma 2.2 *Let* $J_0, J_1, \ldots, J_{n-1}$ *be bounded closed intervals such that they overlap at most at endpoints pairwise. Assume that* $J_0 \longrightarrow J_1 \longrightarrow J_2 \longrightarrow \cdots \longrightarrow J_{n-1} \longrightarrow J_n \equiv J_0$ *holds. Then*

Fig. 2.2 Interval I_1 f-covers I_2

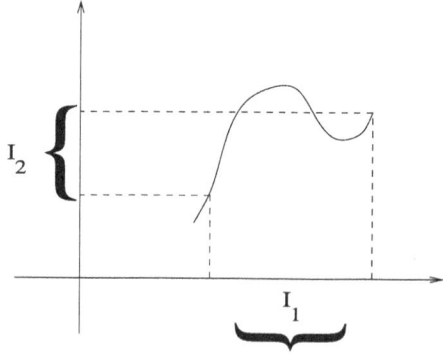

(i) there exists a fixed point x_0 of $f^n: f^n(x_0) = x_0$, such that $f^k(x_0) \in J_k$ for $k = 0, \ldots, n$;

(ii) Further, assume that the loop $J_0 \longrightarrow J_1 \longrightarrow \cdots \longrightarrow J_n$ is not a repetition of a shortened repetitive loop m where $mk = n$ for some integer $k > 0$. If the point x_0 in (i) is in the interior of J_0, then x_0 has prime period n.

Proof Use mathematical induction. □

Theorem 2.3 *Let I be a closed interval and $f: I \to I$ be continuous. Assume that f has two fixed points on I and a pair of period-2 points on I. Then*

$$\lim_{n \to \infty} V_I(f^n) = \infty.$$

Proof Let the two fixed points be x_0 and x_1:

$$f(x_0) = x_0 \quad \text{and} \quad f(x_1) = x_1. \tag{2.2}$$

Let the two period-2 points be p_1 and p_2:

$$f(p_1) = p_2, \quad f(p_2) = p_1. \tag{2.3}$$

Then there are three possibilities:

(i)
$$p_1 < x_0 < x_1 < p_2; \tag{2.4}$$

(ii) $x_0 < p_1 < p_2$;
(iii) $p_1 < p_2 < x_1$.

We consider case (i) only. Cases (ii) and (iii) can be treated in a similar way.
From Fig. 2.3, we see that we have

$$\left. \begin{array}{l} f(I_1) \supseteq I_2 \cup I_3, \text{ i.e., } I_1 \longrightarrow I_2 \cup I_3, \\ f(I_2) \supseteq I_2, \quad \text{ i.e., } I_2 \longrightarrow I_2 \\ f(I_3) \supseteq I_1 \cup I_2, \text{ i.e., } I_3 \longrightarrow I_1 \cup I_2. \end{array} \right\} \tag{2.5}$$

Therefore, we have the covering diagram (Fig. 2.4).

Fig. 2.3 The points and intervals corresponding to (2.2)–(2.4)

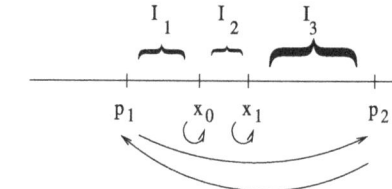

Fig. 2.4 The covering of intervals according to (2.5)

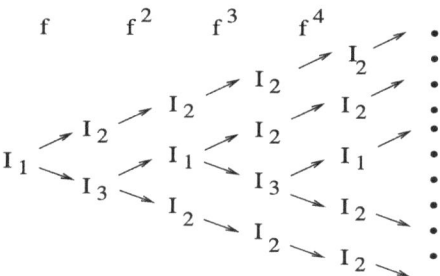

It is easy to verify by mathematical induction that the following statement is true:

For each n, I_1 contains $n+1$ subintervals $\{I_{1,1}^{(n)}; I_{1,2}^{(n)}, \ldots, I_{1,n+1}^{(n)}\}$ such that

$$I_{1,j}^{(n)} \subseteq I_1, \quad I_{1,j_1}^{(n)} \cap I_{1,j_2}^{(n)} \text{ has empty interior if } j_1 \neq j_2,$$
$$f(I_{1,j}^{(n)}) \supseteq I_k \text{ for some } k \in \{1, 2, 3\}.$$

Therefore

$$V_I(f^n) \geq V_{I_1}(f^n) \geq \sum_{j=1}^{n+1} V_{I_{1,j}^{(n)}}(f^n)$$
$$\geq (n+1) \cdot \min\{|I_1|, |I_2|, |I_3|\} \to \infty \text{ as } n \to \infty. \tag{2.6}$$

$\square$

Exercise 2.4 Prove that for the quadratic map

$$f_\mu(x) = \mu x(1-x), \quad x \in I = [0, 1],$$

if $\mu > 3$, then f_μ has two fixed points and at least a pair of period-2 points. Therefore the assumptions of Theorem 2.3 are satisfied and

$$\lim_{n \to \infty} V_I(f_\mu^n) = \infty \text{ for all } \mu: 3 < \mu < 4. \quad \square$$

2.1 The Use of Total Variations as a Measure of Chaos

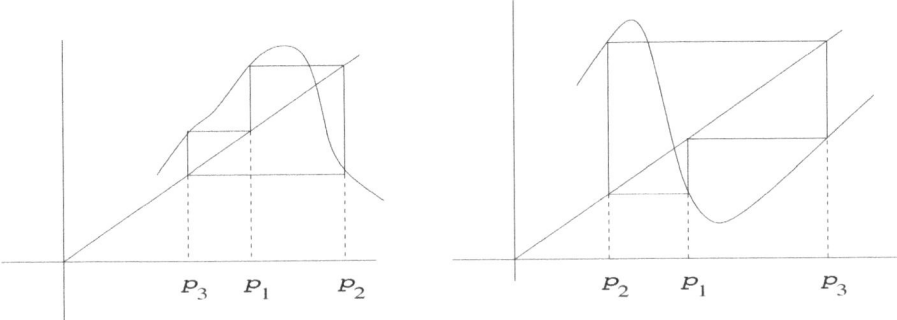

Fig. 2.5 Two period-3 orbits satisfying $f(p_1) = p_2$, $f(p_2) = p_3$, $f(p_3) = p_1$

Theorem 2.5 *Let I be a bounded interval and let $f: I \to I$ be continuous such that f has a period-3 orbit $\{p_1, p_2, p_3\}$ satisfying $f(p_1) = p_2$, $f(p_2) = p_3$ and $f(p_3) = p_1$. Then*

$$\lim_{n \to \infty} V_I(f^n) \geq K e^{\alpha n} \quad \text{for some} \quad K, \alpha > 0, \qquad (2.7)$$

i.e., the total variation of f^n grows exponentially with n.

Proof To help our visualization, we draw the graphs of a period-3 orbit in Fig. 2.5.
We have two possibilities:

$$\text{(i)} \ p_2 < p_1 < p_3, \quad \text{or} \quad \text{(ii)} \ p_3 < p_1 < p_2. \qquad (2.8)$$

Let us treat only case (i). Define

$$I_1 = [p_2, p_1], \qquad I_2 = [p_1, p_3].$$

Then

$$\left.\begin{array}{l} f(I_1) \supseteq I_1 \cup I_2 \ \text{i.e.,} \ I_1 \longrightarrow I_1 \cup I_2 \\ f(I_2) \supseteq I_1, \quad \text{i.e.,} \ I_2 \longrightarrow I_1. \end{array}\right\} \qquad (2.9)$$

Thus we have the following covering diagram in Fig. 2.6.

For each n, one can prove by mathematical induction that if the $(n+1)$th column (after mapping by f^n) contains a_n subintervals of I_1 or I_2, then the following relation is satisfied:

$$\begin{cases} a_{n+1} = a_n + a_{n-1}, \ \text{for} \ n = 2, 3, 4, \ldots \\ a_1 = 2, \quad a_2 = 3. \end{cases} \qquad (2.10)$$

An exact solution to the recurrence relation (2.10) can be determined as follows. Assume that a solution of $a_{n+1} = a_n + a_{n-1}$ can be written in the form

Fig. 2.6 The covering diagram satisfying (2.9)

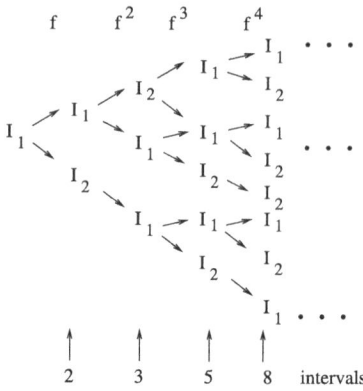

$$a_k = cx^k, \quad \text{for } k = 1, 2, \ldots. \tag{2.11}$$

Then substituting (2.11) into the first equation of (2.10) gives

$$cx^{n+1} = cx^n + cx^{n-1},$$
$$x^2 - x - 1 = 0,$$
$$x = \frac{1 \pm \sqrt{5}}{2}.$$

Therefore, we write the solution of (2.10) as

$$a_n = c_1 \left(\frac{1+\sqrt{5}}{2}\right)^n + c_2 \left(\frac{1-\sqrt{5}}{2}\right)^n, \quad n = 1, 2, \ldots,$$

$$a_1 = 2 = c_1 \left(\frac{1+\sqrt{5}}{2}\right) + c_2 \left(\frac{1-\sqrt{5}}{2}\right),$$

$$a_2 = 3 = c_1 \left(\frac{1+\sqrt{5}}{2}\right)^2 + c_2 \left(\frac{1-\sqrt{5}}{2}\right)^2,$$

and obtain

$$c_1 = \frac{5+3\sqrt{5}}{10}, \quad c_2 = \frac{5-3\sqrt{5}}{10}.$$

It is easy to show that

$$a_n = c_1 \left(\frac{1+\sqrt{5}}{2}\right)^n + c_2 \left(\frac{1-\sqrt{5}}{2}\right)^n$$

$$\geq k_0 \left(\frac{1+\sqrt{5}}{2}\right)^n$$

2.1 The Use of Total Variations as a Measure of Chaos

for some $k_0 > 0$, for all $n = 1, 2, \ldots$. Therefore, using the same arguments as in the proof of Theorem 2.3, we have

$$V_I(f^n) \geq V_{I_1}(f^n) \geq k_0 \left(\frac{1+\sqrt{5}}{2}\right)^n \cdot \min\{|I_1|, |I_2|\} \equiv k e^{\alpha n}, \text{ for } n = 1, 2, \ldots$$

where

$$k \equiv k_0 \cdot \min\{|I_1|, |I_2|\} \quad \text{and} \quad \alpha \equiv \ln\left(\frac{1+\sqrt{5}}{2}\right) > 0. \tag{2.12}$$

□

Using the same arguments as in the proof of Theorem 2.5, we can also establish the following.

Theorem 2.6 *Let I be a bounded closed interval and $f: I \to I$ be continuous. Assume that $I_1, I_2, \ldots, I_n$ are closed subintervals of I which overlap at most at endpoints, and the covering relation*

$$I_1 \longrightarrow I_2 \longrightarrow I_2 \longrightarrow \cdots \longrightarrow I_n \longrightarrow I_1 \cup I_j, \text{ for some } j \neq 1. \tag{2.13}$$

Then for some $K > 0$ and $\alpha > 0$,

$$V_I(f^n) \geq K e^{\alpha n} \longrightarrow \infty, \text{ as } n \to \infty. \tag{2.14}$$

□

Theorem 2.6 motivates us to give the following definition of chaos, the first one of such definitions in this book.

Definition 2.7 Let $f : I \to I$ be an interval map such that there exist $K > 0, \alpha > 0$ such that (2.14) holds. We say that f is *chaotic in the sense of exponential growth of total variations of iterates*. □

In Chap. 9, such a map f will also be said to have *rapid fuctuations of dimension 1*.

Corollary 2.8 *Let $f : I \to I$ be an interval map satisfying (2.14) in the assumption of Theorem 2.6. Then f is chaotic in the sense of exponential growth of variations of iterates.* □

Exercise 2.9 Let $f : [a, b] \to [a, b]$ be a monotonic function. Prove that

$$\forall n \in \mathbb{N}, V_{[a,b]}(f^n) = |f(b) - f(a)|.$$

□

Exercise 2.10 Assume that $f : [a, b] \to [a, b]$ is continuous and piecewise monotonic. Prove that there exists a constant $M \geq 0$ such that

$$\forall n \in \mathbb{N}, V_{[a,b]}(f^n) \leq M,$$

provided that f has no periodic points whose period is larger than 1. □

Exercise 2.11 Let $f : [a, b] \to [a, b]$ be continuous and piecewise monotonic. Suppose f has two distinct fixed points and a periodic point with period 2. Prove that $V_{[a,b]}(f^n) \to +\infty$ as $n \to +\infty$. In addition, show that there exist a constants $c_0 \geq 0$ and an integer k such that

$$\forall n \in \mathbb{N}, \ V_{[a,b]}(f^n) \leq c_0 n^k.$$

That is to say $V_{[a,b]}(f^n)$ grows unbounded but at most polynomially as n goes to infinity. □

Notes for Chapter 2

The total variation of a scalar-valued function on an interval provides a numerical measure of how strong the oscillatory behavior that function has, when the interval is finite. This chapter is based on Chen et al. [1]. It shows that the total variations of iterates of a given map can be bounded, of polynomial growth, and of exponential growth. Only the case of exponential growth of total variations of iterates is classified as chaos (while the case of polynomial growth is associated with the existence of periodic points). This offers a global approach to the study of Chaotic maps.

This chapter will pave the way for the study of chaotic behavior in terms of total variations in higher and fractional dimensions in Chaps. 8 and 9.

Reference

1. G. Chen, T. Huang and Y. Huang, Chaotic behavior of interval maps and total variations of iterates, *Int. J. Bifur. Chaos* **14** (2004), 2161–2186. https://doi.org/10.1142/S0218127404010242.

Ordering Among Periods: The Sharkovski Theorem 3

One of the most beautiful theorems in the theory of dynamical systems is the Sharkovski Theorem. An interval map may have many different periodic points with seemingly unrelated periodicities. What is unexpected and, in fact, amazing is that those periodicities are actually oredered in a certain way, called the Sharkovski ordering. The "top chain of the ordering" consists of all odd integers, with the number 3 at the zenith, and the "bottom chain of the ordering" consists of all decreasing powers of 2, with the number 1 at the nadir.

Here we give the statement of the theorem and provide a sketch of ideas of the proof. We introduce the Sharkovski ordering on the set of all positive integers. The ordering is arranged as in Fig. 3.1.

Theorem 3.1 (Sharkovski's Theorem) *Let I be a bounded closed interval and $f : I \to I$ be continuous. Let $n \triangleright k$ in Sharkovski's ordering. If f has a (prime) period n orbit, then f also has a (prime) period k orbit.* □

The following lemma is key in the proof of Sharkovski's Theorem.

Lemma 3.2 *Let n be an odd integer. Let f have a periodic point of prime period n. Then there exists a periodic orbit $\{x_j \mid j = 1, 2, \ldots n;\ f(x_j) = x_{j+1}$ for $j = 1, 2, \ldots, n-1$; $f(x_n) = x_1\}$ of prime period n such that either*

$$x_n < \cdots < x_5 < x_3 < x_1 < x_2 < x_4 < \cdots < x_{n-1} \tag{3.1}$$

or

$$x_{n-1} < \cdots < x_4 < x_2 < x_1 < x_3 < x_5 < \cdots < x_n. \tag{3.2}$$

□

$$
\begin{array}{ccccc}
3 & 2\cdot 3 & 2^2\cdot 3 & \vdots & 2^n\cdot 3 & \vdots & \vdots \\
\triangledown & \triangledown & \triangledown & & \triangledown & & 2^{m+1} \\
5 & 2\cdot 5 & 2^2\cdot 5 & & 2^n\cdot 5 & & \triangledown \\
\triangledown & \triangledown & \triangledown & & \triangledown & & 2^m \\
7 & 2\cdot 7 & 2^2\cdot 7 & & 2^n\cdot 7 & & \triangledown \\
\triangledown & \triangledown & \triangledown & & \triangledown & & 2^{m-1} \\
9 & 2\cdot 9 & 2^2\cdot 9 & & 2^n\cdot 9 & & \triangledown \\
\triangledown & \triangledown & \triangledown & & \triangledown & & \vdots \\
\vdots & \vdots & \vdots & \vdots & & & \triangledown \\
\triangledown & \triangledown & \triangledown & \triangledown & & & 2^3 \\
2n+1 & 2\cdot(2n+1) & 2^2\cdot(2n+1) & & 2^n\cdot(2m+1) & & \triangledown \\
\triangledown & \triangledown & \triangledown & & \triangledown & & 2^2 \\
2n+3 & 2\cdot(2n+3) & 2^2\cdot(2n+3) & & 2^n\cdot(2m+3) & & \triangledown \\
\triangledown & \triangledown & \triangledown & & \triangledown & & 2 \\
\vdots & \vdots & \vdots & & \vdots & & \triangledown \\
& & & & & & 1.
\end{array}
$$

Fig. 3.1 The Sharkovski ordering

The proof of Lemma 3.2 may be found in Robinson [1, pp. 67–69].

We now consider (3.1) only; (3.2) is a mirror image of (3.1) and can be treated similarly. We define subintervals $I_1, I_2, \ldots, I_n$ according to Fig. 3.2.

We now look at the covering relations of intervals $I_1, I_2, \ldots, I_n$. From

$$f(x_1) = x_2, \quad f(x_2) = x_3,$$

we have

$$f(I_1) \supseteq I_1 \cup I_2. \tag{3.3}$$

From

$$f(x_3) = x_4, \quad f(x_1) = x_2,$$

we have

$$f(I_2) \supseteq I_3. \tag{3.4}$$

Similarly,

$$f(I_3) \supseteq I_4, \ f(I_4) \supseteq I_5, \ldots, f(I_{n-1}) \supseteq I_n. \tag{3.5}$$

However,

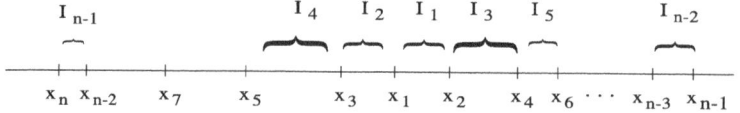

Fig. 3.2 The subintervals $I_1, I_2, \ldots, I_n$ according to (3.1), where $I_{2j-1} = [x_{2j-2}, x_{2j}]$ and $I_{2j} = [x_{2j+1}, x_{2j-1}]$, for $j = 2, \ldots, \frac{n-1}{2}$

Fig. 3.3 Covering relations for intervals $I_1, I_2, \ldots, I_6$ where $n = 7$ in Lemma 3.2. Note that I_1 covers both I_1 and I_2, while I_6 covers all the odd-numbered intervals $I_1, I_3,$ and I_5

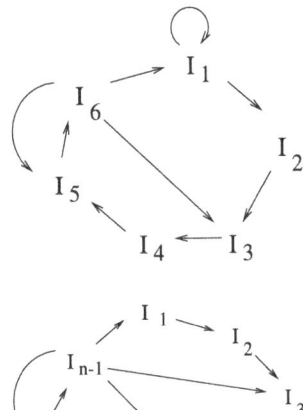

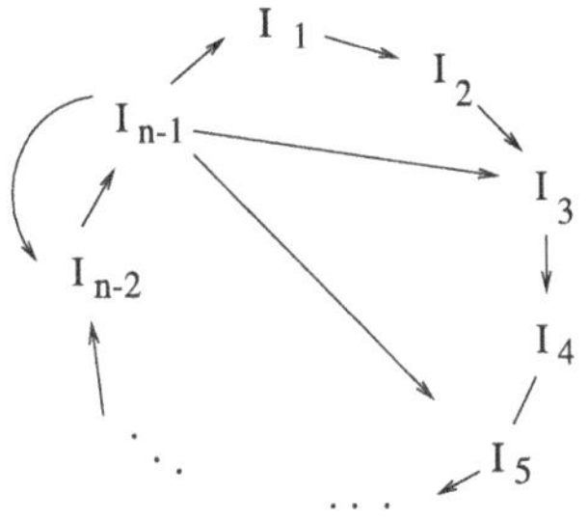

Fig. 3.4 The Stefan cycle corresponding to (3.1)

$$I_{n-1} = [x_n, x_{n-2}], \quad f(x_n) = x_1, \quad f(x_{n-2}) = x_{n-1},$$

and, therefore

$$f(I_{n-1}) \supseteq [x_1, x_{n-1}] = I_1 \cup I_3 \cup I_5 \cup \cdots \cup I_{-2}. \tag{3.6}$$

Example 3.3 In Lemma 3.2, let $n = 7$. Then (3.3)–(3.6) above give us the following diagram in Fig. 3.3. □

For a general odd positive integer n, from (3.3)–(3.6), we can construct the graph in Fig. 3.4, called the *Stefan cycle*.

Proposition 3.4 *Assume that n is odd and k is a positive integer such that $n \triangleright k$ in Sharkovski's ordering. If f has a prime period n, then f also has a prime period k.*

Proof There are two possibilities: (i) k is even and $k < n$; and (ii) $k > n$ and k can be either even or odd.

Consider Case (i) first. We use the loop

$$I_{n-1} \longrightarrow I_{n-k} \longrightarrow I_{n-k+1} \longrightarrow \cdots \longrightarrow I_{n-2} \longrightarrow I_{n-1}. \tag{3.7}$$

↑

note: $n - k$ here is odd.

Then by Lemma 2.2 in Chap. 2, there exists an $x_0 \in I_{n-1}$ with period k. The point x_0 cannot be an endpoint because the endpoints have period n. Therefore, x_0 has prime period k.

Next, we consider Case (ii). We use the following loop of length k:

$$I_1 \longrightarrow I_2 \longrightarrow \cdots \longrightarrow I_{n-1} \longrightarrow I_1 \longrightarrow \underbrace{I_1 \longrightarrow \cdots \longrightarrow I_1}_{k-n+1 \ I_1\text{'s}} \tag{3.8}$$

Thus, there exists an $x_0 \in I_1$ with $f^k(x_0) = x_0$. If x_0 is an endpoint of I_1, then x_0 has period n and, therefore, k is divisible by n, and $k \geq 2n \geq n+3$. Either $x_0 = x_1$ or $x_0 = x_2$ is satisfied for x_1 and x_2 in (3.1). Thus,

(a) if $x_0 = x_1$, then $f^n(x_0) = x_0$ and $f^{n+2}(x_0) = f^{n+2}(x_1) = f^2(x_1) = x_3$;
(b) if $x_0 = x_2$, then $f^n(x_0) = x_0$ and $f^{n+2}(x_0) = f^{n+2}(x_2) = f^2(x_2) = x_4$.

In either case above, $f^{n+2}(x_0) \notin I_1$. This violates our choice of x_0 that it satisfies

$$\begin{array}{ccccccccc}
x_0 \in I_1 & \longrightarrow & I_2 & \longrightarrow & I_3 & \longrightarrow \cdots \longrightarrow & I_{n-1} & \longrightarrow & I_1 & \longrightarrow & I_1 \\
& & \downarrow & & \downarrow & & & & \downarrow & & \downarrow \\
& & f(x_0) & & f^2(x_0) & & & & f^{n-1}(x_0) & & f^n(x_0) \\
\longrightarrow & I_1 & \longrightarrow & I_1 & \longrightarrow \cdots \longrightarrow & I_1, \\
& \downarrow & & \downarrow & & \\
& f^{n+1}(x_0) & & f^{n+2}(x_0) & &
\end{array}$$

from (3.8) because $k - n \geq n + 3$. $\square$

For readers who are interested to see a simple, complete proof of the Sharkovski Theorem, we recommend Du [2–4].

Exercise 3.5 We call a continuous function $f : [a, b] \to \mathbb{R}$ *turbulent* if there exist two nonempty subintervals I_1 and I_2 of $[a, b]$, the cardinal of their intersection is less than or equal to 1, such that $I_1 \cup I_2 \subset f(I_1) \cap f(I_2)$. In particular, we say that f is strictly *turbulent* if $I_1 \cap I_2 = \emptyset$.

1. Assume that a continuous function $f : [a, b] \to \mathbb{R}$ is (strictly) *turbulent*. For every integer n, prove that f has a periodic point with period n.
2. Recall the quadratic map $f_\mu : x \mapsto \mu x(1 - x)$. Determine a range of μ for which f_μ has periodic points with all positive periods. $\square$

Notes for Chapter 3

A.N. Sharkovski (1936–2022) published his paper [5] in the Ukrainian Mathematical Journal in 1964. This paper was ahead of the time before iterations, chaos and nonlinear phenomena became fashionable. Also, the paper was written in Russian. Thus, Sharkovski's results went unnoticed for a decade.

The American mathematicians Tien-Yien Li and James A. Yorke published a famous paper entitled "Period three implies chaos" in the American Mathematical Monthly in 1975. They actually proved a special part of Sharkovski's result besides coining the term chaos. Li and Yorke attended a conference in East Berlin where they met Sharkovski. Although they could not converse in a common language, the meeting led to global recognition of Sharkovski's work.

Sharkovski's Theorem does not hold for multidimensional maps. For circle maps, rotation by one hundred twenty degrees is a map with period three. But it does not have any other periods.

Today, there exist several ways of proving Sharkovski's Theorem: by Stefan [6], Block, Guckenheimer, Misiurewicz and Young [7], Burkart [8], Ho and Morris [9], Ciesielski and Pogoda [10], and Du [2, 3], and others.

References

1. C. Robinson, *Dynamical Systems, Stability, Symbolic Dynamics and Chaos*, CRC Press, Boca Raton, FL, 1995, pp. 67–69.
2. B. Du, A simple proof of Sharkovsky's theorem, *Amer. Math. Monthly* **111** (2004), 595–599. https://doi.org/10.2307/4145161.
3. B. Du, A simple proof of Sharkovsky's theorem revisited, *Amer. Math. Monthly* **114** (2007), 152–155.
4. B. Du, A simple proof of Sharkovsky's theorem re-revisited, preprint. (Version 7, September 28, 2009.) Cited on page(s) 32.
5. A.N. Sharkovskii, Coexistence of cycles of a continuous mapping of a line into itself, *Ukrainian Math. J.* 1964.
6. P. Stefan , A theorem of Sarkovskii on the coexistence of periodic orbits of continuous endomorphisms of the real line, *Comm. Math. Phys.* **54** (1977), 237–248. https://doi.org/10.1007/BF01614086.
7. L. Block, J. Guckenheimer, M. Misiurewicz, and L.-S. Young, Periodic points and topological entropy of one dimensional maps, in *Lecture Notes in Mathematica*, Vol. 819, Springer-Verlag, New York-Heidelberg-Berlin, 1980, 18–34.
8. U. Burkart, Interval mapping graphs and periodic points of continuous functions, *J. Combin. Theory Ser. B* **32** (1982), 57–68. https://doi.org/10.1016/0095-8956(82)90076-4.
9. C.W. Ho and C. Morris, A graph-theoretic proof of Sharkovsky's theorem on the periodic points of continuous functions, *Pacific J. Math.* **96** (1981), 361–370.
10. K. Ciesielski and Z. Pogoda, On ordering the natural numbers or the Sharkovski theorem, *Amer. Math. Monthly* **115** (2008), no. 2, 159–165.

4 Bifurcation Theorems for Maps

Bifurcation means "branching". It is a major nonlinear phenomenon. Bifurcation happens when one or several important system parameters change values in a transition process. After a bifurcation, the system's behavior changes. For example, new equilibrium states emerge, with a different behavior, especially that related to stability.

4.1 The Period-Doubling Bifurcation Theorem

Period doubling is an important *route to chaos*. We have seen from Fig. 1.5 that the (local) diagram looks like what is shown in Fig. 4.1.

In Fig. 4.1, the C_1 branch of fixed points loses its stability at $\mu = \mu_0$ and bifurcates into $C_2 \cup C_3$, which is a curve of period-2 points. We want to analyze this bifurcation. In doing the analysis, there are at least two difficulties involved:

(i) For the iteration $x_{n+1} = f_\mu(x_n) = f(\mu, x_n)$, any period-2 point $\bar{x}$ satisfies

$$\bar{x} = f_\mu^2(\bar{x}) = f(\mu, f(\mu, \bar{x})). \tag{4.1}$$

But if $\tilde{x}$ is a fixed point, $\tilde{x}$ will also satisfy (4.1). How do we pick out those period-2 points $\bar{x}$ which are not fixed points?

(ii) Assume that we can resolve (4.1). Can we also determine the stability (i.e., whether attracting, or repelling) of the period-2 points?

These will be answered in the following theorem.

Fig. 4.1 Period doubling and stability of bifurcated solutions

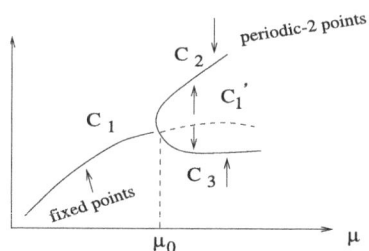

Theorem 4.1 (Period Doubling Bifurcation Theorem) *Consider the map $f(\mu, \cdot): I \to I$ where I is a closed interval and f is C^r for $r \geq 3$. Let the curve C represent a family of fixed points of $f(\mu, \cdot)$, where $C: x = x(\mu)$, and $x(\mu)$ satisfies*

$$x(\mu) = f(\mu, x(\mu)).$$

Assume that at $\mu = \mu_0$,

(i) $\quad \left.\dfrac{\partial f(\mu, x)}{\partial x}\right|_{\substack{\mu=\mu_0 \\ x=x(\mu_0)\equiv x_0}} = -1,$ \hfill (4.2)

(ii) $\quad \left[\dfrac{\partial^2 f(x, \mu)}{\partial \mu \partial x} + \dfrac{1}{2}\dfrac{\partial f(\mu, x)}{\partial \mu}\dfrac{\partial^2 f(\mu, x)}{\partial x^2}\right]_{\substack{\mu=\mu_0 \\ x=x_0}} = \alpha \neq 0,$ \hfill (4.3)

then there is period-doubling bifurcation at $\mu = \mu_0$ and $x = x_0$, i.e., there exists a curve $\gamma: \mu = m(x)$ in a neighborhood of $\mu = \mu_0$ and $x = x_0$ s.t. $\mu_0 = m(x_0)$ and

$$x = f_\mu^2(x)|_{\mu=m(x)} = f(\mu, f(\mu, x))|_{\mu=m(x)}. \tag{4.4}$$

Further, assume that

$$\left\{\dfrac{1}{3!}\dfrac{\partial^3 f(\mu, x)}{\partial x^3} + \left[\dfrac{1}{2!}\dfrac{\partial^2 f(\mu, x)}{\partial x^2}\right]^2\right\}\bigg|_{\substack{\mu=\mu_0 \\ x=x_0}} = \beta \neq 0. \tag{4.5}$$

Then the bifurcated period-2 points on the curve $\mu = m(x)$ are attracting if $\beta > 0$ and repelling if $\beta < 0$.

Proof We follow the proof in [1]. The fixed points x of f_μ satisfy

$$x = f_\mu(x) = f(\mu, x),$$

and, thus,

$$F(\mu, x) \equiv f(\mu, x) - x = 0. \tag{4.6}$$

4.1 The Period-Doubling Bifurcation Theorem

At $x = x(\mu_0) = x_0$ and $\mu = \mu_0$,

$$\frac{\partial}{\partial x} F(\mu, x)\Big|_{\substack{\mu=\mu_0 \\ x=x_0}} = \left[\frac{\partial f(\mu, x)}{\partial x} - 1\right]_{\substack{\mu=\mu_0 \\ x=x_0}} = -1 - 1 = -2 \neq 0.$$

Therefore, by the Implicit Function Theorem, x is solvable (locally near $x = x_0$) in terms of μ:

$$x = x(\mu), \quad \text{s.t.} \quad x_0 = x(\mu_0).$$

This gives us the curve C: $x = x(\mu)$ of fixed points.

Next, we want to capture the bifurcated period-2 points near $x = x_0$ and $\mu = \mu_0$. These points satisfy (4.1). To simplify notation, let us define

$$g(\mu, y) = f(\mu, y + x(\mu)) - x(\mu). \tag{4.7}$$

Then it is easy to check that

$$\frac{\partial^j}{\partial y^j} g(\mu, y)\Big|_{y=0} = \frac{\partial^j}{\partial x^j} f(\mu, x)\Big|_{x=x(\mu)}, \quad \text{for } j = 1, 2, \ldots. \tag{4.8}$$

This change of variable will give us plenty of convenience. We note that $y = y(\mu) \equiv 0$ becomes the curve of fixed points for the map g. Since $g(\mu, y)|_{y=0} = g(\mu, 0) = 0$, we have the Taylor expansion

$$g(\mu, y) = a_1(\mu) y + a_2(\mu) y^2 + a_3(\mu) y^3 + \mathcal{O}(|y|^4).$$

The period-2 points of f_μ now satisfy

$$y = g_\mu^2(y) = g(\mu, g(\mu, y))$$
$$= a_1(\mu)[a_1(\mu) y + a_2(\mu) y^2 + a_3(\mu) y^3 + \mathcal{O}(|y|^4)]$$
$$+ a_2(\mu)[a_1(\mu) y + a_2(\mu) y^2 + a_3(\mu) y^3 + \mathcal{O}(|y|^4)]^2$$
$$+ a_3(\mu)[a_1(\mu) y + a_2(\mu) y^2 + a_3(\mu) y^3 + \mathcal{O}(|y|^4)]^3 + \mathcal{O}(|y|^4),$$
$$y = a_1^2 y + (a_1 a_2 + a_1^2 a_2) y^2 + (a_1 a_3 + 2 a_1 a_2^2 + a_1^3 a_3) y^3 + \mathcal{O}(|y|^4),$$

where in the above, we have omitted the dependence of a_1, a_2 and a_3 on μ. Since $y = y(\mu) = 0$ corresponds to the fixed points of f_μ, we don't want them. Therefore, define

$$M(\mu, y) = \frac{g_\mu^2(y) - y}{y} \quad \text{if} \quad y \neq 0. \tag{4.9}$$

This gives

$$M(\mu, y) = (a_1^2 - 1) + (a_1 a_2 + a_1^2 a_2) y + (a_1 a_3 + 2 a_1 a_2^2 + a_1^3 a_3) y^2 + \mathcal{O}(|y|^3).$$

The above function $M(\mu, y)$ is obviously C^2 even for $y = 0$. Thus, we can extend the definition of $M(\mu, y)$ given in (4.9) by continuity even to $y = 0$. Now, note that period-2 points are determined by the equation $M(\mu, y) = 0$. We have

$$\frac{\partial}{\partial \mu} M(\mu, y)\Big|_{\substack{\mu=\mu_0 \\ y=0}} = 2a_1(\mu_0)a_1'(\mu_0)$$

$$= 2\left[\frac{\partial g(\mu, y)}{\partial y}\Big|_{\substack{\mu=\mu_0 \\ y=0}}\right]\left[\frac{\partial^2 g(\mu, y)}{\partial \mu \partial y}\Big|_{\substack{\mu=\mu_0 \\ y=0}}\right]$$

$$= 2\left[\frac{\partial f(\mu_0, x_0)}{\partial x}\right]\left[\frac{\partial^2 f(\mu_0, x_0)}{\partial \mu \partial x} + \frac{\partial^2 f(\mu_0, x_0)}{\partial x^2}x'(\mu_0)\right]$$

$$= -2\alpha, \tag{4.10}$$

where we have utilized the fact that

$$x(\mu) = f(\mu, x(\mu)),$$

$$x'(\mu) = \frac{\partial f}{\partial \mu} + \frac{\partial f}{\partial x}x'(\mu),$$

$$x'(\mu) = \frac{\partial f/\partial \mu}{1 - \frac{\partial f}{\partial x}},$$

$$x'(\mu_0) = \frac{\partial f(\mu_0, x_0)/\partial \mu}{1 - \frac{\partial f(\mu_0, x_0)}{\partial x}} = \frac{1}{2}\frac{\partial f(\mu_0, x_0)}{\partial \mu}.$$

From (4.3) and (4.10), we thus have

$$\frac{\partial}{\partial \mu} M(\mu, y)\Big|_{\substack{\mu=\mu_0 \\ y=0}} = -2\alpha \neq 0. \tag{4.11}$$

On the other hand,

$$\frac{\partial}{\partial y} M(\mu, y)\Big|_{\substack{\mu=\mu_0 \\ y=0}} = a_1(\mu_0)a_2(\mu_0) + a_1(\mu_0)^2 a_2(\mu_0)$$

$$= -a_2(\mu_0) + a_2(\mu_0)$$

$$= 0. \tag{4.12}$$

From (4.11) and the Implicit Function theorem, we thus conclude that near $\mu = \mu_0$ and $y = 0$, there exists a curve $\gamma: \mu = m(y)$ s.t.

$$M(m(y), y) = 0, \tag{4.13}$$

i.e., $\mu = m(y)$ represents period-2 points of the map f_μ. We have

4.1 The Period-Doubling Bifurcation Theorem

$$m(0) = \mu_0,$$

and $m'(0)$, $m''(0)$ may be computed from (4.13):

$$0 = \left[\frac{\partial M}{\partial \mu}m'(y) + \frac{\partial M}{\partial y}\right]\bigg|_{\substack{y=0 \\ \mu=m(0)}} = -2\alpha m'(0) + 0 \quad \text{(by (4.11) and (4.12))} \quad (4.14)$$

i.e., $m'(0) = 0$.

$$\left[\frac{\partial M}{\partial \mu}m''(y) + \frac{\partial^2 M}{\partial \mu^2}[m'(y)]^2 + 2\frac{\partial M}{\partial \mu \partial y}m'(y) + \frac{\partial^2 M}{\partial y^2}\right]\bigg|_{\substack{y=0 \\ \mu=m(0)}} = 0,$$

which implies

$$-2\alpha m''(0) + \frac{\partial^2 M(m(0), 0)}{\partial y^2} = -2\alpha m''(0) + 2[a_1(\mu_0)a_3(\mu_0)$$
$$+ 2a_1(\mu_0)a_2^2(\mu_0) + a_1^3(\mu_0)a_3(\mu_0)]$$
$$= -2\alpha m''(0) + 2 \cdot (-1)[2(a_3(\mu_0) + a_2^2(\mu_0))] = 0,$$

$$m''(0) = -\frac{2[a_3(\mu_0) + a_2^2(\mu_0)]}{\alpha} = -\frac{2}{\alpha}\left[\frac{1}{3!}\frac{\partial^3 g}{\partial y^3} + \left(\frac{1}{2!}\frac{\partial^2 g}{\partial y^2}\right)^2\right]\bigg|_{\substack{y=0 \\ \mu=m(0)}}$$

$$= -\frac{2}{\alpha}\left[\frac{1}{3!}\frac{\partial^3 f}{\partial x^3} + \left(\frac{1}{2!}\frac{\partial^2 f}{\partial x^2}\right)^2\right]\bigg|_{\substack{x=x_0 \\ \mu=\mu_0}}$$

$$= -\frac{2\beta}{\alpha} \neq 0.$$

Therefore, near $y = 0$, the function $\mu = m(y)$ has an expansion

$$\mu = m(0) + m'(0)y + \frac{m''(0)}{2!}y^2 + \mathcal{O}(|y|^3) \quad (4.15)$$

$$= \mu_0 - \frac{\beta}{\alpha}y^2 + \mathcal{O}(|y|^3). \quad (4.16)$$

Exercise 4.2 Verify that $\frac{\partial^3 g^2(\mu_0, 0)}{\partial y^3} = 3\frac{\partial^2 M(\mu_0, 0)}{\partial y^2} = -12\beta \neq 0$. □

We now check the stability of the period-2 points by computing $\frac{\partial(g^2)}{\partial y}$ about $y = 0$ and $\mu = \mu_0$:

$$\frac{\partial (g^2)(\mu, y)}{\partial y} = \frac{\partial (g^2)(\mu_0, 0)}{\partial y} + \frac{\partial^2 (g^2)(\mu_0, 0)}{\partial y^2} y + \frac{\partial^2 (g^2)(\mu_0, 0)}{\partial \mu \partial y}(\mu - \mu_0)$$
$$+ \frac{1}{2} \frac{\partial^3 (g^2)(\mu_0, 0)}{\partial y^3} y^2 + \cdots . \tag{4.17}$$

But

$$\frac{\partial (g^2)(\mu_0, 0)}{\partial y} = a_1^2(\mu_0) = (-1)^2 = 1, \tag{4.18}$$

$$\frac{\partial^2 (g^2)(\mu_0, 0)}{\partial y^2} = a_1(\mu_0) a_2(\mu_0) + a_1^2(\mu_0) a(\mu_0)$$

$$= -a_2(\mu_0) + a_2(\mu_0) = 0, \tag{4.19}$$

$$\frac{\partial^2 (g^2)(\mu_0, 0)}{\partial \mu \partial y} = 2a_1(\mu_0) a_1'(\mu_0) = -2\alpha \quad \text{(by (4.10))} \tag{4.20}$$

and

$$\frac{1}{2} \frac{\partial^3 (g^2)(\mu_0, 0)}{\partial y^3} = \frac{1}{2}(-12\beta) = -6\beta, \quad \text{(by Exercise)}. \tag{4.21}$$

Substituting (4.12) and (4.18)–(4.21) into (4.17), we obtain

$$\frac{\partial (g^2)(m(y), y)}{\partial y} = 1 + (-2\alpha) \cdot \left(-\frac{\beta}{\alpha} y^2\right) + (-6\beta) y^2 + \mathcal{O}(|y|^3)$$
$$= 1 - 4\beta y^2 + \mathcal{O}(|y|^3).$$

Therefore,

$$|1 - 4\beta y^2| = 1 - 4\beta y^2 < 1, \quad \text{if } \beta > 0, \quad \text{and so the period-2 orbit is } \textit{attracting};$$
$$|1 - 4\beta y^2| = 1 - 4\beta y^2 > 1, \quad \text{if } \beta < 0, \quad \text{and so the period-2 orbit is } \textit{repelling}.$$

□

Exercise 4.3 Consider the iteration

$$x_{n+1} = \mu \sin(\pi x_n), \quad 0 \le x_n \le 1, \quad 0 < \mu < 1.$$

(1) Use the Period Doubling Bifurcation Theorem to determine when (i.e., the value of μ) the first period doubling happens.
(2) Determine the stability of the bifurcated period-2 solution.
(3) Plot an orbit diagram to confirm your answers in (1) and (2). □

4.2 Saddle-Node Bifurcation

Next, we consider a different type of bifurcation.

Example 4.4 Let $f(\mu, x) = \mu e^x$, $x \in \mathbb{R}$. If $\mu > e^{-1}$, then the graph of $f(\mu, x)$ looks like Fig. 4.2a, where there is no intersection between the curve $f(\mu, x)$ and the diagonal line $y = x$. However, when $\mu = e^{-1}$, $f(\mu, x)$ is tangent to the diagonal line at $x = 1$. See Fig. 4.2b. For $\mu < e^{-1}$, the graph of $f(\mu, x)$ intersects $y = x$ at two points, both of which are then fixed points. Among these two fixed points, one is stable and the other is unstable. See Fig. 4.2c.

In Fig. 4.2b, we see that the slope $\frac{\partial}{\partial x} f(\mu, x)$ is equal to 1 at the point of tangency where $x = 1$. There is a bifurcation of fixed points when $\mu = e^{-1}$ because $f(\mu, x)$ has changed behavior from having no fixed points in Fig. 4.2a to having two fixed points in Fig. 4.2c. This bifurcation is now analyzed in the following theorem.

Theorem 4.5 (Saddle-Node or Tangent Bifurcation) *Assume that $f(\cdot, \cdot): \mathbb{R}^2 \to \mathbb{R}$ is C^2 satisfying the following conditions:*

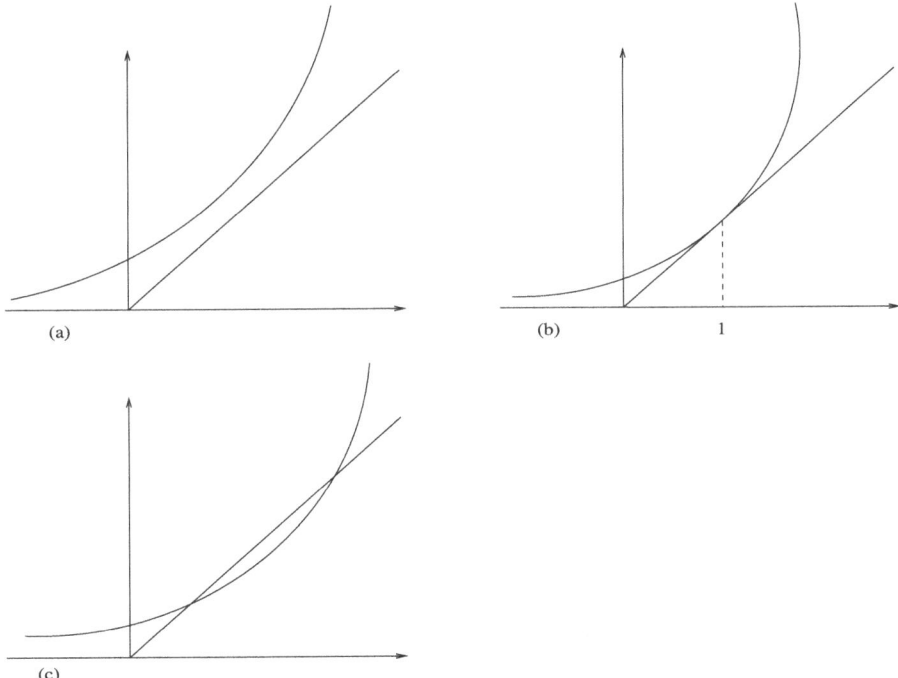

Fig. 4.2 The graphs of $y = f(\mu, x) = \mu e^x$. **a** $\mu > e^{-1}$; **b** $\mu = e^{-1}$; (c) $0 < \mu < e^{-1}$

(i) $f(\mu_0, x_0) = x_0$,
(ii) $\frac{\partial}{\partial x} f(\mu, x)\big|_{\substack{\mu=\mu_0 \\ x=x_0}} = 1$,
(iii) $\frac{\partial^2}{\partial x^2} f(\mu, x)\big|_{\substack{\mu=\mu_0 \\ x=x_0}} \neq 0$,
(iv) $\frac{\partial f}{\partial \mu}(\mu, x)\big|_{\substack{\mu=\mu_0 \\ x=x_0}} \neq 0$.

Then there exists a curve $C \colon \mu = m(x)$ *of fixed points defined in a neighborhood of* x_0 *such that* $\mu_0 = m(x_0)$, $m'(x_0) = 0$, *and*

$$f(m(x), x) = x, \qquad m''(x_0) = -\frac{\partial^2 f/\partial x^2}{\partial f/\partial \mu}\bigg|_{\substack{x=x_0 \\ \mu=\mu_0}} \neq 0. \qquad (4.22)$$

The fixed points on C are either (i) stable for $x > x_0$ *and unstable for* $x < x_0$, *or (ii) stable for* $x < x_0$ *and unstable for* $x < x_0$. *See Fig. 4.3.*

Proof We follow Robinson [2, pp. 212 and 213]. To determine the fixed points, we define the function

$$G(\mu, x) = f(\mu, x) - x = 0.$$

Fig. 4.3 Saddle-node bifurcation

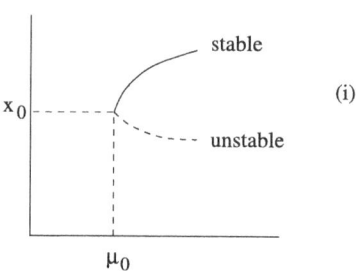

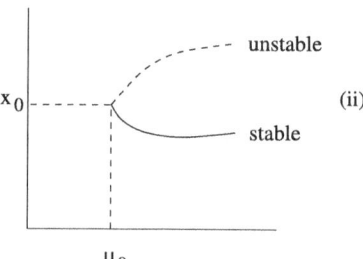

4.2 Saddle-Node Bifurcation

Then $G(\mu_0, x_0) = 0$. We have

$$\frac{\partial}{\partial x} G(\mu, x) \bigg|_{\substack{\mu=\mu_0 \\ x=x_0}} = \left(\frac{\partial f}{\partial x} - 1\right) \bigg|_{\substack{\mu=\mu_0 \\ x=x_0}} = 0,$$

so near $x = x_0$, x may not be solved in terms of y. However,

$$\frac{\partial}{\partial \mu} G(\mu, x) \bigg|_{\substack{\mu=\mu_0 \\ x=x_0}} = \frac{\partial f(\mu, x)}{\partial x} \bigg|_{\substack{\mu=\mu_0 \\ x=x_0}} \neq 0.$$

Thus, there exists a function $\mu = m(x)$ defined in a neighborhood of $x = x_0$ such that $\mu_0 = m(x_0)$ and

$$f(m(x), x) = x.$$

This function describes a curve C of fixed points near (μ_0, x_0). We have

$$0 = \frac{d}{dx} G(m(x), x) = \frac{\partial f}{\partial \mu} m'(x) + \frac{\partial f}{\partial x} - 1 = 0.$$

At $(\mu, x) = (\mu_0, x_0)$,

$$0 = \frac{\partial f(\mu_0, x_0)}{\partial \mu} m'(x_0) + \frac{\partial f(\mu_0, x_0)}{\partial x} - 1 = \frac{\partial f(\mu_0, x_0)}{\partial \mu} m'(x_0)$$

and so $m'(x_0) = 0$.

Also,

$$0 = \frac{d^2}{dx^2} G(m(x), x) = \frac{\partial^2 f}{\partial \mu^2} [m'(x)]^2 + 2 \frac{\partial^2 f}{\partial \mu \partial x} m'(x) + \frac{\partial f}{\partial \mu} m''(x) + \frac{\partial^2 f}{\partial x^2} = 0. \quad (4.23)$$

At $(\mu, x) = (\mu_0, x_0)$, using the fact that $m'(x_0) = 0$, $\frac{\partial f}{\partial \mu} \neq 0$, we obtain from (4.23) the second equation of (4.22).

Finally, let us analyze stability of points on C near $(\mu, x) = (\mu_0, x_0)$. The stability is determined by whether

$$\left| \frac{\partial}{\partial x} f(\mu, x) \right|_{\substack{\mu=m(x) \\ x}} \quad \text{is less than 1 or greater than 1.}$$

We have

$$\frac{\partial}{\partial x}f(\mu,x)\Big|_{\mu=m(x)} = \frac{\partial}{\partial x}f(\mu_0,x_0) + \frac{\partial^2 f(\mu_0,x_0)}{\partial x^2}(x-x_0) + \frac{\partial^2 f(\mu_0,x_0)}{\partial \mu \partial x}(m(x)-\mu_0)$$
$$+ \frac{1}{2!}\frac{\partial^3 f(\mu_0,x_0)}{\partial x^3}(x-x_0)^2 + \frac{\partial^3 f(\mu_0,x_0)}{\partial \mu \partial x^2}(m(x)-\mu_0)(x-x_0)$$
$$+ \frac{1}{2!}\frac{\partial^3 f(\mu_0,x_0)}{\partial \mu^2 \partial x}(m(x)-\mu_0)^2 + \cdots$$
$$= 1 + \frac{\partial^2 f(\mu_0,x_0)}{\partial x^2}(x-x_0) + \mathcal{O}(|x-x_0|^2).$$

If $\frac{\partial^2 f(\mu_0,x_0)}{\partial x^2} > 0$, then

$$1 + \frac{\partial^2 f(\mu_0,x_0)}{\partial x^2}(x-x_0) \begin{cases} > 1 & x > x_0, \\ \text{if} & \\ < 1 & x < x_0. \end{cases}$$

So the stability can be determined. A similar conclusion can be obtained if $\frac{\partial^2 f(\mu_0,x_0)}{\partial x^2} < 0$. □

The Saddle-Node Bifurcation Theorem does not apply to the map $f(\mu,x) = \mu \sin(\pi x)$ we studied in the second homework set. To analyze the bifurcation behavior of $f(\mu,x)$ near $x_0 = 0$ and $\mu_0 = 1/\pi$, we note that even though

$$\frac{\partial}{\partial x}f(\mu,x)\Big|_{\substack{x=x_0=0 \\ \mu=\mu_0=1/\pi}} = \mu\pi \cos 0 = 1,$$

a different bifurcation analysis must be done, as

$$\frac{\partial}{\partial \mu}f(\mu,x)\Big|_{\substack{x=0 \\ \mu=1/\pi}} = \sin 0 = 0.$$

This will be investigated in the next section.

4.3 The Pitchfork Bifurcation

The following theorem applies to the map $f(\mu,x) = \mu \sin(\pi x)$.

Theorem 4.6 (Pitchfork Bifurcation) *Let $f(\mu,x) = xg(\mu,x)$ and $f(\cdot,\cdot): \mathbb{R}^2 \to \mathbb{R}$ is C^3. Then the curve $C_1: x \equiv 0$ represents a curve of fixed points on the (μ,x)-plane. Assume that at $(\mu,x) = (\mu_0, 0)$, we have*

4.3 The Pitchfork Bifurcation

$$\left.\frac{\partial}{\partial x}f(\mu, x)\right|_{\substack{\mu=\mu_0 \\ x=0}} = 1,$$

$$\left.\frac{\partial}{\partial \mu}g(\mu, x)\right|_{\substack{\mu=\mu_0 \\ x=0}} \neq 0,$$

$$\left.\frac{\partial^2}{\partial x^2}f(\mu, x)\right|_{\substack{\mu=\mu_0 \\ x=0}} = 0,$$

$$\left.\frac{\partial^3}{\partial x^3}f(\mu, x)\right|_{\substack{\mu=\mu_0 \\ x=0}} \neq 0.$$

Then at $(\mu, x) = (\mu_0, 0)$, there is a new curve C_2: $\mu = m(x)$ of bifurcated fixed points such that

$$\mu_0 = m(0), \quad m'(0) = 0, \quad m''(0) \neq 0,$$
$$x = f(m(x), x).$$

The stability of the points on C_2 near $(\mu, x) = (\mu_0, 0)$ is attracting if $\partial^2 g(\mu_0, 0)/\partial x^2 < 0$ and repelling if $\partial^2 g(\mu_0, 0)/\partial x^2 > 0$ (Fig. 4.4).

Proof Define the implicit relation

$$G(\mu, x) = g(\mu, x) - 1 = 0.$$

Note that if $x \neq 0$ satisfies

$$G(\mu, x) = g(\mu, x) - 1 = 0,$$

then $xg(\mu, x) - x = 0 = f(\mu, x) - x$. Therefore x is a fixed point of the map $f(\mu, x)$. Since

$$\left.\frac{\partial}{\partial x}G(\mu, x)\right|_{\substack{\mu=\mu_0 \\ x=0}} = \left.\frac{\partial}{\partial x}g(\mu, x)\right|_{\substack{\mu=\mu_0 \\ x=0}} = \frac{1}{2}\left[\left.\frac{\partial^2}{\partial x^2}f(\mu, x)\right|_{\substack{\mu=\mu_0 \\ x=0}}\right] = 0,$$

we may not be able to solve x in terms of μ locally near $\mu = \mu_0$. However,

Fig. 4.4 Pitchfork bifurcation

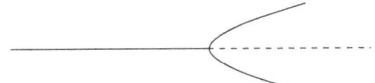

$$\left.\frac{\partial}{\partial \mu} G(\mu, x)\right|_{\substack{x=0 \\ \mu=\mu_0}} = \left.\frac{\partial g(\mu, x)}{\partial \mu}\right|_{\substack{\mu=\mu_0 \\ x=0}} \neq 0,$$

so we have a curve C_2: $\mu = m(x)$ of fixed points near $(\mu, x) = (\mu_0, 0)$, such that $m(0) = \mu_0$. Let us compute $m'(0)$ and $m''(0)$. We have

$$\frac{d}{dx} G(m(x), x) = 0 = \frac{\partial g}{\partial \mu} m'(x) + \frac{\partial g}{\partial x} = 0.$$

Since $\partial g(\mu_0, 0)/\partial \mu \neq 0$,

$$m'(0) = -\frac{\partial g(\mu_0, 0)/\partial x}{\partial g(\mu_0, 0)/\partial \mu} = 0. \tag{4.24}$$

Differentiating again,

$$\frac{d^2}{dx^2} G(m(x), x) = 0 = \frac{\partial^2 g}{\partial \mu^2}[m'(x)]^2 + 2\frac{\partial^2 g}{\partial \mu \partial x} m'(x)$$
$$+ \frac{\partial g}{\partial \mu} m''(x) + \frac{\partial^2 g}{\partial x^2},$$

we obtain

$$m''(0) = -\frac{\partial^2 g(\mu_0, 0)/\partial x^2}{\partial g(\mu_0, 0)/\partial \mu}. \tag{4.25}$$

But

$$\frac{\partial^3}{\partial x^3} f(\mu, x) = \frac{\partial^3}{\partial x^3}[xg(\mu, x)] = 3\frac{\partial^2 g(\mu, x)}{\partial x^2} + x\frac{\partial^3 g(\mu, x)}{\partial x^3}$$

and at $x = 0$, $\mu = \mu_0$,

$$\frac{\partial^2 g(\mu_0, 0)}{\partial x^2} = \frac{1}{3}\frac{\partial^3}{\partial x^3} f(\mu_0, 0) \neq 0. \tag{4.26}$$

From (4.25) and (4.26), we thus have

$$m''(0) \neq 0. \tag{4.27}$$

Combining (4.24) and (4.27), we see that the curve C_2 looks like a parabola near $(\mu_0, 0)$ on the (μ, x)-plane. C_2 opens to the left if $m''(0) < 0$, and to the right if $m''(0) > 0$.

Finally, let us analyze stability of the bifurcated fixed points on C_2 near $(\mu, x) = (\mu_0, 0)$. We have

4.3 The Pitchfork Bifurcation

$$\left.\frac{\partial f(\mu, x)}{\partial x}\right|_{\mu=m(x)} = \frac{\partial f(\mu_0, 0)}{\partial x} + \frac{\partial^2 f(\mu_0, 0)}{\partial x^2}(x - 0) + \frac{\partial^2 f(\mu_0, 0)}{\partial \mu \partial x}(m(x) - \mu_0)$$

$$+ \frac{1}{2!}\frac{\partial^3 f(\mu_0, 0)}{\partial x^3}(x - 0)^2 + \frac{\partial^3 f(\mu_0, 0)}{\partial \mu \partial x^2}(x - 0)(m(x) - \mu_0)$$

$$+ \frac{1}{2!}\frac{\partial^3 f(\mu_0, 0)}{\partial \mu^2 \partial x}(m(x) - \mu_0)^2 + \cdots$$

$$= 1 + 0 \cdot x + \left[\frac{\partial g(\mu_0, 0)}{\partial \mu} \cdot m''(0) x^2 + \mathcal{O}(x^3)\right]$$

$$+ \frac{1}{2}\left[3 \frac{\partial^2 g(\mu_0, 0)}{\partial x^2} x^2\right] + \mathcal{O}(x^3). \quad (4.28)$$

But, by (4.25),

$$\frac{\partial g(\mu_0, 0)}{\partial \mu} m''(0) = -\frac{\partial^2 g(\mu_0, 0)}{\partial x^2}, \quad (4.29)$$

and by substituting (4.29) into (4.28), we obtain

$$\left.\frac{\partial f(\mu, x)}{\partial x}\right|_{\mu=m(x)} = 1 + \frac{1}{3}\frac{\partial^2 g(\mu_0, 0)}{\partial x^2} x^2 + \mathcal{O}(x^3).$$

Therefore, fixed points on C_2 near $(\mu, x) = (\mu_0, 0)$ are attracting if $\partial^2 g(\mu_0, 0)/\partial x^2 < 0$, and repelling if $\partial^2 g(\mu_0, 0)/\partial x^2 > 0$. □

Example 4.7 For the map $f(\mu, x) = \mu \sin(\pi x)$, we have

$$f(\mu, x) = x g(\mu, x) \text{ where}$$

$$g(\mu, x) = \mu \left[\pi - \frac{\pi(\pi x)^2}{3!} + \frac{\pi(\pi x)^4}{5!} - \cdots + (-1)^n \frac{\pi(\pi x)^{2n}}{(2n+1)!} \pm \cdots\right].$$

At $(\mu, x) = \left(\frac{1}{\pi}, 0\right)$, we have

$$\frac{\partial f(\mu, x)}{\partial x} = 1, \frac{\partial g(\mu, x)}{\partial \mu} = \pi \neq 0, \frac{\partial^2 f(\mu, x)}{\partial x^2} = 0, \frac{\partial^3 f(\mu, x)}{\partial x^3} = -\mu \pi^3 \neq 0,$$

$$\frac{\partial^2 g(\mu, x)}{\partial x^2} = -\frac{1}{3}\pi^2 < 0.$$

So the bifurcated fixed points are stable, as can be seen in Fig. 4.5. □

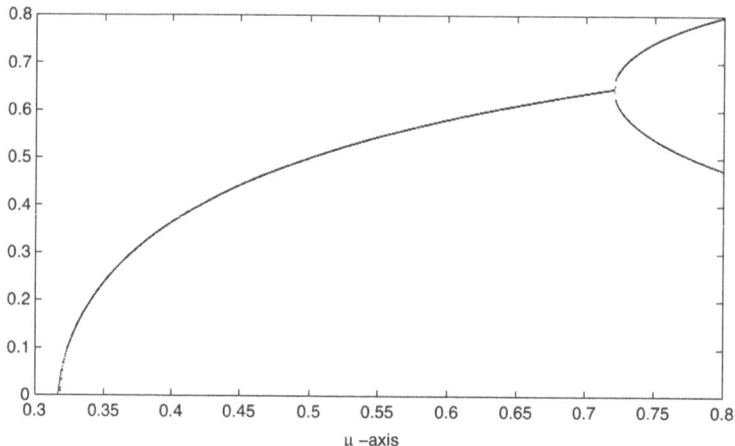

Fig. 4.5 Bifurcation of the fixed points $x \equiv 0$ into a new curve C_2 of fixed points near $(\mu, x) = \left(\frac{1}{\pi}, 0\right)$, for the map $f(\mu, x) = \mu \sin(\pi x), 0 \le x \le 1$

Exercise 4.8 For the quadratic map

$$f(\mu, x) = \mu x(1 - x),$$

at $(\mu, x) = (1, 0)$, we have

$$\left. \frac{\partial}{\partial x} f(\mu, x) \right|_{\substack{\mu=1 \\ x=0}} = 1.$$

Analyze the bifurcation of fixed points near $(\mu, x) = (1, 0)$ by stating and proving a theorem similar to Theorem 4.6. □

4.4 Hopf Bifurcation

To begin with, we offer as an example a 2-dimensional map $F : \mathbb{R}^2 \to \mathbb{R}^2$ defined by

$$F(x_1, x_2) = \begin{bmatrix} \cos\theta & -\sin\theta \\ \sin\theta & \cos\theta \end{bmatrix} \left\{ (1+\alpha) \begin{bmatrix} x_1 \\ x_2 \end{bmatrix} + (x_1^2 + x_2^2) \begin{bmatrix} a & -b \\ b & a \end{bmatrix} \begin{bmatrix} x_1 \\ x_2 \end{bmatrix} \right\} \quad (4.30)$$

where α is a parameter; $\theta = \theta(\alpha)$, $a = a(\alpha)$, $b = b(\alpha)$ are smooth functions of α, satisfying $0 < \theta(0) < \pi$, $a(0) \ne 0$, for $\alpha = 0$.

It is easy to check that the origin, $(x_1, x_2) = (0, 0)$, is a fixed point of F for all α. At $(0, 0)$, the Jacobian matrix of the map F is

$$A \equiv (1+\alpha) \begin{bmatrix} \cos\theta & -\sin\theta \\ \sin\theta & \cos\theta \end{bmatrix}$$

4.4 Hopf Bifurcation

The two eigenvalues of matrix A are $\mu_{1,2} \equiv (1+\alpha)e^{\pm i\theta}$. In particular, when $\alpha = 0$, we have $|\mu_{1,2}| = 1$. Thus, the origin is not a hyperbolic fixed point; cf. Definition 1.4. To facilitate the study of bifurcation of the system when α passes $\alpha = 0$, we rewrite F as a map of the complex plane:

for $z = x_1 + ix_2$, $x_1, x_2 \in \mathbb{R}$

$$F(z) = e^{i\theta} z \left(1 + \alpha + d|z|^2\right) = \mu z + cz|z|^2 \qquad (4.31)$$

$$c = c(\alpha) \equiv e^{i\theta(\alpha)} d(\alpha), \quad d(\alpha) \equiv a(\alpha) + ib(\alpha), \quad \mu = \mu(\alpha) \equiv (1+\alpha)e^{i\theta(\alpha)}$$

We look at the phase relation of (4.31): letting $z = \rho e^{i\phi}$ with $\rho = |z|$, we have

$$F(z) = e^{i\theta} \left(\rho e^{i\phi}\right) \left[1 + \alpha + (a+ib)\rho^2\right]$$

$$= \rho \left[\left(1 + \alpha + a\rho^2\right)^2 + b^2 \rho^4\right]^{1/2} e^{i(\theta + \phi + \psi)}$$

where

$$\psi = \sin^{-1} \frac{b\rho^2}{\left[\left(1 + \alpha + a\rho^2\right)^2 + b^2 \rho^4\right]^{1/2}}$$

Thus, in polar coordinates, system (4.31) becomes

$$\begin{bmatrix} \rho \\ \varphi \end{bmatrix} \longmapsto G(\rho, \varphi) = \begin{bmatrix} \rho\left[1 + \alpha + a(\alpha)\rho^2\right] + \rho^4 R_\alpha(\rho) \\ \varphi + \theta(\alpha) + \rho^2 Q_\alpha(\rho) \end{bmatrix} \qquad (4.32)$$

where $R_\alpha(\cdot)$ and $Q_\alpha(\cdot)$ are smooth functions of (ρ, α). From (4.32), we know that the first component on the RHS of (4.32) is independent of φ. Thus, we have achieved decoupling between ρ and φ, making the subsequent discussions on bifurcation more intuitive. With regard to the ρ-variable, the transformation (4.31) actually constitutes a 1-dimensional dynamical system:

$$\widetilde{G}(\rho) \equiv \rho \left[1 + \alpha + a(\alpha)\rho^2\right] + \rho^4 R_\alpha(\rho).$$

For this dynamical system, $\rho = 0$ is a fixed point for any parameter value α. When $\alpha > 0$, the fixed point $\rho = 0$ is unstable. When $\alpha = 0$, the stability of $\rho = 0$ is determined by the sign of $a(0)$:

(i) if $a(0) < 0$, then $\rho = 0$ is (nonlinear) stable;
(ii) if $\alpha > 0$ and $a(0) < 0$, then in addition to the fixed point $\rho = 0$, there is another stable fixed point

$$\rho(\alpha) = \sqrt{-\frac{\alpha}{a(\alpha)}} + \mathcal{O}(\alpha). \qquad (4.33)$$

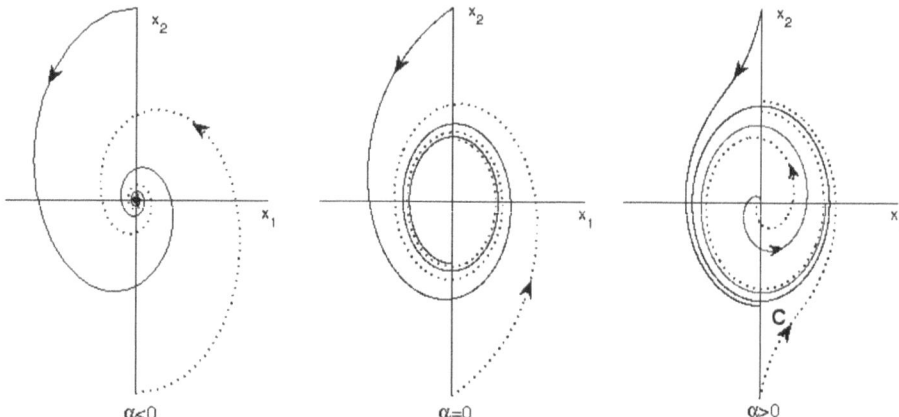

Fig. 4.6 Hopf bifurcation

With respect to the phase angle φ, the second component of the RHS of (4.32) shows that the action of the map is similar to a rotation by an angle $\theta(\alpha)$ (but it depends on both ρ and α).

Summarizing the above, we see that for the 2-dimensional dynamical system (4.30), assuming that $a(0) < 0$, then when the parameter α passes 0, we have the following bifurcation phenomena:

(1) When $\alpha < 0$, the origin $(0, 0)$ is a stable fixed point. The phase plot is a stable focus; see Fig. 4.6a.
(2) When $\alpha > 0$, the origin $(0, 0)$ is an unstable focus in a small neighborhood of the origin. There is a closed curve C with approximate radius $\rho(\alpha)$ (see Eq. 4.33), which is a stable curve such that the trajectories of all points starting from either within C or outside C will be attracted to C; see Fig. 4.6c.
(3) When $\alpha = 0$, the origin $(0, 0)$ is nonlinear stable; cf. Fig. 4.6b.

The above bifurcation phenomena are called the Hopf bifurcation (or the Neimark-Sacker bifurcation). Its special feature is the occurrence of the closed curve C, which is invariant under the map F. Similarly, one may consider the case $a(0) > 0$. For this case, there is an unstable closed curve C when $\alpha > 0$, Hopf bifurcation happens when α crosses 0, and the curve C disappears.

Next, we consider the following two-dimensional map:

$$F(x_1, x_2) = \begin{bmatrix} \cos\theta & -\sin\theta \\ \sin\theta & \cos\theta \end{bmatrix} \left\{ (1+\alpha) \begin{bmatrix} x_1 \\ x_2 \end{bmatrix} + (x_1^2 + x_2^2) \begin{bmatrix} a & -b \\ b & a \end{bmatrix} \begin{bmatrix} x_1 \\ x_2 \end{bmatrix} \right\}$$
$$+ \mathcal{O}\left(|x|^4\right) \tag{4.34}$$

4.4 Hopf Bifurcation

Similarly, to the conversion from (4.30) to (4.31), one can assert that (4.34) can be converted to the form

$$z \longmapsto F(z) = \mu z + cz|z|^2 + \mathcal{O}\left(|z|^4\right). \tag{4.35}$$

In comparison, system (4.34) contains some higher order terms than those of system (4.30). Even though (4.34) is not locally topologically conjugate to (4.30), and the higher order terms in (4.34) do affect bifurcation phenomena, some special characters of Hopf bifurcation are preserved.

Lemma 4.9 *The higher order term $\mathcal{O}\left(|x|^4\right)$ in (4.34) does not affect the occurrence (or disappearance) of the invariant closed close C and its stability. The local stability of the origin $(0, 0)$ and the bifurcation patterns remain the same.*

Proof The justification is lengthy. We refer to Kuznetsov [3, pp. 131–136], for example. □

Now, we consider a general planar map

$$f : \mathbb{R}^2 \to \mathbb{R}^2, \quad x \longmapsto f(x, \alpha), \quad x = (x_1, x_2) \in \mathbb{R}^2, \quad \alpha \in \mathbb{R}. \tag{4.36}$$

We will prove that any planar map with the Hopf bifurcation property can be transformed into the form (4.34).

Assume that f is smooth, and at $\alpha = 0$, $f(x, \alpha)$ has a fixed point $x = 0$, i.e., $f(0, 0) = 0$. The eigenvalues of the Jacobian matrix at $(x, \alpha) = (0, 0)$ are $\mu_{1,2} = e^{\pm i\theta_0}$, for some $\theta_0 : 0 < \theta_0 < \pi$. By the Implicit Function Theorem, for sufficiently small $|\alpha|$, f has a one-parameter family of unique fixed points $x(\alpha)$, and the map f is invertible. By using translation, the fixed points $x(\alpha)$ can be relocated to the origin $\mathbf{0}$. Thus, without loss of generality, we assume that for $|\alpha|$ small, $x = \mathbf{0}$ is a fixed point of the system. Thus, the map can be written as

$$x \longmapsto A(\alpha)x + F(x, \alpha), \tag{4.37}$$

where $A(\alpha)$ is a 2×2 matrix depending on α, and $F(x, \alpha)$ is a vector valued function with components F_1 and F_2 such that their leading terms begin quadratically with respect to x_1 and x_2 in the Taylor expansions of F_1 and F_2, and $F(0, \alpha) = 0$ for sufficiently small $|\alpha|$. The Jacobian matrix $A(\alpha)$ has eigenvalues

$$\mu_{1,2}(\alpha) = r(\alpha)e^{\pm i\varphi(\alpha)}, \text{ with } r(0) = 1, \varphi(0) = \theta_0. \tag{4.38}$$

Set $\beta(\alpha) = r(\alpha) - 1$. Then $\beta(\alpha)$ is smooth and $\beta(0) = 0$. Assume that $\beta'(0) \neq 0$, then (locally) we can use β in lieu of α for the parametrization. Thus, we have

$$\mu_1(\beta) = \mu(\beta) \equiv (1 + \beta)e^{i\theta(\beta)}, \quad \mu_2(\beta) = \bar{\mu}(\beta), \tag{4.39}$$

where $\theta(\beta)$ is a smooth function of β and $\theta(0) = \theta_0$.

Lemma 4.10 *Under the assumptions of (4.38) and (4.39) for $|\alpha|$ small, the map (4.37) can be rewritten as*

$$z \longmapsto \mu(\beta)z + g(z, \bar{z}, \beta), \quad z \in \mathbb{C}, \quad \beta \in \mathbb{R}, \tag{4.40}$$

where g is a smooth function with a (local) Taylor's expansion

$$g(z, \bar{z}, \beta) = \sum_{k+\ell \geq 2} \frac{1}{k!\ell!} g_{k\ell}(\beta) z^k \bar{z}^\ell, \quad k, \ell = 0, 1, 2, \ldots$$

Proof Let $q(\beta)$ be the eigenvector corresponding to eigenvalue $\mu(\beta)$:

$$A(\beta)q(\beta) = \mu(\beta)q(\beta). \tag{4.41}$$

Then $A^T(\beta)$, the transpose of $A(\beta)$, has an eigenvector $p(\beta)$ corresponding to eigenvalue $\bar{\mu}(\beta)$:

$$A^T(\beta)p(\beta) = \bar{\mu}(\beta)p(\beta)$$

First, we prove that

$$\langle p, \bar{q} \rangle \equiv \langle p, \bar{q} \rangle_{\mathbb{C}^2} = 0$$

where $\langle p, q \rangle$ is defined as

$$\langle p, q \rangle = \bar{p}_1 q_1 + \bar{p}_2 q_2 \quad \text{for} \quad p = (p_1, p_2)^T, \quad q = (q_1, q_2)^T$$

In fact, since $Aq = \mu q$ and A is a real matrix, we have $A\bar{q} = \bar{\mu}\bar{q}$. Thus,

$$\langle p, \bar{q} \rangle = \left\langle p, \frac{1}{\bar{\mu}} A\bar{q} \right\rangle$$
$$= \frac{1}{\mu} \langle p, A\bar{q} \rangle = \frac{1}{\mu} \left\langle A^T p, \bar{q} \right\rangle$$
$$= \frac{1}{\mu} \langle \bar{\mu} p, \bar{q} \rangle = \frac{\mu}{\bar{\mu}} \langle p, \bar{q} \rangle.$$

Therefore,

$$\left(1 - \frac{\mu}{\bar{\mu}}\right) \langle p, \bar{q} \rangle = 0.$$

But $\mu \neq \bar{\mu}$ because for sufficiently small $|\beta|$, we have $0 < \theta(\beta) < \pi$. So we have

$$\langle p, \bar{q} \rangle = 0.$$

Second, since A is a real matrix and the imaginary part of μ is nonzero, the imaginary part of q is also nonzero. By the above equality, we have

4.4 Hopf Bifurcation

$$\langle p, q \rangle \neq 0.$$

By normalization, we can assume that

$$\langle p, q \rangle = 1.$$

For any sufficiently small $|\beta|$, any $x \in \mathbb{R}^2$, there exists a unique $z \in \mathbb{C}$ such that

$$x = zq(\beta) + \bar{z}\bar{q}(\beta). \tag{4.42}$$

(Due to the fact that $\langle p(\beta), \bar{q}(\beta) \rangle_{\mathbb{C}^2} = 0$, we can simply choose $z = \langle p(\beta), x \rangle_{\mathbb{C}^2}$.) From (4.37) and (4.41) through (4.42), for the complex variable z, we have

$$z \longmapsto \mu(\beta)z + \langle p(\beta), F(zq(\beta) + \bar{z}\bar{q}(\beta), \beta) \rangle_{\mathbb{C}^2}. \tag{4.43}$$

Denote the very last term in (4.43) as $g(z, \bar{z}, \beta)$. Then we obtain

$$g(z, \bar{z}, \beta) = \sum_{k+\ell \geq 2} \frac{1}{k!\ell!} g_{k\ell}(\beta) z^k \bar{z}^\ell,$$

where

$$g_{k\ell}(\beta) = \frac{\partial^{k+\ell}}{\partial z^k \partial \bar{z}^\ell} \langle p(\beta), F(zq(\beta) + \bar{z}\bar{q}(\beta), \beta) \rangle \bigg|_{z=0}$$

for $k + \ell \geq 2, k, \ell = 0, 1, 2, \ldots$. Hence, Eq. (4.40) is obtained. □

The following three lemmas show that under proper conditions, we can convert the map from the form (4.40) to the standard form (4.36).

Lemma 4.11 *Assume that $e^{i\theta_0} \neq 1$, $e^{3i\theta_0} \neq 1$. Consider the map*

$$z \longmapsto \mu z + \frac{g_{20}}{2} z^2 + g_{11} z \bar{z} + \frac{g_{02}}{2} \bar{z}^2 + \mathcal{O}\left(|z|^3\right), \quad z \in \mathbb{C}, \tag{4.44}$$

where $\mu = \mu(\beta) = (1 + \beta)e^{i\theta(\beta)}$, $g_{ij} = g_{ij}(\beta)$. Then for $|\beta|$ sufficiently small, there exists a (locally) invertible transformation

$$z = w + \frac{h_{20}}{2} w^2 + h_{11} w \bar{w} + \frac{h_{02}}{2} \bar{w}^2, \tag{4.45}$$

such that (4.44) is transformed to

$$w \longmapsto \mu w + \mathcal{O}\left(|w|^3\right), \quad w \in \mathbb{C}$$

i.e., the quadratic terms $\mathcal{O}\left(|z|^2\right)$ in (4.44) are eliminated.

Proof It is easy to check that (4.45) is invertible near the origin, as

$$w = z - \left(\frac{h_{20}}{2}z^2 + h_{11}z\bar{z} + \frac{h_{02}}{2}\bar{z}^2\right) + \mathcal{O}\left(|z|^3\right).$$

With respect to the new complex variable w, Eq. (4.45) becomes

$$\tilde{w} \equiv \mu w + \frac{1}{2}\left[g_{20} + \left(\mu - \mu^2\right)h_{20}\right]w^2 = \left[g_{11} + \left(\mu - |\mu|^2\right)h_{11}\right]w\bar{w}$$

$$+ \frac{1}{2}\left[g_{02} + (\mu - \bar{\mu})^2 h_{02}\right]\bar{w}^2 + \mathcal{O}\left(|w|^3\right). \tag{4.46}$$

As

$$\mu^2(0) - \mu(0) = e^{i\theta_0}\left(e^{i\theta_0} - 1\right) \neq 0$$

$$|\mu(0)|^2 - \mu(0) = 1 - e^{i\theta_0} \neq 0$$

$$\bar{\mu}(0)^2 - \mu(0) = e^{i\theta_0}\left(e^{-i3\theta_0} - 1\right) \neq 0$$

for $|\beta|$ sufficiently small, we thus can let

$$h_{20} = \frac{g_{20}}{\mu^2 - \mu}, \quad h_{11} = \frac{g_{11}}{|\mu|^2 - \mu}, \quad h_{02} = \frac{g_{02}}{\bar{\mu}^2 - \mu},$$

Hence, all the quadratic terms in (4.46) disappear. The proof is complete. □

Remark 4.12 We here give some remarks as follow:

(i) Denote $\mu_0 = \mu(0)$. Then the conditions $e^{i\theta_0} \neq 1$ and $e^{3i\theta_0} \neq 1$ in Lemma 4.11 mean that $\mu_0 \neq 1$, $\mu_0^3 \neq 1$. The condition $\mu_0 \neq 1$ is automatically satisfied as $\mu_0 = e^{i\theta_0}$ and $0 < \theta_0 < \pi$.
(ii) From the transformation (4.45), we see that in the neighborhood of the origin, it is nearly an identity transformation.
(iii) The transformation (4.45) generally alters the coefficients of the cubic terms.

□

Lemma 4.13 *Assume that* $e^{2i\theta_0} \neq 1$, $e^{4i\theta_0} \neq 1$. *Consider the map*

$$z \mapsto \mu z + \left[\frac{g_{30}}{6}z^3 + \frac{g_{21}}{2}z^2\bar{z} + \frac{g_{12}}{2}z\bar{z}^2 + \frac{g_{03}}{6}\bar{z}^3\right] + \mathcal{O}\left(|z|^4\right), \tag{4.47}$$

where $\mu = \mu(\beta) = (1 + \beta)e^{i\theta(\beta)}$, $g_{ij} = g_{ij}(\beta)$. *For* $|\beta|$ *small, the following transformation*

$$z = w + \left[\frac{h_{30}}{6}w^3 + \frac{h_{21}}{2}w^2\bar{w} + \frac{h_{12}}{2}w\bar{w}^2 + \frac{h_{03}}{6}\bar{w}^3\right] \tag{4.48}$$

4.4 Hopf Bifurcation

converts (4.47) to
$$w \mapsto \mu w + \frac{g_{21}}{2} w^2 \bar{w} + \mathcal{O}|w|^4, \tag{4.49}$$

i.e., only one cubic term is retained in (4.49).

Proof The map is locally invertible near the origin:
$$w = z - \left[\frac{h_{30}}{6}z^3 + \frac{h_{21}}{2}z^2\bar{z} + \frac{h_{12}}{2}z\bar{z}^2 + \frac{h_{03}}{6}\bar{z}^3\right] + \mathcal{O}\left(|z|^4\right).$$

Substituting (4.48) into (4.47), we obtain
$$\tilde{w} \equiv \mu w + \left\{\frac{1}{6}\left[g_{30} + (\mu - \mu^3) h_{30}\right] w^3 + \frac{1}{2}\left[g_{21} + (\mu - \mu|\mu|^2) h_{21}\right] w^2\bar{w}\right.$$
$$\left. + \frac{1}{2}\left[g_{12} + (\mu - \bar{\mu}|\mu|^2) h_{12}\right] w\bar{w}^2 + \frac{1}{6}\left[g_{03} + (\mu - \bar{\mu}^3) h_{03}\right] \bar{w}^3\right\} + \mathcal{O}\left(|w|^4\right).$$

If we set
$$h_{30} = \frac{g_{30}}{\mu^3 - \mu}, \quad h_{12} = \frac{g_{12}}{\bar{\mu}|\mu|^2 - \mu}, \quad h_{03} = \frac{g_{03}}{\bar{\mu}^3 - \mu},$$
which is viable as the denominators are nonzero by assumption, as well as $h_{21} = 0$, then we obtain (4.49). □

The terms $\frac{g_{21}}{2} w^2 \bar{w}$ in (4.49) is called the resonance terms. Note that its coefficient, $g_{21}/2$, is the same as the corresponding term in (4.47).

Lemma 4.14 (Normal form of Hopf bifurcation) *Assume that $e^{ik\theta_0} \neq 1$ for $k = 1, 2, 3, 4$. Consider the map*
$$z \mapsto \mu z + \left(\frac{g_{20}}{2}z^2 + g_{11}z\bar{z} + \frac{g_{02}}{2}\bar{z}^2\right) + \left(\frac{g_{30}}{6}z^3 + \frac{g_{21}}{2}z^2\bar{z} + \frac{g_{12}}{2}z\bar{z}^2 + \frac{g_{03}}{2}\bar{z}^3\right)$$
$$+ \mathcal{O}\left(|z|^4\right), \tag{4.50}$$

where $\mu = \mu(\beta) = (1 + \beta)e^{i\theta(\beta)}$, $g_{ij} = g_{ij}(\beta)$, $\theta_0 = \theta(\beta)|_{\beta=0}$. Then there exists a locally invertible transformation near the origin:
$$z = w + \left(\frac{h_{20}}{2}w^2 + h_{11}w\bar{w} + \frac{h_{02}}{2}\bar{w}^2\right) + \left(\frac{h_{30}}{6}w^3 + \frac{h_{12}}{2}w\bar{w}^2 + \frac{h_{03}}{6}\bar{w}^3\right)$$

such that for $|\beta|$ sufficiently small, Eq.(4.50) is transformed to
$$w \mapsto \mu w + c_1 w^2 \bar{w} + \mathcal{O}\left(|w|^4\right),$$

where
$$c_1 = c_1(\beta) = \frac{g_{20}g_{11}(\bar{\mu} - 3 + 2\mu)}{2(\mu^2 - \mu)(\bar{\mu} - 1)} + \frac{|g_{11}|^2}{1 - \bar{\mu}} + \frac{|g_{02}|^2}{2(\mu^2 - \bar{\mu})} + \frac{g_{21}}{2}. \quad (4.51)$$

Proof This follows as a corollary to Lemmas 4.11 and 4.13. The value c_1 in (4.51) can be obtained by straightforward calculations. □

We can now summarize all of the preceding discussions in this section. As the map (4.34) is (4.35) in essence, from Lemma 4.9, we obtain the Hopf bifurcation theorem for general planar maps as follows.

Theorem 4.15 (Hopf-(Neimark-Sacker) Bifurcation Theorem) *For the 1-parameter family of planar maps*
$$x \longmapsto f(x, \alpha)$$
assume that

(i) *When $\alpha = 0$, the system has a fixed point $x_0 = \mathbf{0}$, and the Jacobian matrix has eigenvalues*
$$\mu_{1,2} = e^{\pm i\theta_0}, \quad 0 < \theta_0 < \pi$$

(ii) $r'(0) \neq 0$, *where $r(\alpha)$ is defined through (4.38)*.

(iii) $e^{ik\theta_0} \neq 1$, *for $k = 1, 2, 3, 4$*.

(iv) $a(0) \neq 1$, *where $a(0) = \text{Re}\left(e^{-i\theta_0} c_1(0)\right)$, with c_1 as given in (4.51)*. □

Then when α passes through 0, the system has a closed invariant curve C bifurcating from the fixed point $x_0 = \mathbf{0}$.

In applications, we often want to obtain the actual value of $a(0)$, which, from (4.51):
$$c_1(0) = \frac{g_{20}(0)g_{11}(0)(1 - 2\mu_0)}{2(\mu_0^2 - \mu_0)} + \frac{|g_{11}(0)|^2}{1 - \bar{\mu}_0} + \frac{|g_{02}(0)|^2}{2(\mu^2 - \bar{\mu}_0)} + \frac{g_{21}(0)}{2}$$

is
$$a(0) = \text{Re}\left(\frac{e^{-i\theta_0} g_{21}(0)}{2}\right) - \text{Re}\left[\frac{(1 - 2e^{i\theta_0})e^{-2i\theta_0}}{2(1 - e^{i\theta_0})} g_{20}(0)g_{11}(0)\right] - \frac{1}{2}|g_{11}(0)|^2$$
$$- \frac{1}{4}|g_{02}(0)|^2.$$

Exercise 4.16 Let μ be real number and consider a real-valued function
$$f_\mu : x \mapsto \mu + x - x^2.$$

4.4 Hopf Bifurcation

(a) Determine, if any the fixed points of f_μ.
(b) Prove that there is a *saddle-node bifurcation* for some μ.
(c) Apply the Saddle-Node Bifurcation theorem to prove (b).
(d) Sketch the bifurcation diagram of the fixed points versus μ. □

Exercise 4.17 Do likewise as in Exercise 4.16 for the following:

(a)
$$f_\mu : x \mapsto 1 - \mu x + x^2;$$

(b)
$$g_\mu : x \mapsto \mu + x - ch(x);$$

(c)
$$h_\mu : x \mapsto \mu + \frac{2}{3}x - \frac{x}{1+x};$$

(d)
$$k_\mu : x \mapsto \mu + 2x - \ln(1+x).$$
□

Exercise 4.18 Let μ be real number and set
$$f_\mu(x) = (\mu + 1)x - x^3.$$

(a) Determine, if any the fixed points of f_μ.
(b) Prove that there is a *Pitchfork bifurcation* for some μ.
(c) Apply the Pitchfork Bifurcation Theorem to prove (b).
(d) Sketch the bifurcation diagram of the fixed points versus μ. □

Exercise 4.19 Do likewise for the following:

(a)
$$f_\mu : x \mapsto (\mu + 1)x + x^2;$$

(b)
$$g_\mu : x \mapsto \mu x - \ln(x+1);$$

(c)
$$h_\mu : x \mapsto x(\mu - e^x);$$

(d)
$$k_\mu : x \mapsto 2x - \mu x(1-x).$$
□

Exercise 4.20 We revisit the quadratic map

$$f_\mu : x \mapsto \mu x(1-x).$$

(a) For $\mu \in [0, 4]$, determine, if any nonzero fixed point x of f_μ. Denote such x as $x = x(\mu)$.
(b) Analyze the stability of the fixed point $x(\mu)$, and determine a region of μ for which $x(\mu)$ is stable.
(c) Prove that f_μ undergoes a period-doubling bifurcation for some μ where $f'_\mu(x(\mu)) = -1$.
(d) Sketch the bifurcation diagram of the periodic points versus μ.
(e) Consider a function g_μ defined by

$$g_\mu(x) = \begin{cases} \frac{\mu}{2}x, & 0 \le x \le \frac{1}{2}, \\ \frac{\mu}{2}(1-x), & \frac{1}{2} < x \le 1. \end{cases}$$

  (i) Sketch the bifurcation diagram of the periodic points versus μ.
  (ii) Does g undergo a period-doubling bifurcation, even though g_μ and f_μ have the same monotonicity? Why? □

Exercise 4.21 Let μ be a real number. Consider a two-dimensional map f_μ given by

$$f_\mu : (x, y) \mapsto ((\mu+1)y - x^2 y - y^3, (\mu+1)x + x^3 + xy^3).$$

1. Determine all fixed points of f_μ.
2. Study the stability of the fixed points.
3. Prove that f_μ undergoes a Hopf-bifurcation at $(x, y) = 0$ as μ varies.
4. Using a computer, plot the phase portrait and determine whether the bifurcated limit circle is stable. □

Exercise 4.22 Let μ be a real number. Consider a two-dimensional map g_μ given by

$$g_\mu : (x, y) \mapsto ((\mu+1)x - y + xy^2, x + (\mu+1)y + y^3).$$

1. Determine all fixed points of g_μ.
2. Study the stability of the fixed points.
3. Show that a Hopf-bifucation occurs at the origin as μ varies.
4. Using a computer, plot the phase portrait and determine whether the bifurcated limit circle is stable. □

Exercise 4.23 Define a 2D map as follow:

$$f_\lambda(x, y) = (\lambda x(1 - y), x).$$

1. Determine all fixed points of f_λ.
2. Let $\lambda > 1$. Prove that f_λ has only one fixed point whose coordinates are positive. Denote X_0 as that fixed point.
3. Determine the Jacobian matrix of f at X_0 and its eigenvalues.
4. Prove that when λ passes through $\lambda_0 = 2$, f_λ undergoes a Hopf-bifurcation at X_0. □

Notes for Chapter 4

The word bifurcation or Abzweigung (German) seems to have been first introduced by the celebrated German mathematician Carl Jacobi (1804–1851) [4] in 1834 in his study of the bifurcation of the McLaurin spheroidal figures of equilibrium of self-gravitating rotating bodies (Abraham and Shaw [5, p. 19], Iooss and Joseph [6, p. 11]). Poincare introduced the French word bifurcation in [7] in 1885. The bifurcation theorems studied in this chapter are of the local character, namely, local bifurcations, which analyze changes in the local stability properties of equilibrium points, periodic points or orbits or other invariant sets as system parameters cross through certain critical thresholds. The analysis of change of stability and bifurcation is almost always technical. No more so than the case of maps when the governing system consists of ordinary differential equations or even partial differential equations.

A partial list of reference sources for the study of bifurcations of maps, ordinary and partial differential equations are Hale and Kocak [8], Iooss and Joseph [6], Guckenheimer and Holmes [9], Robinson [2], and Wiggins [10, 11].

References

1. R.L. Devaney, *An Introduction to Chaotic Dynamical Systems*, 2nd ed., Addison-Wesley, New York, 1989. Cited on page(s) 19, 36, 75, 86, 209, 211.
2. C. Robinson, *Dynamical Systems, Stability, Symbolic Dynamics and Chaos*, CRC Press, Boca Raton, FL, 1995, pp. 67–69.
3. Y.A. Kuznetsov, *Elements of Applied Bifurcation Theory* (Applied Mathematical Sciences), Springer, New York, 2010.
4. C.G.J. Jacobi, Über die Figur des Gleichgewichts, *Poggendorff Annalen der Physik und Chemie* **32** (229), 1834. https://doi.org/10.1002/andp.18341090808
5. R.H. Abraham and C.D. Shaw, *Dynamics–The Geometry of Behavior, Part 4: Bifurcation Behavior*, Aerial Press, Inc., Santa Cruz, CA, 1988.
6. G. Iooss and D.D. Joseph, *Elementary Stability and Bifurcation Theory*, Springer Verlag, New York-Heidelberg-Berlin, 1980.

7. H. Poincaré, Sur l'équilibre d'une masse fluide animée d'un mouvement de rotation, *Acta Math.* **7** (1885), 259–380. https://doi.org/10.1007/BF02402204
8. J.K. Hale and H. Kocak, *Dynamics and Bifurcations*, Springer Verlag, New York-Heidelberg-Berlin, 1991.
9. C.W. Ho and C. Morris, A graph-theoretic proof of Sharkovsky's theorem on the periodic points of continuous functions, *Pacific J. Math.* **96** (1981), 361–370.
10. S. Wiggins, *Global Bifurcations and Chaos: Analytical Methods*, Springer Verlag, New York-Heidelberg-Berlin, 1988.
11. S. Wiggins, *Introduction to Applied Nonlinear Dynamical Systems and Chaos*, 2nd ed., Springer, New York, 2003.

5 Homoclinicity, Lyapunov Exponents

5.1 Homoclinic Orbits

There is a very important *geometric* concept, called *homoclinic orbits*, that leads to chaos.

Let p be a fixed point of a C^1-map f:

$$f(p) = p.$$

Assume that p is repelling so that $|f'(p)| > 1$. Since p is repelling, there is a neighborhood $N(p)$ of p such that

$$|f(x) - f(p)| = |f(x) - p| > |x - p|, \quad \forall x \in N(p). \tag{5.1}$$

We denote $W_{\text{loc}}^u(p)$ the largest open neighborhood of p such that (5.1) is satisfied. $W_{\text{loc}}^u(p)$ is called the *local unstable set of* p.

Definition 5.1 Let p be a repelling fixed point of a continuous map f, and let $W_{\text{loc}}^u(p)$ be the local unstable set of p. Let $x_0 \in W_{\text{loc}}^u(p)$. We say that x_0 is *homoclinic* to p if there exists a positive integer n such that

$$f^n(x_0) = p.$$

We say that x_0 is *heteroclinic* to p if there exists another different periodic point q such that

$$f^m(x_0) = q.$$

□

See some illustrations in Fig. 5.1a and b.

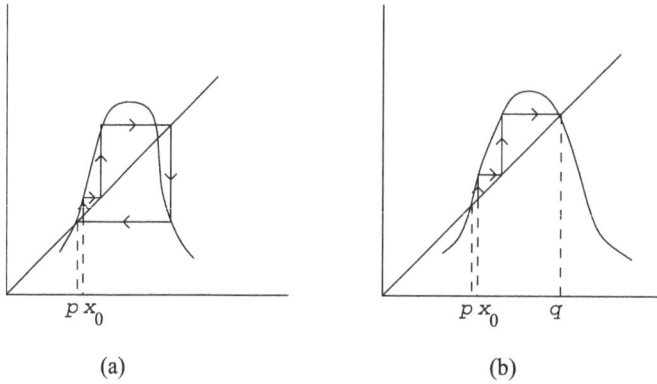

(a)   (b)

Fig. 5.1 a p is a fixed point and $x_0 \in W^u_{loc}(p)$. But $f^4(x_0) = p$. So x_0 is homoclinic to p. **b** p and q are two different fixed points and $x_0 \in W^u_{loc}(p)$. But $f^3(x_0) = q$. So x_0 is heteroclinic to p

Fig. 5.2 The orbit of x_0 is a degenerate homoclinic orbit because $f'(x_2) = 0$

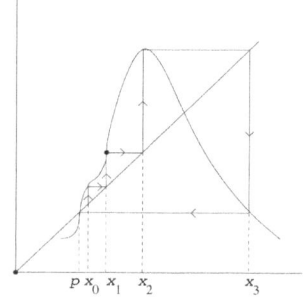

Definition 5.2 A homoclinic orbit is said to be *nondegenerate* if $f'(x) \neq 0$ for all x on the orbit. Otherwise it is said to be *degenerate* (see Fig. 5.2). □

A nondegenerate homoclinic orbit will lead to chaos, as the following theorem shows.

Theorem 5.3 *Let I be a bounded closed interval and $f : I \to I$ is C^1. Assume that p is a repelling fixed point of f, and p has a nondegenerate homoclinic orbit. Then*

$$V_I(f^n) \geq K e^{\alpha n} \to \infty \quad \text{as } n \to \infty, \tag{5.2}$$

for some K and $\alpha > 0$.

Proof Let $x_0 \in W^u_{loc}(p)$ such that $f^n(x_0) = p$, for some positive integer n. Since x_0 is nondegenerate,

$$[f^n(x)]'_{x=x_0} = f'(f^{n-1}(x_0)) \cdot f'(f^{n-2}(x_0)) \cdot \;\cdots\; \cdot f'(f(x_0)) \cdot f'(x_0) \neq 0. \tag{5.3}$$

5.1 Homoclinic Orbits

But assumption, $|f'(p)| > 1$. We can choose an open set $W \subseteq W^u_{\text{loc}}(p)$ such that

$$|f'(x)| \geq d > 1, \quad \forall x \in W, \quad \text{and} \quad p \in W. \tag{5.4}$$

Now, choose an open interval $V \ni x_0$ such that $p \notin V$. Then by (5.3), if we choose V sufficiently small, we have

$$(f^n)'(x) \neq 0 \quad \forall x \in V. \tag{5.5}$$

This implies that $f, f^2, \ldots, f^n$ are 1–1 on V and, therefore,

$$V, f(V), f^2(V), \ldots, f^n(V)$$

are all open intervals, with $x_0 \in V$ and $p \in f^n(V)$. Furthermore, by choosing V sufficiently small, we may assume that

$$f^j(V) \cap f^k(V) = \emptyset \quad \text{for} \quad j \neq k, \quad j, k = 0, 1, 2 \ldots, n. \tag{5.6}$$

If V is sufficiently small, then

$$f^n(V) \subseteq W. \tag{5.7}$$

By the fact that the orbit of x_0 is nondegenerate, we have

$$|(f^j)'(x)| \geq \varepsilon \quad \text{for all} \quad j = 1, 2, \ldots, n, \quad \forall x \in V. \tag{5.8}$$

From (5.4), (5.6) and (5.7), we now have

$$|(f^{n+k})'(x)| = |f'(f^{n+k-1}(x))| \cdot |f'(f^{n+k-2}(x))| \cdots |f'(f(x))||f'(x)|$$
$$\geq d^k \varepsilon^n \tag{5.9}$$

$\forall x \in V$, provided that $f^{n+j}(x) \in W$ for $j = 1, 2, \ldots, k$.

Since n in (5.9) is fixed, by choosing k sufficiently large, we have some k such that

$$d^k \varepsilon^n \geq M \tag{5.10}$$

for any given $M > 1$. This implies that

$$f^{n+k}(V) \supseteq V$$

if k is chosen sufficiently large.

We now choose an open interval $V_1 \ni p$, $V_1 \subseteq W$, $V_1 \cap V = \emptyset$ and, by (5.10), we can choose k sufficiently large such that

$$f^{n+k}(V_1) \supseteq V.$$

Then by choosing V_1 sufficiently small, we have

$$f^{n+k}(V) \supseteq V_1 \cup V.$$

Therefore, we obtain

$$\overline{V}_1 \xrightarrow{f^{n+k}} \overline{V} \xrightarrow{f^{n+k}} \overline{V}_1 \cup \overline{V}. \tag{5.11}$$

This gives the growth of total variations

$$V_{\overline{V}_1}((f^{n+k})^j) \geq K' e^{\alpha' j} \to \infty \text{ as } j \to \infty, \text{ for some } K', \alpha' > 0,$$

where $V_{\overline{V}_1}$ denotes the total variation over the set $\overline{V}_1$. Using the above, we can further show that

$$V_{\overline{V}_1}((f^{n+k})^j \circ f^\ell) \geq K e^{\alpha j}, \text{ for } \ell = 1, 2, \ldots, n+k-1,$$

for some $K, \alpha > 0$.

Therefore, Eq. (5.1) has been proven. □

Usually, the covering-interval sequence is much "stronger" than what (5.11) indicates (see Fig. 5.3).

Theorem 5.3 tells us that chaos occur when there is a nondegenerate homoclinic orbit. What happens if, instead, a homoclinic orbit is *degenerate*? This happens, e.g., for the quadratic map $f_\mu(x) = \mu x(1-x)$ when $\mu = 4$, and $x = 1/2$ lies on a degenerate homoclinic orbit.

In this case, the map has rather complex bifurcation behavior. For example, for the quadratic map f_μ mentioned above, near μ (actually, for $\mu > 4$) there are μ-values where there are infinitely many distinct homoclinic orbits. The maps f_μ also have sadddle-node or period-doubling bifurcations [1], which means that these bifurcations are accumulation points of simple bifurcations. This phenomenon is called *homoclinic bifurcation*; see also Afraimovich and Hsu [2, pp. 195–208].

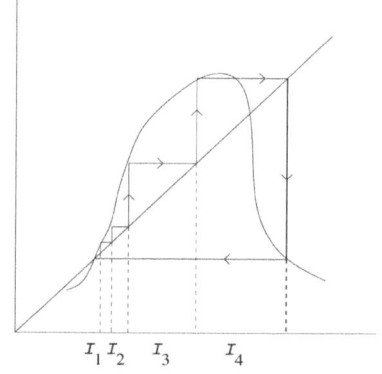

Fig. 5.3 There is a homoclinic orbit showing the interval covering relation
$I_1 \longrightarrow I_2 \longrightarrow I_3 \longrightarrow I_4 \longrightarrow I_1 \cup I_2 \cup I_3$

5.2 Lyapunov Exponents

Let $f: I \to I$ be C^1 everywhere except at finitely many points on I. The *Lyapunov exponent* of f at $x_0 \in \mathbb{R}$ is defined by

$$\lambda(x_0) = \limsup_{n \to \infty} \frac{1}{n} [\ln |(f^n)'(x_0)|]$$

if it exists. Because

$$(f^n)'(x_0) = f'(x_{n-1}) f'(x_{n-2}) \cdots f'(x_1) f'(x_0), \quad \text{where} \quad x_j = f^j(x_0),$$

we have

$$\lambda(x_0) = \limsup_{n \to \infty} \frac{1}{n} \left[\sum_{j=0}^{n-1} \ln |f'(f^j(x_0))| \right].$$

Notation. Given x_0, denote

$$\mathcal{O}^+(x_0) = \text{the forward orbit of } x_0$$
$$= \{f^j(x_0) \mid j = 0, 1, 2, \ldots\};$$
$$\mathcal{O}^-(x_0) = \text{a backward orbit of } x_0$$
$$= \{f^{-j}(x_0) \mid j = 0, 1, 2, \ldots\}.$$

□

Example 5.4 The roof (or tent) function; see Fig. 5.4.

At $x = 1/2$, $T(x)$ is not differentiable. So remove the set $\mathcal{O}^-(1/2)$. Choose any $y_0 \notin \mathcal{O}^-(1/2)$. Then $T(x)$ is differentiable at any $T^j(y_0)$. We have the Lyapunoff exponent

Fig. 5.4 The roof function defined by $T(x) = \{2x, 0 \le x \le 1/2, 2(1-x), 1/2 < x \le 1$

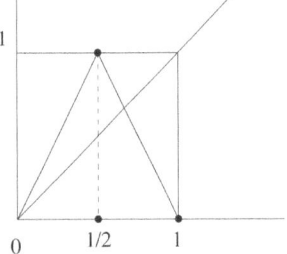

$$\lambda(y_0) = \limsup_{n\to\infty} \frac{1}{n} \left[\sum_{j=0}^{n-1} \ln |T'(T^j(y_0))| \right]$$

$$= \limsup_{n\to\infty} \frac{1}{n} \underbrace{[\ln |2| + \ln |2| + \cdots + \ln |2|]}_{n \text{ terms}} = \ln 2.$$

This suggests that when chaos occurs, $\lambda(y_0) > 0$. □

Example 5.5 The quadratic map

$$f_\mu(x) = \mu x(1-x), \quad 3 < \mu < 3 + \delta_0,$$

where $3 + \delta_0$ is the μ-value where the second period-doubling bifurcation happens (see Fig. 5.5).

At $x = 1/2$, $f'_\mu\left(\frac{1}{2}\right) = 0$ and $\ln \left| f'_\mu\left(\frac{1}{2}\right) \right| = -\infty$. So we need to exclude $\mathcal{O}^-(1/2)$. If $y_0 \notin \mathcal{O}^-(1/2)$, then $f'(f_\mu^j(y_0)) \neq 0$ for any $j = 0, 1, 2, \ldots$. Note that if $y_0 \notin \{0\} \cup \mathcal{O}^-(1)$, then y_0 will be attracted to the period-2 orbit $\{z_0, f_\mu(z_0)\}$, which is globally attracting, i.e., either

$$\lim_{n\to\infty} f_\mu^{2n}(y_0) = z_0 \ \left(\text{and } \lim_{n\to\infty} f_\mu^{2n+1}(y_0) = f(z_0)\right) \quad (5.12)$$

or

$$\lim_{n\to\infty} f_\mu^{2n+1}(y_0) = z_0 \ \left(\text{and } \lim_{n\to\infty} f_\mu^{2n}(y_0) = f(z_0)\right). \quad (5.13)$$

Then we have (check!)

$$\lambda(y_0) = \max \left\{ \limsup_{n\to\infty} \frac{1}{2n+1} \{\ln |(f_\mu^{2n+1})'(y_0)|\}, \right.$$
$$\left. \limsup_{n\to\infty} \frac{1}{2n} \{\ln |(f_\mu^{2n})'(y_0)|\} \right\}. \quad (5.14)$$

Let $g_\mu = f_\mu^2$. Then

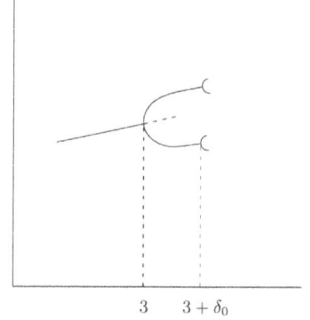

Fig. 5.5 Second period-doubling bifurcation of the period-two points

5.2 Lyapunov Exponents

$$\limsup_{n\to\infty} \frac{1}{2n}\{\ln |(f_\mu^{2n})'(y_0)|\}$$
$$= \frac{1}{2}\limsup_{n\to\infty} \frac{1}{n}\{\ln |(g_\mu^n)'(y_0)|\}$$
$$= \frac{1}{2}\limsup_{n\to\infty} \frac{1}{n}\{\ln |(g_\mu^n)'(z_0)|\} \quad \text{(if (5.12) holds)}$$
$$< 0,$$

because we know that z_0 is an attracting fixed point of g_μ^2 and, thus, $|(g_\mu^2)'(z_0)| < 1$. Similarly,

$$\limsup_{n\to\infty} \frac{1}{2n+1}\{\ln |(f_\mu^{2n+1})'(y_0)|\} = \limsup_{n\to\infty} \frac{1}{2n}\{\ln |(f_\mu^{2n})'(y_0)|\} < 0.$$

□

The argument in Example 5.5 can easily be generalized to the following result, which is left as an exercise.

Exercise 5.6 Prove the following:

Let $f: I \to I$ be C^1 except at finitely many points. Assume that f has a globally attracting period-n orbit $\mathcal{O} = \{y_0, y_1, \ldots, y_{n-1}\}$ such that $f'(y_j) \neq 0$ for $j = 0, 1, \ldots, n-1$. Then there are infinitely many x_0 such that

$$\lambda(x_0) < 0.$$

□

We thus see that a *negative Lyapunov exponent* is a sign that there is *no chaos*.

Definition 5.7 Let X and Y be two topological spaces. A map $h: X \to Y$ is said to be a *homeomorphism* if

(i) h is 1–1 and onto;
(ii) h is continuous; and
(iii) h^{-1} is also continuous. □

Definition 5.8 Let X and Y be two topological spaces and let $f: X \to X$ and $g: Y \to Y$ be continuous. We say that f and g are *topologically conjugate* if there exists a homeomorphism $h: X \to Y$ such that

$$h \circ f(x) = g \circ h(x) \quad \forall x \in X.$$

$$\begin{array}{ccc} X & \xrightarrow{f} & X \\ h \downarrow & & \downarrow h \\ Y & \longrightarrow & Y \end{array}$$

We call h a *topological conjugacy*. □

Example 5.9 The quadratic map $f_4(x) = 4x(1-x)$ is topologically conjugate to the roof function $T(x)$ in Example 5.4. Let us verify this by defining

$$h(y) = \sin^2\left(\frac{\pi y}{2}\right), \quad y \in I = [0, 1].$$

Then

$$h \circ T(y) = \begin{cases} \sin^2\left[\frac{\pi}{2}(2y)\right], & \text{if } 0 \le y \le 1/2; \\ \sin^2\left[\frac{\pi}{2} \cdot 2(1-y)\right], & \text{if } 1/2 < y \le 1, \end{cases}$$

$$= \sin^2(\pi y), \quad \forall y \in [0, 1].$$

On the other hand,

$$f_4 \circ h(y) = 4\sin^2\left(\frac{\pi y}{2}\right)\left[1 - \sin^2\left(\frac{\pi y}{2}\right)\right]$$

$$= 4\sin^2\left(\frac{\pi y}{2}\right)\cos^2\left(\frac{\pi y}{2}\right)$$

$$= \sin^2(\pi y), \quad \forall y \in [0, 1].$$

Therefore

$$h \circ T = f_4 \circ h$$

and

$$f_4 = h \circ T \circ h^{-1}$$
$$f_4^n = (h \circ T \circ h^{-1}) \circ (h \circ T \circ h^{-1}) \circ \cdots \circ (h \circ T \circ h^{-1})$$
$$= h \circ T^n \circ h^{-1}. \tag{5.15}$$

□

Exercise 5.10 For $f_4(x) = 4x(1-x)$, use computer to calculate the Lyapunoff exponent

$$\lambda(x_0), \quad \text{for} \quad x_0 = \frac{1}{n}, \quad n = 3, 4, 5, \ldots, 10,$$

by approximating

$$\lambda(x_0) \approx \frac{1}{1000} \ln |(f^{1000})'(x_0)|. \quad \square$$

5.2 Lyapunov Exponents

Using the topological conjugacy that we established in Example 5.9, we can now compute the Lyapunoff exponent of the quadratic map $f_\mu(x) = \mu x(1-x)$ when $\mu = 4$, as follows.

Example 5.11 For the quadratic map $f_4(x) = 4x(1-x)$, assume that $x_0 \notin \mathcal{O}^-(1/2) \cup \mathcal{O}^{-1}(0) = \mathcal{O}^{-1}(0)$. We claim that for such x_0,

$$\lambda(x_0) = \text{the Lyapunoff exponent of } x_0$$
$$= \ln 2 > 0. \tag{5.16}$$

First, note that $h(x) = \sin^2\left(\frac{\pi}{2}x\right)$ is (C^∞) differentiable, so

$$|h'(x)| \leq K \quad \forall x \in [0, 1] \tag{5.17}$$

for some positive constant $K > 0$. Also, note that if x is bounded away from 0 and 1, i.e.,

$$x \in [\delta, 1-\delta] \text{ for some } \delta > 0,$$

then $\sin\left(\frac{\pi x}{2}\right)$ and $\cos\left(\frac{\pi x}{2}\right)$ are bounded away from 0, so

$$|h'(x)| = \left|2 \cdot \frac{\pi}{2} \cos\left(\frac{\pi x}{2}\right) \sin\left(\frac{\pi x}{2}\right)\right| \geq K_\delta, \quad \forall x \in [\delta, 1-\delta], \tag{5.18}$$

for some constant $K_\delta > 0$.

Now, for $x \notin \mathcal{O}^-(0)$, we have

$$\lambda(x_0) = \limsup_{n \to \infty} \frac{1}{n}[\ln |(f_4^n)'(x_0)|]$$

$$= \limsup_{n \to \infty} \frac{1}{n}[\ln |[h \circ T^n \circ h^{-1}]'(x_0)|]$$

$$= \limsup_{n \to \infty} \frac{1}{n}[\ln |h'(y_n)| + \ln |(T^n)'(y_0)| + \ln |(h^{-1})'(y_0)|]$$

(where $y_0 = h^{-1}(x_0)$, and $y_n = T^n(y_0)$)

$$\leq \lim_{n \to \infty} \frac{1}{n}[\ln K + n \ln 2 + \ln |(h^{-1})'(y_0)|]$$

(by (5.17) and Example 5.4)

$$= \ln 2. \tag{5.19}$$

On the other hand, if we choose a subsequence $\{n_j \mid j = 1, 2, \ldots\} \subseteq \{0, 1, 2, \ldots\}$ such that

$$y_{n_j} = T^{n_j}(y_0) \in [\delta, 1-\delta],$$

then for $x_0 \notin \mathcal{O}^{-1}(0)$, $x_0 = h(y_0)$, we have

$$\lambda(x_0) = \limsup_{n\to\infty} \frac{1}{n}[\ln|(h\circ T^n \circ h^{-1})'(x_0)|]$$

$$\geq \limsup_{j\to\infty} \frac{1}{n_j}[\ln|(h\circ T^{n_j} \circ h^{-1})'(x_0)|]$$

$$= \lim_{j\to\infty} \frac{1}{n_j}[\ln|h'(y_n)| + n_j \ln 2 + \ln|(h^{-1})'(x_0)|]$$

$$\geq \lim_{j\to\infty} \frac{1}{n_j}[\ln K_\delta + n_j \ln 2 + \ln|(h^{-1})'(x_0)|] \quad \text{(by (5.18))}$$

$$= \ln 2. \tag{5.20}$$

From (5.19) and (5.20), we conclude (5.16). □

Next, we study a different kind of chaos involving "fractal" structure. We begin by introducing the concept of a *Cantor set*.

Example 5.12 (The Cantor no-middle-third set) Consider the unit interval $I = [0, 1]$. We first remove the middle-third section of the interval I, i.e.,

$$I - \left(\frac{1}{3}, \frac{2}{3}\right) = [0, 1/3] \cup [2/3, 1].$$

Then for the two closed intervals $[0, 1/3]$ and $[2/3, 1]$, we again remove their respective middle-third sections:

$$[0, 1/3] - (1/9, 2/9) = [0, 1/9] \cup [2/9, 1/3], [2/3, 1] - (2/3 + 1/3^2, 2/3 + 2/3^2)$$
$$= [2/3, 1] - (7/9, 8/9) = [2/3, 7/9] \cup [8/9, 1].$$

This process is continued indefinitely, as the following diagram shows in Fig. 5.6.

The final outcome of this process is called the *Cantor ternary set*. It has a "*fractal*" structure for if we put it under a microscope, we see that the set looks the same no matter what magnification scale we use, i.e., it is self-similar.

Denote this set by $\mathcal{C}$. Elements in $\mathcal{C}$ are best described from the way $\mathcal{C}$ is created using *ternary representation* of numbers:

Fig. 5.6 The process of removing the middle-third open segments of subintervals

5.2 Lyapunov Exponents

$$\mathcal{C} = \left\{ a \in [0,1] \mid a = 0.a_1 a_2 \ldots a_n \ldots = \sum_{n=1}^{\infty} \frac{a_n}{3^n}, a_j \in \{0,1,2\} \right.$$

$$\left. \text{satisfying (5.21)} \right\},$$

where either

$$\left. \begin{array}{l} \text{(i)} \ \sum_{j=1}^{\infty} |a_j| = \infty, \quad a_j \neq 1 \ \forall j, \\ \text{or} \\ \text{(ii)} \ \sum_{j=1}^{\infty} |a_j| < \infty, \ \text{i.e.,} \ a = 0.a_1 a_2 \ldots a_n 000 \ldots, \quad a_n \in \{1,2\}; \\ a_j \neq 1 \ \forall j: \ 1 \leq j \leq n-1. \end{array} \right\} \quad (5.21)$$

$\square$

Definition 5.13 Let $S \subseteq \mathbb{R}^N$. A point $x_0 \in \mathbb{R}^N$ is said to be an *accumulation point* of S if every open set containing x_0 also contains at least one point of $S \setminus \{x_0\}$. $\square$

Definition 5.14 A set S in $\mathbb{R}$ is said to be *totally disconnected* if S does not contain any interval. $\square$

Definition 5.15 A set $S \subseteq \mathbb{R}^n$ is said to be a *perfect set* if every point is an accumulation point of S. $\square$

Theorem 5.16 *The Cantor ternary set $\mathcal{C}$ is closed, totally disconnected and perfect.*

Proof The set $\mathcal{C}$ is closed because

$$\mathcal{C} = I - \bigcup_{j=1}^{\infty} I_j \quad (5.22)$$

where I_j's are the middle-third open intervals that are removed in the construction process of $\mathcal{C}$. Since $\bigcup_{j=1}^{\infty} I_j$ is a union of open intervals and any union of open intervals is open, $\mathcal{C}$ as given by (5.22) is closed.

Assume that $\mathcal{C}$ is not totally disconnected. Then $\mathcal{C}$ contains some intervals. By the construction process of $\mathcal{C}$, this interval will have a subinterval whose middle-third will be removed. So this middle-third interval does not belong to $\mathcal{C}$, a contradiction.

To show that $\mathcal{C}$ is perfect, we need to show that every point $a \in \mathcal{C}$ is an accumulation point. We use (5.21). If

(i) $a = 0.a_1a_2\ldots a_n\ldots$ such that $\sum_{j=1}^{\infty}|a_j| = \infty$ and $a_j \neq 1 \,\forall j$, then a is an accumulation point of

$$\{y \mid y = 0.a_1a_2\ldots a_m, \text{ for } m = 1, 2, \ldots, k, k+1, \ldots\} \subseteq \mathcal{C};$$

(ii) $a = 0.a_1a_2\ldots a_n$ where $a_n \in \{1, 2\}$ and $a_j \neq 1$ for $j = 1, 2, \ldots, n-1$, then a is an accumulation point of

$$\{y \mid y = 0.a_1a_2\ldots a_{n-1}b_nb_{n+1}\ldots b_{n+k}, \text{ for } k = 0, 1, 2, \ldots,$$
$$b_j = 2 \text{ for } j = n, n+1, \ldots, n+k\}.$$

□

Exercise 5.17 Let μ be a real number. Define a real-valued function $f : x \mapsto \mu x$. Calculate the Lyapunoff exponent of f at every point $x \in \mathbb{R}$. □

Exercise 5.18 Recall the tent map given by

$$g_\mu(x) = \begin{cases} \mu x, & 0 \leq x \leq \frac{1}{2}, \\ \mu(1-x), & \frac{1}{2} < x \leq 1. \end{cases}$$

(a) Prove that $[0, 1]$ is a stable interval of g_μ if and only if $0 \leq \mu \leq 2$.
(b) Let $1 < \mu \leq 2$. Prove that g_μ has at least one *homoclinic orbit*.
(c) Calculate the Lyapunoff exponent for g_μ at every point $x \in [0, 1]$.
(d) Explain why there is no period-doubling bifurcation as the parameter μ varies. □

Notes for Chapter 5

The term "homoclinic" or "homoclinicity" cannot not even be found in most major English dictionaries. Basically, here it means "of the same orbit", while "heteroclinic" is its antonym, meaning "of the different orbit".

Homoclinicity is a global concept. In this chapter, we have only discussed it for interval maps. A point x_0 in a neighborhood of a repelling fixed point p will move away from p after a few iterations at first. But some actually return exactly to p. This is *totally astonishing*. We did not discuss homoclinic orbits of *higher-dimensional maps* in this book. They may have highly complex behaviors. We refer the readers to Robinson [3] and Wiggins [4], for example.

The definition of Lyapunoff exponents for smooth interval maps can be traced back to Lyapunoff's dissertation [5] in 1907. We did not discuss that for multidimensional diffeomorphisms, but they can easily be found in many dynamical systems books. Lyapunoff exponents provide an easily computable quantity, using commercial or open source algorithms and softwares (see [6–8], Maple, Matlab, etc.). For example, the Maple software computer codes written by Rucklidge [9] for computing the Lyapunoff exponents of the quadratic map consists of only 21 lines. Many papers in the engineering literature, where rigorous proofs are not required, claim that the systems under study are chaotic once the authors are able to compute that the Lyapunoff exponents are >1.

References

1. R.L. Devaney, *An Introduction to Chaotic Dynamical Systems*, 2nd ed., Addison-Wesley, New York, 1989. Cited on page(s) 19, 36, 75, 86, 209, 211.
2. V.S. Afraimovich and S.-B. Hsu, *Lectures on chaotic dynamical systems* (AMS/IP Studies in Advanced Mathematics), *American Mathematical Society, Providence, R.I.*, 2002.
3. C. Robinson, *Dynamical Systems, Stability, Symbolic Dynamics and Chaos*, CRC Press, Boca Raton, FL, 1995, pp. 67–69.
4. S. Wiggins, *Introduction to Applied Nonlinear Dynamical Systems and Chaos*, 2nd ed., Springer, New York, 2003.
5. A. Lyapunov, Problemes general de la stabilitede mouvement, *Ann. Fac. Sci. Univ. Toulouse* **9** (1907), 203–475.
6. J.-P. Eckmann, S.O. Kamphorst, D. Ruelle, and S. Ciliberto, Lyapunov exponent from time series, *Phys. Rev. A* **34** (1986), 4971–4979 http://mpej.unige.ch/~eckmann/ps_files/eckmannkamphorstruelle.pdf/. https://doi.org/10.1103/PhysRevA.34.4971.
7. R. Hegger, H. Kantz and T. Schreiber, *Nonlinear Time Series Analysis*, TISEAN 3.0.1 (2007), http://en.wikipedia.org/wiki/Tisean/.
8. A. Wolf, J.B. Swift, H.L. Swinney, and J.A. Vastanoa, Determining Lyapunov exponents from a time series, *Phys. D* **16** (1985), 285–317. https://doi.org/10.1016/0167-2789(85)90011-9.
9. A.M. Rucklidge, http://www.maths.leeds.ac.uk/~alastair/MATH3395/examples_2.pdf/.

Symbolic Dynamics, Conjugacy, and Shift Invariant Sets

6.1 The Itinerary of an Orbit

Return to the quadratic map $f_\mu(x) = \mu x(1-x)$ for $\mu > 4$. When $\mu > 4$, $I = [0, 1]$ is no longer invariant under f_μ. Part of I will be mapped outside I, as can be seen in Fig. 6.1. Also, from Fig. 6.1, we see that there are two subintervals I_0 and I_1 of I, such that

$$f_\mu(I_0) = I, \quad f_\mu(I_1) = I. \tag{6.1}$$

Apply f_μ^2 to I_0 and I_1 respectively, we see that I_0 (resp., I_1) will have a subinterval mapped out of I. So we remove that (open) subinterval from I_0 (resp., I_1). This process can be continued indefinitely. We see that it is analogous to the process of constructing the Cantor Ternary Set C. The outcome of this process, which is obtained by removing all the points which are eventually mapped out of I, is denoted as

$$\Lambda = \{x \in I \mid f_\mu^n(x) \in I \ \forall n = 0, 1, 2, \ldots\}. \tag{6.2}$$

This Λ is invariant under f_μ:

$$f_\mu : \Lambda \longrightarrow \Lambda.$$

Also, if $x \in I \setminus \Lambda$, then $\lim_{n \to \infty} f_\mu^n(x) = -\infty$.

© The Author(s), under exclusive license to Springer Nature Switzerland AG 2025
L. Li et al., *Chaotic Maps, Fractals, and Rapid Fluctuations*, Synthesis Lectures
on Mathematics & Statistics, https://doi.org/10.1007/978-3-031-84828-5_6

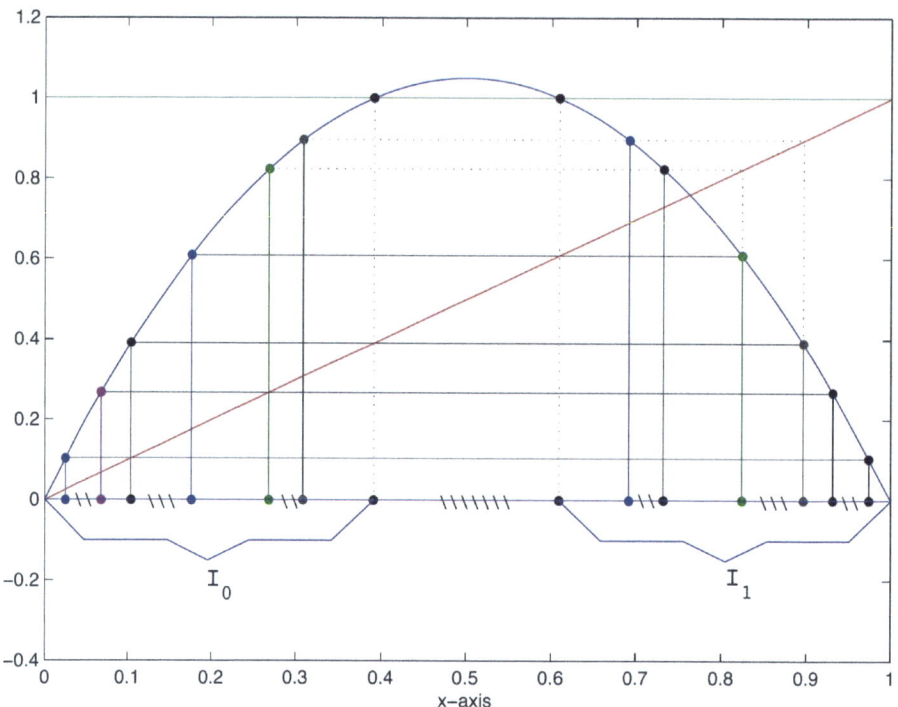

Fig. 6.1 The graph of $f_\mu(x) = \mu x(1-x)$ has a portion that exceeds the height of 1. Here, we show that the interval $I = [0, 1]$ has $1 + 2 + 2^2 = 7$ open segments removed

For each $x \in \Lambda$, define its *itinerary* as follows:

$$S(x) = \text{itinerary of } x$$
$$= (s_0 s_1 s_2 \cdots s_n \cdots), \quad s_j \in \{0, 1\} \; \forall j$$
$$s_j = \begin{cases} 0 & f_\mu^j(x) \in I_0; \\ & \text{if} \\ 1 & f_\mu^j(x) \in I_1. \end{cases} \quad (6.3)$$

We collect all such binary strings (6.3) together and define

$$\sum\nolimits_2 = \text{the space of all binary symbols}$$
$$= \{s \,|\, s = (s_0 s_1 \cdots s_n \cdots), \text{ where } s_j \in \{0, 1\}, j = 0, 1, 2, \ldots\}.$$

For any $s, t \in \sum_2$, where

$$s = (s_0 s_1 s_2 \cdots s_n \cdots), \quad t = (t_0 t_1 t_2 \cdots t_n \cdots),$$

define

$$d(s,t) \equiv \text{the distance between } s \text{ and } t$$

$$= \sum_{j=0}^{\infty} \frac{|s_j - t_j|}{2^j}.$$

Then for any $s, t, u \in \Sigma_2$, we can easily verify that the following triangle inequality is satisfied:

$$d(s,t) \leq d(s,u) + d(u,t). \tag{6.4}$$

We say the (Σ_2, d) forms a metric space, and d is a metric on Σ_2.
On Σ_2, we now define a "left-shift" map

$$\sigma : \Sigma_2 \longrightarrow \Sigma_2,$$
$$\sigma(s) = \sigma(s_0 s_1 \cdots s_n \cdots) = (s_1 s_2 \cdots s_n \cdots).$$

This map σ looks simple and innocent. Yet it is totally surprising to learn that σ is actually chaotic!

6.2   Properties of the Shift Map σ

We prove some basic properties of σ in this section.

Lemma 6.1 *Let $s, \tilde{s} \in \Sigma_2$, where*

$$s = (s_0 s_1 \cdots s_n s_{n+1} s_{n+2} \cdots), \qquad \tilde{s} = (s_0 s_1 \cdots s_n \tilde{s}_{n+1} \tilde{s}_{n+2} \cdots),$$

i.e., s and $\tilde{s}$ are identical up to the first $n+1$ bits. Then

$$d(s, \tilde{s}) \leq 2^{-n}.$$

Conversely, if $s, \tilde{s} \in \Sigma_2$ such that $d(s, \tilde{s}) < 2^{-n}$, then s and $\tilde{s}$ must agree up to at least the first $n+1$ bits.

Proof We have

$$d(s, \tilde{s}) = \sum_{j=0}^{n} \frac{|s_j - s_j|}{2^j} + \sum_{j=n+1}^{\infty} \frac{|s_j - \tilde{s}_j|}{2^j}$$

$$\leq \sum_{j=n+1}^{\infty} \frac{1}{2^j} = 2^{-n}.$$

The proof of the converse follows also in a similar way. □

Theorem 6.2 *The map* $\sigma : \Sigma_2 \longrightarrow \Sigma_2$ *is continuous with respect to the metric (6.4).*

Proof We need to prove the following: for any given $\tilde{s} \in \Sigma_2$, for any $\varepsilon > 0$, there exists a $\delta > 0$ such that if $d(\tilde{s}, t) < \delta$, then $d(\sigma(\tilde{s}), \sigma(t)) < \varepsilon$.

Find a positive integer n sufficiently large such that

$$\frac{1}{2^{n-1}} < \varepsilon, \quad \text{and choose} \quad \delta = \frac{1}{2^n}.$$

Write

$$\tilde{s} = (s_0 s_1 \cdots s_n \cdots).$$

Consider any $t \in \Sigma_2$ such that $d(\tilde{s}, t) < \delta = 2^{-n}$. Then by Lemma 6.1 we know that

$$t = (s_0 s_1 \cdots s_n t_{n+1} t_{n+2} \cdots),$$

i.e., t must agree with $\tilde{s}$ in the first $n + 1$ bits. Thus

$$\sigma(\tilde{s}) = (s_1 s_2 \cdots s_n s_{n+1} s_{n+2} \cdots), \quad \sigma(t) = (s_1 s_2 \cdots s_n t_{n+1} t_{n+2} \cdots)$$

and

$$d(\sigma(\tilde{s}), \sigma(t)) = \sum_{j=1}^{n} \frac{|s_j - s_j|}{2^{j-1}} + \sum_{j=n+1}^{\infty} \frac{|s_j - t_j|}{2^{j-1}}$$

$$= \sum_{j=n+1}^{\infty} \frac{|s_j - t_j|}{2^{j-1}} \leq \sum_{j=n+1}^{\infty} \frac{1}{2^{j-1}} = 2^{-(n-1)} < \varepsilon. \quad (6.5)$$

□

Let A and B be two sets in a metric space X. We say that A is *dense* in B if $A \subset B$ and for every $y \in B$, there exists a sequence $\{x_n\}_{n=1}^{\infty} \subseteq A$ such that $y = \lim_{n \to \infty} x_n$. Thus, $\bar{A} = B$, where $\bar{A}$ is the closure of A.

Theorem 6.3 *Consider* $\sigma : \Sigma_2 \longrightarrow \Sigma_2$. *Denote*

$$Per_n(\sigma) = \text{the set of all points in } \Sigma_2 \text{ with period } n$$

and

$$Per(\sigma) = \bigcup_{n=1}^{\infty} Per_n(\sigma).$$

6.2 Properties of the Shift Map σ

Then

(i) *the cardinality of* $\text{Per}_n(\sigma)$ *is* 2^n, *and*
(ii) $\text{Per}(\sigma)$ *is dense in* Σ_2.

Proof (i) If $s \in \text{Per}_n(\sigma)$, then it is easy to see that

$$s = (\overline{s_0 s_1 \cdots s_{n-1}}) \equiv (s_0 s_1 s_2 \cdots s_{n-1} s_0 s_1 \cdots s_{n-1} s_0 s_1 \cdots s_{n-1} \cdots),$$

and vice versa. There are $\overbrace{2 \times 2 \times \cdots \times 2}^{n \text{ times}} = 2^n$ different combinations of $\text{Per}_n(\sigma)$. Therefore the cardinality of $\text{Per}_n(\sigma)$ is 2^n.

(ii) For any given

$$\tilde{s} = (\tilde{s}_0 \tilde{s}_1 \tilde{s}_2 \cdots \tilde{s}_n \tilde{s}_{n+1} \cdots),$$

define

$$\tilde{s}_k = (\overline{\tilde{s}_0 \tilde{s}_1 \cdots \tilde{s}_{k-1}}), \text{ for } k = 1, 2, 3, \ldots.$$

Then $\tilde{s}_k \in \text{Per}_k(\sigma)$. By Lemma 6.1, we have

$$d(\tilde{s}, \tilde{s}_k) \leq \frac{1}{2^{k-1}} \longrightarrow 0 \text{ as } k \to \infty.$$

Therefore $\text{Per}(\sigma)$ is dense in Σ_2. □

Theorem 6.4 *There exists a* $\tau_0 \in \Sigma_2$ *such that* $\overline{\mathcal{O}(\tau_0)} = \Sigma_2$.

Proof We construct τ_0 as follows:

$$\tau_0 = (\underbrace{0\ 1}_{\text{block 1}}\ \underbrace{00\ 01\ 10\ 11}_{\text{block 2}}\ \underbrace{000\ 001\ 010\ 011\ 100\ 101\ 110\ 111}_{\text{block 3}}\ \underbrace{\cdots}_{\text{block 4}}$$

$$\cdots \underbrace{\cdots}_{\text{block } n}\ \underbrace{\cdots}_{\text{block } n+1}\ \cdots)$$

where the n block consists of all n-bit strings arranged sequentially in ascending order.

For any $s \in \Sigma_2$, write $s = (s_0 s_1 \cdots s_k \cdots)$. The first k bits of s, i.e., $s_0 s_1 \cdots s_{k-1}$, appear somewhere in the kth block of τ_0. Thus, there exists some $n(k)$ such that

$$\sigma^{n(k)}(\tau_0) = (s_0 s_1 \cdots s_{k-1} t_k t_{k+1} \cdots)$$

where $t_{k+j} \in \{0, 1\}$ for $j = 0, 1, 2, \ldots$. Then by Lemma 6.1, we have

$$d(s, \sigma^{n(k)}(\tau_0)) \leq 1/2^{k-1}.$$

Since k is arbitrary, we see that
$$\lim_{k \to \infty} d(s, \sigma^{n(k)}(\tau_0)) = 0.$$
Therefore $\overline{\mathcal{O}(\tau_0)} = \Sigma_2$. □

We now introduce the important concept of *topological transitivity* in the following.

Definition 6.5 Let X be a topological space and $f: X \to X$ be continuous. We say that f is *topologically transitive* if for every pair of nonempty open sets U and V of X, there exists an $n > 0$ such that $f^n(U) \cap V \neq \emptyset$. □

The topological transitivity of a map causes the "mixing" of any two open sets by that map.

Exercise 6.6 Let X be a compact metric space and $f : X \to X$ be continuous and onto. Show the following statements are equivalent:

(i) f is topologically transitive.
(ii) If Λ is a closed subset of X and $f(\Lambda) \subset \Lambda$, then either $\Lambda = X$ or Λ is nowhere dense in X.
(iii) If E is a open subset of X and $f^{-1}(E) \subset E$, then either $E = \emptyset$ or E is everywhere dense in X. □

Theorem 6.7 *The map* $\sigma: \Sigma_2 \longrightarrow \Sigma_2$ *is topologically transitive.*

Proof Let U and V be any two nonempty open sets in Σ_2. Since $\overline{\mathcal{O}(\tau_0)} = \Sigma_2$ by Theorem 6.4, we have
$$\mathcal{O}(\tau_0) \cap U \neq \emptyset.$$
Thus there exists an $\alpha_0 \in \mathcal{O}(\tau_0) \cap U$. We have
$$\alpha_0 = \sigma^{\ell_1}(\tau_0) \quad \text{for some positive integer } \ell_1.$$
But then from the proof of Theorem 6.4, we easily see that
$$\overline{\mathcal{O}(\alpha_0)} = \Sigma_2.$$
Thus $\mathcal{O}(\alpha_0) \cap V \neq \emptyset$. Choose $\beta_0 \in \mathcal{O}(\alpha_0) \cap V$. Then
$$\beta_0 = \sigma^{\ell_2}(\alpha_0) \in V.$$
Therefore $\beta_0 \in \sigma^{\ell_2}(U) \cap V \neq \emptyset$. □

6.2 Properties of the Shift Map σ

An important property for many genuinely nonlinear maps is that of sensitive dependence on initial data.

Definition 6.8 Let (X, d) be a metric space and $f: X \to X$ be continuous. We say that f has *sensitive dependence on initial data* if there exists a $\delta > 0$ such that for every $x \in X$ and for every open neighborhood $\mathcal{N}(x)$ of x, there exist a $y \in \mathcal{N}(x)$ and an n (depending on y) such that
$$d(f^n(x), f^n(y)) \geq \delta.$$
Here δ is called the sensitivity constant of f. □

Theorem 6.9 *The map* $\sigma: \Sigma_2 \longrightarrow \Sigma_2$ *is sensitively dependent on initial data.*

Proof Let $\eta_0 = (s_0 s_1 \cdots s_n \cdots) \in \Sigma_2$ and let $\mathcal{N}(\eta_0)$ be an open neighborhood of η_0 in Σ_2. Define, for a positive integer m_0,
$$A = \left\{ s \in \Sigma_2 \mid s \text{ agrees with } \eta_0 \text{ for the first } m_0 \text{ bits} \right\}.$$
If $\zeta_0 \in A$, then
$$d(\eta_0, \zeta_0) \leq \frac{1}{2^{m_0}}, \quad \text{by Lemma 6.1.}$$
We choose m_0 sufficiently large such that $A \subseteq \mathcal{N}(\eta_0)$. Also, choose $\delta = 1$ and
$$\eta_1 = (s_0 s_1 \cdots s_{m_0} t_{m_0+1} t_{m_0+2} \cdots)$$
where
$$t_{m_0+1} = s_{m_0+1} + 1 \pmod{2}$$
but
$$t_{m_0+j} \in \{0, 1\} \text{ is arbitrary for } j > 1.$$
Then $\eta_1 \in A \subseteq \mathcal{N}(\eta_0)$ and
$$d(\sigma^{m_0+1}(\eta_0), \sigma^{m_0+1}(\eta_1)) = d((s_{m_0+1} s_{m_0+2} \cdots s_{m_0+k} \cdots),$$
$$(t_{m_0+1} t_{m_0+2} \cdots t_{m_0+k} \cdots)) \geq 1 = \delta. \quad (6.6)$$
□

Summarizing Theorems 6.3, 6.7, and 6.9, we see that the shift map $\sigma: \Sigma_2 \to \Sigma_2$ satisfies the following three properties:

$$\left.\begin{array}{l} \text{(i) The set of all periodic points is dense;} \\ \text{(ii) It is topologically transitive; and} \\ \text{(iii) It has sensitive dependence on initial data.} \end{array}\right\} \quad (6.7)$$

Properties (i)–(iii) above are known to be (individually) independent of each other. Devaney [1] used these three important properties to define chaos as follows:

Let (X, d) be a metric space and $f: X \to X$ be continuous. We say that the map f is *chaotic* on X if f satisfies (i)–(iii) above.

However, a paper by Banks et al. [2] points out that conditions (i) and (ii) actually imply (iii). Thus, condition (iii) is redundant in Devaney's definition. See Theorem 6.26.

Exercise 6.10 For the quadratic map $f_\mu(x) = \mu x(1-x)$, let $\mu > 2 + \sqrt{5}$. Define

$$I_0 = \{x \in [0, 1] \mid 0 \leq x < 1/2, \ f_\mu(x) \leq 1\},$$
$$I_1 = \{x \in [0, 1] \mid 1/2 \leq x \leq 1, \ f_\mu(x) \leq 1\}.$$

Prove that $|f'_\mu(x)| \geq 1 + \delta$ for some $\delta > 0 \ \forall x \in I_0 \cup I_1$. ☐

Let us explore the properties of $S: \Lambda \longrightarrow \Sigma_2$ and $\sigma: \Sigma_2 \longrightarrow \Sigma_2$ and their relationship.

Theorem 6.11 *Assume that $\mu > 2 + \sqrt{5}$. Then we have*

$$S \circ f_\mu = \sigma \circ S \tag{6.8}$$

as indicated in the commutative diagram in Fig. 6.2.

Proof Let $x \in \Lambda$ such that

$$S(x) = \text{the itinerary of } x = (x_0 s_1 \cdots s_n \cdots).$$

Denote $y = f_\mu(x) \in \Lambda$, with $S(y) = (t_0 t_1 \cdots t_n \cdots)$. Then by the definition of $S(x)$ and $S(y)$, we have

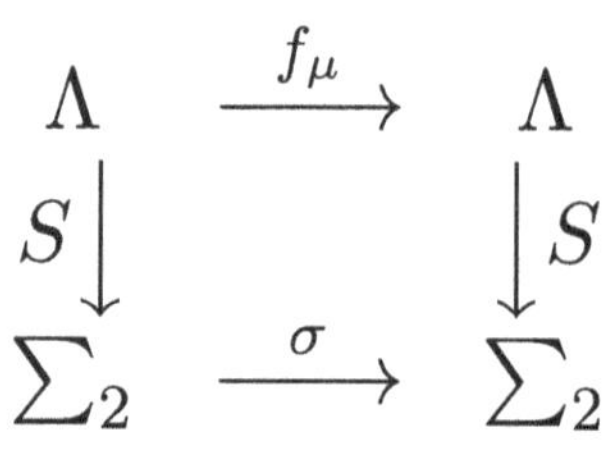

Fig. 6.2 Commutative diagram

$$\begin{array}{ccc} \Lambda & \xrightarrow{f_\mu} & \Lambda \\ S \downarrow & & \downarrow S \\ \Sigma_2 & \xrightarrow{\sigma} & \Sigma_2 \end{array}$$

6.2 Properties of the Shift Map σ

$$x \in I_{s_0}, \quad f_\mu(x) \in I_{s_1}, \quad f_\mu^2(x) \in I_{s_2}, \ldots, f_\mu^n(x) \in I_{s_n}, \ldots,$$
$$y \in I_{t_0}, \quad f_\mu(x) \in I_{t_1}, \quad f_\mu^2(y) \in I_{t_2}, \ldots, f_\mu^n(y) \in I_{t_n}, \ldots.$$

But $f_\mu^n(y) = f_\mu^n(f_\mu(x)) = f_\mu^{n+1}(x)$ for any $n = 0, 1, 2, \ldots$. Therefore

$$f_\mu^n(y) \in I_{t_n} \text{ and } f_\mu^n(y) = f_\mu^{n+1}(x) \in I_{s_{n+1}}, \text{ implying } t_n = s_{n+1},$$
$$\forall n = 0, 1, 2, \ldots,$$

i.e., $S(y) = (s_1 s_2 \cdots s_n s_{n+1} \cdots) = S(f_\mu(x)) = \sigma(s_0 s_1 \cdots s_n \cdots) = \sigma(S(x))$. □

Theorem 6.12 *Assume that $\mu > 2 + \sqrt{5}$. Then $S : \Lambda \to \Sigma_2$ is a homeomorphism.*

Proof We need to prove that S is 1–1, onto, continuous, and that S^{-1} is also continuous.

(i) 1–1: Let $x, y \in \Lambda$ such that $S(x) = S(y)$. This means that x and y have the same itinerary, i.e., $f_\mu^n(x)$ and $f_\mu^n(y)$ belong to the same I_j for $j \in \{0, 1\}$ for any $n = 0, 1, 2, \ldots$. Since f_μ is monotonic on both I_0 and I_1, every point $z \in (x, y)$ satisfy the same itinerary as that of x and y. Therefore, f_μ^n maps the closed interval $[x, y] \equiv J$ to either I_0 or I_1, for any $n = 0, 1, 2, \ldots$. But, by the fact that $\mu > 2 + \sqrt{5}$, we have

$$|f_\mu'(x)| \geq 1 + \delta \quad \text{for some} \quad \delta > 0, \forall x \in I_0 \cup I_1.$$

Using the mean value theorem, we therefore have

$$\text{length of } f_\mu^n(J) \geq (1 + \delta)^n \cdot (y - x) \longrightarrow \infty \text{ as } n \to \infty,$$

a contradiction.

(ii) Onto: Choose any $s = (s_0 s_1 \cdots s_n \cdots) \in \Sigma_2$. We want to find an $x \in \Lambda$ such that $S(x) = s$. Define, for any $n = 0, 1, 2, \ldots$

$$I_{s_0 s_1 \cdots s_n} = \{x \in I = [0, 1] \mid x \in I_{s_0} f_\mu(x) \in I_{s_1}, \ldots, f_\mu^n(x) \in I_{s_n}\} \quad (6.9)$$
$$= I_{s_0} \cap f_\mu^{-1}(I_{s_1}) \cap f_\mu^{-2}(I_{s_2}) \cap \cdots \cap f_\mu^{-n}(I_{s_n}).$$

Then

$$I_{s_0 s_1 \cdots s_n} = [I_{s_0} \cap f_\mu^{-1}(I_1) \cap \cdots \cap f_\mu^{-(n-1)}(I_{s_{n-1}})] \cap f_\mu^{-n}(I_{s_n})$$
$$= I_{s_0 s_1 \cdots s_{n-1}} \cap f_\mu^{-n}(I_{s_n}).$$

Thus $I_{s_0 s_1 \cdots s_n} \subset I_{s_0 s_1 \cdots s_{n-1}}$ and the closed intervals $I_{s_0 s_1 \cdots s_n}$ form a nested sequence of nonempty closed sets. From topology, we know that

$$\bigcap_{n=0}^{\infty} I_{s_0 s_1 \cdots s_n} \neq \emptyset.$$

Therefore, there exists some $x \in \bigcup_{n=0}^{\infty} I_{s_0 s_1 \cdots s_n}$. By definition, $x \in \Lambda$, and $S(x) = (s_0 s_1 \cdots s_n \cdots)$. This x is actually unique by part (i) of the proof.

(iii) S is continuous: For any given $\varepsilon > 0$, at any point $x \in \Lambda$, we want to choose a $\delta > 0$ such that

$$d(S(x), S(y)) < \varepsilon \quad \text{provided that} \quad |x - y| < \delta, \quad \forall y \in \Lambda.$$

First, let n be a positive integer such that $2^{-n} < \varepsilon$. Let $I_{s_0 \cdots s_n}$ be defined as in (6.9) such that $x \in I_{s_0 s_1 \cdots s_n}$. Choose $\delta > 0$ sufficiently small such that if $|y - x| < \delta$ and $y \in \Lambda$, we have $y \in I_{s_0 s_1 \cdots s_n}$. Then

$$S(y) = (s_0 s_1 \cdots s_n t_{n+1} t_{n+2} \cdots t_{n+k} \cdots)$$

where $t_{n+k} \in \{0, 1\}$ for any $k = 1, 2, \ldots$. This gives

$$d(S(x), S(y)) \leq \frac{1}{2^n} < \varepsilon.$$

(iv) S^{-1} is continuous from Σ_n to Λ: This is left as an exercise. $\square$

Remark 6.13 Theorem 6.12 is also true as long as $\mu > 4$, but the proof is somewhat more involved. In the proof of Theorem 6.12, we have utilized the property that

$$|f'_\mu(x)| \geq 1 + \delta \quad \forall x \in I_0 \cup I_1, \quad (\mu > 2 + \sqrt{5}) \tag{6.10}$$

which is not true if $4 < \mu \leq 2 + \sqrt{5}$. However, instead of (3), we can utilize the property that for any μ s.t. $4 < \mu \leq 2 + \sqrt{5}$, we have

$$|(f_\mu^n)'(x)| \geq 1 + \delta \quad \forall x \in I_0 \cup I_1,$$

for some positive integer n. Thus, the proof goes through.

Since $S: \Lambda \to \Sigma_2$ is a homeomorphism, by Theorem 6.11 we can write

$$f_\mu = S^{-1} \circ \sigma \circ S.$$

Therefore, many useful properties of σ pass on to f_μ, as stated below. $\square$

Corollary 6.14 *Let $f_\mu(x) = \mu x(1-x)$ with $\mu > 4$. Then the map $f_\mu: \Lambda \to \Lambda$ has the following properties:*

(i) *The cardinality of $Per_n(f_\mu)$ is 2^n.*
(ii) *$Per(f_\mu)$ is dense in Λ.*
(iii) *The map f_μ has a dense orbit, i.e., there exists an $x_0 \in \Lambda$ such that $\mathcal{O}(x_0)$ is dense in Λ.* □

6.3 Symbolic Dynamic Systems Σ_k and Σ_k^+

Consider the set consisting of k symbols,

$$S(k) = \{0, 1, \ldots, k-1\}. \tag{6.11}$$

Endowed with the discrete topology, this set becomes topological space. All the subsets in $S(k)$ are open. The set $S(k)$ is metrizable, a compatible metric is

$$\delta(a, b) = \begin{cases} 1, & \text{if } a \neq b \\ 0, & \text{if } a = b. \end{cases} \tag{6.12}$$

Obviously $S(k)$ is compact, so the topological products

$$\Sigma_k = \prod_{j=-\infty}^{+\infty} S_j, \quad S_j = S(k),$$

or

$$\Sigma_k^+ = \prod_{j=0}^{+\infty} S_j, \quad S_j = S(k),$$

is also compact space by the well-known Tychonov's Theorem. Σ_k and Σ_k^+ is called, respectively, two-side or one-side symbol spaces with k symbols. any element s in Σ_k is a two-sided symbol sequence

$$s = (\ldots, s_{-2}, s_{-1}; s_0, s_1, s_2, \ldots),$$

and one-side symbol sequence in Σ_k^+

$$s = (s_0, s_1, s_2, \ldots),$$

respectively. Σ_k and Σ_k^+ can be metrizable. A usual metric is, for two-side case,

$$d(s, t) = \sum_{j=-\infty}^{+\infty} \frac{\delta(s_j, t_j)}{2^{|j|}}, \tag{6.13}$$

for
$$\begin{cases} s = (\ldots, s_{-2}, s_{-1}; s_0, s_1, s_2, \ldots), \\ t = (\ldots, t_{-2}, t_{-1}; t_0, t_1, t_2, \ldots), \end{cases} \quad (6.14)$$

and for the one-side case
$$d(s,t) = \sum_{j=0}^{+\infty} \frac{\delta(s_j, t_j)}{2^j}, \quad (6.15)$$

for
$$s = (s_0, s_1, s_2, \ldots),$$
$$t = (t_0, t_1, t_2, \ldots).$$

Exercise 6.15 Prove that $d(\cdot, \cdot)$ defined by (6.13) and (6.15) is a metric on, respectively, Σ_k and Σ_k^+. Furthermore, (Σ_k, d) and (Σ_k^+, d) are compact metric spaces. □

Exercise 6.16 In lieu of (6.13), define
$$d(s,t) = \sum_{j=-\infty}^{\infty} \frac{1}{2^{|j|}} \frac{|s_j - t_j|}{1 + |s_j - t_j|}.$$

Prove that $d(\cdot, \cdot)$ is a metric on Σ_k equivalent to (6.15).

On Σ_k, we now define a "left-shift" map
$$\sigma : \Sigma_k \longrightarrow \Sigma_k$$
$$\sigma(s) = \sigma(\ldots, s_{-2}, s_{-1}, \dot{s}_0, s_1, s_2, \ldots) = (\ldots, s_{-2}, s_{-1}, s_0, \dot{s}_1, s_2, \ldots).$$

Similarly, on Σ_k^+, we define a "left-shift" map
$$\sigma^+ : \Sigma_k^+ \longrightarrow \Sigma_k^+$$
$$\sigma^+(s) = \sigma(s_0, s_1, s_2, \ldots) = (s_1, s_2, \ldots).$$
□

Lemma 6.17 (1) σ is a homeomorphism from Σ_k to itself.
(2) σ^+ is continuous from Σ_k^+ to itself.

Proof Let
$$s = (\ldots, s_{-2}, s_{-1}, \dot{s}_0, s_1, s_2, \ldots),$$
$$t = (\ldots, t_{-2}, t_{-1}, \dot{t}_0, t_1, t_2, \ldots).$$

6.4 The Dynamics of $\left(\Sigma_k^+, \sigma^+\right)$ and Chaos

By definition, we have the following implication relations

$$d(s,t) < 2^{-n} \Rightarrow s_j = t_j, \ j = 0, \pm 1, \pm 2, \ldots, \pm n \Rightarrow d(s,t) \leq 2^{-(n-1)}. \quad (6.16)$$

For the continuity of σ, we need to prove the following: for any given $s \in \Sigma_k$, for any $\varepsilon > 0$, there exists a $\delta > 0$ such that if $d(s,t) < \delta$, then $d(\sigma(s), \sigma(t)) < \varepsilon$.

Find a positive integer n sufficiently large such that $2^{-(n-1)} < \varepsilon$, and choose $\delta = 2^{-(n+1)}$. When $d(s,t) < \delta$, $s_j = t_j$, $j = 0, \pm 1, \pm 2, \ldots, \pm(n+1)$ by (6.16), and so $(\sigma(s))_j = (\sigma(t))_j$, $j = 0, \pm 1, \pm 2, \ldots, \pm n$. Again by (6.16), we have

$$d(\sigma(s), \sigma(t)) \leq 2^{-(n-1)} < \varepsilon.$$

Obviously the map σ is 1–1 and onto. The continuity of its inverse σ^{-1} can be proved similarly. Therefore σ is a homeomorphism.

With the similar method, we have (2). □

We have two dynamical systems (Σ_k, σ) and (Σ_k^+, σ^+) which are called two-side shift and one-side shift, respectively. The dynamics between them are very similar. For simplicity, in the following, we just discuss the dynamics of the one-side shift. Actually all the results on (Σ_k^+, σ^+) are also true for the two-side shift (Σ_k, σ).

6.4 The Dynamics of $\left(\Sigma_k^+, \sigma^+\right)$ and Chaos

Throughout this subsection, we assume $k \geq 2$, i.e., the symbol space Σ_k^+ with the symbols not <2.

Theorem 6.18 *Consider (Σ_k^+, σ^+). Denote*

$$Per_n(\sigma^+) = \text{the set of all points in } \Sigma_k^+ \text{ with period less than or equal to } n$$

and

$$Per(\sigma^+) = \bigcup_{n=1}^{+\infty} Per_n(\sigma^+).$$

Then

(1) the cardinality of $Per_n(\sigma^+)$ is k^n, and
(2) $Per(\sigma^+)$ is dense in Σ_k^+.

Proof

(1) If $s \in Per_n(\sigma^+)$, then it is easy to show (Exercise!) that

$$s = (\overline{s_0 s_1 \cdots s_{n-1}}) \equiv (s_0 s_1 \cdots s_{n-1} s_0 s_1 \cdots s_{n-1} \cdots),$$

and vice verse. There are $\overbrace{k \times k \times \cdots k}^{} = k^n$ different combinations of $Per_n(\sigma^+)$. Therefore the cardinality of $Per_n(\sigma^+)$ is k^n.

(2) For any given

$$s = (s_0 s_1 \cdots s_n s_{n+1} \cdots) \in \Sigma_k^+,$$

define

$$\tilde{s}^n = (\overline{s_0 s_1 \cdots s_{n-1}}), \qquad n = 1, 2, \ldots.$$

Then $\tilde{s}^n \in Per_n(\sigma^+)$. By (6.16), we have

$$d(s, \tilde{s}^n) \le \frac{1}{2^{n-1}} \to 0 \qquad (n \to \infty).$$

Therefore $Per(\sigma^+)$ is dense in Σ_k^+. $\square$

Theorem 6.19 *The shift map σ^+ is topological transitive. That is, there exists a $s \in \Sigma_k^+$ such that*

$$\overline{orb(s)} = \Sigma_k^+,$$

where $\overline{orb(s)}$ denotes the orbit $\{s, \sigma^+(s), \ldots, (\sigma^+)^n(s), \ldots\}$ starting from s.

Proof For any positive integer n, taking n symbols from the k symbols each time, we have totally k^n different order n-blocks. Such k^n n-blocks are arranged in any order to form a nk^n-block, which is denoted by P_{nk^n}. Construct s as

$$s = (P_k P_{2k^2} \cdots P_{nk^n} \cdots) \in \Sigma_k^+.$$

We claim that

$$\overline{orb(s)} = \Sigma_k^+.$$

For any $t \in \Sigma_k^+$ and for any $n \ge 1$, it follows from the construction of s that there exists $l_n \ge 1$ such that $\sigma^+(s)$ and t are identical up to the first $n+1$ bits. Then by (6.16), we have

$$d((\sigma^+)^{l_n}(s), t) \le \frac{1}{2^{n-1}} \to 0, \qquad (n \to \infty).$$

Therefore $\overline{orb(s)} = \Sigma_k^+$. $\square$

From Theorem 6.19, the one side shift (Σ_m^+, σ^+) is topological transitive. There exists a point $s = (s_0 s_1 \cdots) \in \Sigma_m^+$ such that the orbit $\{(\sigma^+)^n(s) : n = 0, 1, \ldots\}$ is dense in Σ_m^+.

6.4 The Dynamics of $\left(\Sigma_k^+, \sigma^+\right)$ and Chaos

Denote by Σ_m^* the set of all such points. The following Proposition shows that the set Σ_m^* is almost equal to the whole symbol space Σ_m^+. See [3, Theorem 14.3.20]

Proposition 6.20 *Let* $\{p_0, \ldots, p_{m-1}\}$ *be a system of weights, with* $0 < p_i < 1$ *for all* $0 \le i \le m - 1$ *and* $p_0 + \cdots + p_{m-1} = 1$.

(i) *Then, there is a probability measure μ_p defined on Σ_m^+ that is invariant by σ^+ and that satisfies the following: given* $a_0, \ldots, a_k \in \{0, 1, \ldots m - 1\}$, *then*

$$\mu_p(s : s_i = a_0) = p_{a_0},$$

and

$$\mu_p(s : s_i = a_i, \ i = 0, \ldots, n) = \prod_{i=0}^{n} p_{a_i}.$$

(ii) $\mu_p(\Sigma_m^*) = 1$. □

Let X be a compact metric space and $f : X \to X$ be continuous. Define

$$f \times f : X \times X \to X \times X,$$
$$f \times f(x, y) = (f(x), f(y)).$$

Then we obtain a topological system $(X \times X, f \times f)$ with $f \times f$ being continuous on the topological product space $X \times X$.

Definition 6.21 Let f be continuous from a compact metric space X into itself. We say that f is weakly topological mixing if $f \times f$ is topological transitive. □

It is easy to see from the definition that weakly topological mixing implies topological mixing.

Theorem 6.22 *The map* $\sigma^+ : \Sigma_k^+ \to \Sigma_k^+$ *is weakly topological mixing.*

Proof It suffices to prove the following: there exist $u, v \in \Sigma_k^+$ such that for any $s, t \in \Sigma_k^+$, there exists an increasing integer sequences l_n with

$$\lim_{n \to \infty} (\sigma^+ \times \sigma^+)^{l_n}(u, v) = (s, t),$$

i.e.

$$\lim_{n \to \infty} (\sigma^+)^{l_n}(u) = s, \quad \lim_{n \to \infty} (\sigma^+)^{l_n}(v) = t.$$

Let $n \geq 1$ and P_{nk^n} be any nk^n-block which was constructed in the proof of Theorem 6.19. There are totally $k^n!$ different nk^n-blocks. Arranging them in any order, we have a $k^n!nk^n$-block, which is denoted by $Q_{k^n!nk^n}$.

Let
$$P_{k^n n!k^n} = \overbrace{P_{nk^n} \ldots P_{nk^n}}^{k^n! \text{ times}}.$$

(In total $k^n!$ nk^n-blocks P_{nk^n}).

Define
$$u = (Q_{k!k} Q_{k^2!2k^2} \cdots Q_{k^n!nk^n} \cdots) \in \Sigma_k^+,$$
$$v = (P_{k!k} P_{k^2!2k^2} \cdots P_{k^n!nk^n} \cdots) \in \Sigma_k^+.$$

For any $n > 0$, we can see from the construction of u and v that there exists $l_n > 0$ such that $(\sigma^+)^{l_n}(u)$ and s, $(\sigma^+)^{l_n}(v)$ and t, respectively, are identical up to the first $n+1$ bits. Then by (6.16), we have

$$d((\sigma^+)^{l_n}(u), s) \leq \frac{1}{2^n} \to 0, \quad (n \to \infty),$$
$$d((\sigma^+)^{l_n}(v), t) \leq \frac{1}{2^n} \to 0, \quad (n \to \infty).$$

Thus
$$\lim_{n \to \infty} (\sigma^+)^{l_n}(u) = s, \quad \lim_{n \to \infty} (\sigma^+)^{l_n}(v) = t. \tageq (6.17)$$

□

Definition 6.23 Let f be continuous from a compact metric space X into itself. We say that f is topological mixing if for any nonempty open sets $U, V \in X$, there exists a positive integer N such that
$$f^n(U) \cap V \neq \phi, \quad \forall n > N.$$

□

Exercise 6.24 Verify the following implications [4]: Topological mixing $\Rightarrow$ Weakly topological mixing $\Rightarrow$ Topological transitivity. □

We remark that there are the following implication relations.

Theorem 6.25 *The map $\sigma^+ : \Sigma_k^+ \to \Sigma_k^+$ is topological mixing.*

Proof Let $U, V \in \Sigma_k^+$ be any two nonempty open sets, and
$$s = (s_0 s_1 \cdots) \in U.$$

6.4 The Dynamics of $\left(\Sigma_k^+, \sigma^+\right)$ and Chaos

Then there exists a $\varepsilon_0 > 0$ such that

$$O_{\varepsilon_0}(s) = \{t \in \Sigma_k^+ \mid d(s,t) < \varepsilon_0\} \subset U.$$

Let N be a sufficiently large integer such that $2^{-N} < \varepsilon_0$. Thus by (6.16), we have

$$_0[s_0 s_1 \cdots s_{N-1}] \equiv \{t = (t_0 t_1 \cdots) \in \Sigma_k^+ \mid t_i = s_i,\ 0 \le i \le N-1\} \subset O_{\varepsilon_0}(s) \subset U.$$

It is easy to see that $(\sigma^+)^n({}_0[s_0 s_1 \cdots s_{N-1}]) = \Sigma_k^+$ when $n \ge N$. Therefore,

$$(\sigma^+)^n(U) \cap V \supset (\sigma^+)^n({}_0[s_0 s_1 \cdots s_{N-1}]) \cap V \supset \Sigma_k^+ \cap V = V \neq \phi,$$

if $n \ge N$. □

Actually, we know that topological mixing implies weakly topological mixing and the latter implies topological transitive. Thus Theorem 6.25 implies 6.22 and Theorem 6.22 implies 6.19. Here we give the independent proofs of the theorems just as to help the readers to have a deep understanding for the symbol dynamical system, in particular, how to construct a particular symbol sequence with a desired property.

Theorem 6.26 *Let $f : X \to X$ satisfy (6.7) that*

(i) *the set of all f's periodic points is dense,*
(ii) *f is topologically transitive,*
(iii) *f has sensitive dependence on initial data.*

Then conditions (i) and (ii) imply condition (iii).

Proof When X is a finite set, it must be a periodic orbit of f under the conditions (i) and (ii). Thus f satisfies condition (iii) obviously. In the following, we assume that X is a infinite set.

We first claim that there exists a $\delta_0 > 0$ such that for every $x \in X$ there exists a periodic point p such that

$$d(orb(p), x) \ge \frac{\delta_0}{2}.$$

In fact, taking two different periodic points p_1 and p_2 which are on different periodic orbits, respectively. Let

$$\delta_0 = d(orb(p_1), orb(p_2)) > 0.$$

The triangle inequality

$$\delta_0 = d(orb(p_1), orb(p_2)) \le d(orb(p_1), x) + d(orb(p_2), x),$$

implies either
$$d(orb(p_1), x) \geq \frac{\delta_0}{2}$$
or
$$d(orb(p_2), x) \geq \frac{\delta_0}{2}.$$

This proves our claim.

Next, we shall show that $\delta \triangleq \frac{\delta_0}{8}$ is a sensitive constant of f. Let x be a given point and $0 < \varepsilon < \delta$. Then there exists a periodic point $p \in V(x, \varepsilon)$ by the denseness of periodic points, where $V(x, \varepsilon)$ is the ε-neighborhood od x. Denote by n the period of p.

From the claim above, there exists a periodic point q such that
$$d(orb(q), x) \geq 4\delta.$$

Let
$$U = \cap_{i=1}^{n} f^{-1}(V(f^i(q)), \delta).$$

Then U is a nonempty open set containing q. It follows from the topological transitivity of f that there exists $y \in V(x, \varepsilon)$ and $k > 0$ such that $f^k(y) \in U$.

Let $j = [\frac{k}{n} + 1]$, where $[a]$ denotes the largest integer that is not larger than a. Then $1 \leq nj - k \leq n$. We have
$$f^{nj}(y) = f^{nj-k}(f^k(y)) \in f^{nj-k}(U) \subset V(f^{nj-k}(q), \delta).$$

Since $f^{nj}(p) = p$, we have
$$d(f^{nj}(p), f^n j(y)) = d(p, f^{nj}(y))$$
$$\geq d(x, f^{nj-k}(q)) - d(f^{nj-k}(q), f^{nj}(y)) - d(p, x)$$
$$> 4\delta - \delta - \delta = 2\delta.$$

again by the triangle inequality
$$2\delta < d(f^{nj}(p), f^n j(y)) = d(f^{nj}(x), f^n j(y)) + d(f^{nj}(x), f^{nj}(p))$$

implies either
$$d(f^{nj}(x), f^{nj}(y)) > \delta,$$
or
$$d(f^{nj}(x), f^{nj}(p)) > \delta. \qquad (6.18)$$

□

Based on what Devaney [1] first gave in his book [1, Definition 8.5, p. 50] and Theorem 6.26, we now give the following definition of chaos.

6.4 The Dynamics of $\left(\Sigma_k^+, \sigma^+\right)$ and Chaos

Definition 6.27 (*Devaney*) Let (X, d) be a metric space and $f : X \to X$ be continuous. We say that f is chaotic in the sense of Devaney if f satisfies

(i) f is topologically transitive;
(ii) The set of all periodic points of f is dense. □

Summarizing Theorems 6.18, 6.19, and 6.26, we see that the one-side shift map σ^+ is Devaney's Chaos.

Later in this section, we show that σ^+ is also chaotic in the sense of Li-Yorke. To this end, we need some notations and lemmas. Without loss of generality, in the following, we assume that $k = 2$, i.e., 2-symbol dynamics.

We recall a famous result on chaos of interval maps. In 1975, Li and Yorke [5] obtained that for an interval map, period three implies chaos. More precisely, they proved the following.

Theorem 6.28 ([5]) *Let $f : I \to I$ be continuous. Assume that there exists an $a \in I$ such that*
$$f^3(a) \leq a < f(a) < f^2(a), \text{ or } f^3(a) \geq a > f(a) > f^2(a).$$

Then we have

(1) for every positive integer n, f has a periodic point with period n;
(2) There exists a uncountable set $C \subset I - P(f)$ with the following properties

    (i) $\limsup_{n\to\infty} |f^n(x) - f^n(y)| > 0,$ $\forall x, y \in C, \; x \neq y;$
    (ii) $\liminf_{n\to\infty} |f^n(x) - f^n(y)| = 0,$ $\forall x, y \in C;$
    (iii) $\limsup_{n\to\infty} |f^n(x) - f^n(p)| > 0,$ $\forall x \in C, \; \forall p \in P(f).$ □

The theorem of Li and Yorke motivates the following definition.

Definition 6.29 Let (X, d) be a metric space and $f : X \to X$ be continuous. We say that f is Li-Yorke's chaos if

(1) for every positive integer n, f has a periodic point with period n;
(2) There exists a uncountable set $C \subset X - P(f)$ with the following properties

    (i) $\limsup_{n\to\infty} d(f^n(x), f^n(y)) > 0,$ $\forall x, y \in C, \; x \neq y;$
    (ii) $\liminf_{n\to\infty} d(f^n(x), f^n(y)) = 0,$ $\forall x, y \in C;$
    (iii) $\limsup_{n\to\infty} d(f^n(x), f^n(p)) > 0,$ $\forall x \in C, \; \forall p \in P(f).$ □

For every $s = (s_0 s_1 \cdots) \in \Sigma_2^+$, define

$$r(s, l) = \#\{i \mid s_i = 0, \ i = 0, 1, \ldots, l\},$$

where #A denotes the cardinality of the set A.

Lemma 6.30 *For every $0 < \eta < 1$, there exists $s^\eta \in \Sigma_2^+$, such that*

$$\lim_{l \to \infty} \frac{r(s^\eta, l^2)}{l} = \eta.$$

Proof Let $l_0 > 0$ be the smallest integer such that $[l_0 \eta] = 1$. Define

$$s^\eta = (s_0^\eta s_1^\eta \cdots s_l^\eta \cdots) \in \Sigma_2^+$$

by the following induction way: For $0 \leq i \leq l_0^2$,

$$s_i^\eta = \begin{cases} 1, & 0 \leq i < l_0^2 \\ 0, & i = l_0^2, \end{cases}$$

For $l_0^2 < i \leq (l_0 + 1)^2$,

$$s_i^\eta = \begin{cases} 1, & l_0^2 < i < (l_0 + 1)^2 \\ 0, & \text{if } i = (l_0 + 1)^2 \text{ and } [(l_0 + 1)\eta] - [l_0 \eta] = 1 \\ 1, & \text{if } i = (l_0 + 1)^2 \text{ and } [(l_0 + 1)\eta] - [l_0 \eta] = 0. \end{cases}$$

Continuing by induction, for $(l_0 + k)^2 < i \leq (l_0 + k + 1)^2$, $k = 0, 1, \ldots$,

$$s_i^\eta = \begin{cases} 1, & (l_0 + k)^2 < i < (l_0 + k + 1)^2 \\ 0, & \text{if } i = (l_0 + k + 1)^2 \text{ and } [(l_0 + k + 1)\eta] - [(l_0 + k)\eta] = 1 \\ 1, & \text{if } i = (l_0 + k + 1)^2 \text{ and } [(l_0 + k + 1)\eta] - [(l_0 + k)\eta] = 0. \end{cases}$$

By the construction of s^η, it is easy to see that

$$r(s^\eta, l^2) = [l\eta], \qquad \forall l > 0.$$

Since $l\eta \leq [l\eta] \leq l\eta + 1$, we have

$$\lim_{l \to \infty} \frac{r(s^\eta, l^2)}{l} = \lim_{l \to \infty} \frac{[l\eta]}{l} = \eta. \tag{6.19}$$

$\square$

6.4 The Dynamics of $\left(\Sigma_k^+, \sigma^+\right)$ and Chaos

Lemma 6.31 *Let s^η be defined as above. Then we have*

(1) $s_i^\eta = 0$ *if and only if there exists $l > 0$ such that $i = l^2$ and $[l\eta] - [(l-1)\eta] = 1$.*
(2) *For any $l \geq 1$, we have*
$$s_{l^2+j}^\eta = 1, \qquad 0 < j < 2l+1.$$
(3) *There exist infinite many integers $l > 0$ such that $s_{l^2}^\eta = 0$.*
(4) *Let $0 < \eta < \theta < 1$. Then for any $N > 0$, there exists $l > N$ such that $s_{l^2}^\theta \neq s_{l^2}^\eta$.*

Proof (1) and (2) follow from the construction of s^η.

For (3), suppose that there are only finite many integers $l > 0$ such that $s_{l^2}^\eta = 0$. Then it is easy to see that $r(s^\eta, l^2)$ is bounded by the construction of s^η. From Lemma 6.30, we have
$$\eta = \lim_{l \to \infty} \frac{r(s^\eta, l^2)}{l} = 0.$$
This is a contradiction.

For (4), suppose that it is not true. Then $r(s^\eta, l^2) - r(s^\theta, l^2)$ is bounded, which implies $\eta = \theta$ by Lemma 6.30. A contradiction. □

Let (X, d) be a metric space and $f : X \to X$ be continuous. Recall that for $x \in X$, $y \in X$ is said to be a ω-limit point of x if there exists a increasing positive integer sequence n_i such that
$$\lim_{i \to \infty} f^{n_i}(x) = y.$$

Definition 6.32 Denote by $\omega(x, f)$ the set of all ω-limit points of x and call it ω-limit set of x. □

Lemma 6.33 *Let (X, d) be a compact metric space and $f : X \to X$ be continuous. We have*

(1) $\omega(x, f)$ *is a nonempty closed set for every $x \in X$;*
(2) *For every $x \in X$,*
$$f(\omega(x, f)) = \omega(x, f) = \omega(f^n(x), f), \qquad \forall n > 0.$$

□

We left the proof as an exercise.

Lemma 6.34 *Let*

$$C_0 = \{s^\eta \in \Sigma_2^+ \mid \eta \in (0, 1)\}, \quad C = \cup_{i=0}^\infty (\sigma^+)^i (C_0),$$

where s^η is defined in Lemma 6.30. Then we have

$$\omega(s, \sigma^+) = \{e, e_i \mid i = 0, 1, 2, \ldots\}, \quad \forall s \in C, \tag{6.20}$$

where $e, e_i \in \Sigma_2^+$ are given by

$$\begin{aligned} e &= (1, 1, 1, \ldots, 1, \ldots), \\ e_i &= (1, 1, \ldots, 1, 0, 1, 1, \ldots). \end{aligned} \tag{6.21}$$

Proof By Lemma 6.33, it suffices to prove (6.20) for $s \in C_0$. For $s^\eta \in C_0$, by (2) in Lemma 6.31, it is easy to see that

$$\lim_{l \to \infty} d((\sigma^+)^{l^2+1}(s^\eta), e) = 0.$$

Thus $e \in \omega(s^\eta, \sigma^+)$.

By (3) in Lemma 6.31, there exists an increasing sequence

$$l_1 < l_2 < \cdots < l_j < \cdots,$$

such the first bit in $(\sigma^+)^{l_j^2}(s^\eta)$ is 0. From (1) and (2) in Lemma 6.31, we have, for any $i > 0$, the first i bits are 1 and the $i+1$th is 0 in $(\sigma^+)^{l_j^2 - i}(s^\eta)$ when j is larger enough. At the same time, there are at least $(l_j + 1)^2 - l_j^2 - 1 - (i+1) = 2l_j - (i+1)$ number bits with 1 following the 0 bit. Thus

$$\lim_{j \to \infty} d((\sigma^+)^{l_j - i}(s^\eta), e_i) = 0.$$

So we have $e_i \in \omega(s^\eta, \sigma^+), \forall i > 0$.

To complete our proof, it suffices to show that if $t \in \Sigma_2^+$ has at least two bits with symbols 0, say $t_h = t_m = 0, \ h < m$, then

$$t \notin \omega(s^\eta, \sigma^+), \quad \forall s^\eta \in C_0.$$

In fact, suppose that there exists a $\eta \in (0, 1)$ such that $t \in \omega(s^\eta, \sigma^+)$. Then by definition there exists a increasing sequence $\{l_j\}$ such that

$$\lim_{j \to \infty} (\sigma^+)^{l_j}(s^\eta) = t.$$

6.4 The Dynamics of $\left(\Sigma_k^+, \sigma^+\right)$ and Chaos

This implies that the $(h+1)$th and $(m+1)$th bits in $(\sigma^+)^{l_j}(s^\eta)$ are 0 when j is larger enough. This contradicts to the fact that in $(\sigma^+)^{l_j}(s^\eta)$, there are infinite many 1 bits between any two 0 bits. □

Theorem 6.35 *Let C be defined as Lemma 6.34. We have*

(1) for every positive integer n, σ^+ has a periodic point with period n;
(2) The set $C \subset \Sigma_2^+ - P(\sigma^+)$ is uncountable, $\sigma^+(C) \subset C$ and has the following properties

  (i) $\limsup_{n\to\infty} d((\sigma^+)^n(s), (\sigma^+)^n(t)) > 0$, $\forall s, t \in C, \; s \neq t$;
  (ii) $\liminf_{n\to\infty} d((\sigma^+)^n(s), (\sigma^+)^n(t)) = 0$, $\forall s, t \in C$;
  (iii) $\liminf_{n\to\infty} d((\sigma^+)^n(s), (\sigma^+)^n(p)) > 0$, $\forall s \in C, \; \forall p \in P(\sigma^+) - \{e\}$, where e is given by Lemma 6.34.

Thus, σ^+ has Li-Yorke's chaos. □

Comparing (iii) in this theorem and that in Definition 6.29, we can see that σ^+ has chaotic dynamics more stronger than Li-Yorke's chaos. Furthermore, the chaos set C for σ^+ can be taken to be invariant.

Proof of Theorem 6.35 (1) follows from (1) in Theorem 6.18.

For (2), first of all, since $s^\eta \neq s^\theta$ if $\eta \neq \theta$ from (4) in Lemma 6.31, there is one-to-one corresponding between $(0, 1)$ and the set C_0. Thus C_0 is uncountable, so is C. Obviously $\sigma^+(C) \subset C$. From the construction of s^η it is easy to see that for any $l > 0$, s^η and $(\sigma^+)^l(s^\eta)$ are not the periodic points of σ^+. So $C \subset \Sigma_2^+ - Per(\sigma^+)$.

Next, we are about to prove (i)–(iii) in (2).

For (i), let $s, t \in C, \; s \neq t$, then there exist $s^\eta, s^\theta \in C_0$ with $0 < \eta \leq \theta < 1$ and two negative integers h, m such that

$$s = (\sigma^+)^h(s^\eta), \qquad t = (\sigma^+)^h(s^\theta).$$

Assume that $0 \leq h \leq m$. ($h < m$ if $\eta = \theta$.)

When $h = m$, we have $0 < \eta < \theta < 1$ and by (4) in Lemma 6.31 there exists $l_1 < l_2 < \cdots < l_j < \cdots$ such that $s^\eta_{l_j^2} \neq s^\theta_{l_j^2}$, for $j = 1, 2, \ldots$. Thus

$$d((\sigma^+)^{l_j^2 - h}((\sigma^+)^h(s^\eta)), (\sigma^+)^{l_j^2 - h}((\sigma^+)^h(s^\theta))) = d((\sigma^+)^{l_j^2}(s^\eta), (\sigma^+)^{l_j^2}(s^\theta)) \geq 1.$$

Therefore

$$\limsup_{n\to\infty} d((\sigma^+)^n(s), (\sigma^+)^n(t)) \geq 1.$$

We now consider the case that $h < m$. From (3) in Lemma 6.31 it follows that there exist $l_1 < l_2 < \cdots < l_j < \cdots$ such that $s^\eta_{l^2_j} = 0$. Thus for any $j > 0$, the first bit in $(\sigma^+)^{l^2_j - h}((\sigma^+)^h(s^\eta))$ is 0. On the other hand, there exists j_0 such that for $j > j_0$, we have
$$l^2_j < l^2_j + m - h < (l_j + 1)^2 - 1.$$
Thus if $j > j_0$, the $(l^2_j + m - h + 1)$-th bit in s^θ must be 1 by (1) in Lemma 6.31. This implies that
$$\limsup_{n \to \infty} d((\sigma^+)^n(s), (\sigma^+)^n(t))$$
$$\geq \limsup_{j \to \infty} d((\sigma^+)^{l^2_j - h}((\sigma^+)^h(s^\eta)), (\sigma^+)^{l^2_j - h}((\sigma^+)^m(s^\theta))) \geq 1.$$

This proves (i).

For (ii), let s, t be given as in (i). Let l_0 be such that
$$l^2_0 \leq m - h + 1 \leq (l_0 + 1)^2.$$

For $j > j_0$, we consider
$$(\sigma^+)^{l^2 - h + 1}((\sigma^+)^h(s^\eta)) = (\sigma^+)^{l^2 + 1}(s^\eta),$$
$$(\sigma^+)^{l^2 - h + 1}((\sigma^+)^m(s^\theta)) = (\sigma^+)^{l^2 - h + m + 1}(s^\eta).$$

From (2) in Lemma 6.31, the above two elements must agree with each other in the first $(l+1)^2 - l^2 - (m - h + 1) = 2l - (m - h)$ bits. Thus
$$\liminf_{n \to \infty} d((\sigma^+)^n(s), (\sigma^+)^n(t)) = 0.$$

Finally for (iii), let $s = (\sigma^+)^h(s^\eta) \in C$ for some $s^\eta \in C_0$. Assume, on the contrary, that there exists an $p \in Per(\sigma^+)$ such that
$$\liminf_{n \to \infty} d((\sigma^+)^n((\sigma^+)^h(s^\eta)), (\sigma^+)^n(p)) = 0.$$

It follows the periodic property of p that $p \in \omega(s^\eta, \sigma^+)$. But by Lemma 6.34, we know that e_i is not a periodic point of σ^+ for any $i > 0$. Thus $p = e$. This completes the proof of (iii) □

Exercise 6.36 Define $d_1(\cdot, \cdot)$ on Σ^+_k by
$$d_1(s, t) = \max_n \left\{ \frac{1}{n+1} \mid s_n \neq t_n \right\},$$

for $s = (s_0 s_1 \cdots)$, $t = (t_0 t_1 \cdots \in \Sigma_k^+)$. Show that d_1 is a metric and is equivalent to d. □

Exercise 6.37 Show that the shift map $\sigma^+ : \Sigma_k^+ \to \Sigma_k^+$ is k to one. □

Exercise 6.38 Let $A = (a_{ij})$ be a k by k matrix with a_{ij} equal to 0 or 1 and there is at least an entry 1 in each row and each column. Define a subset Λ_A of Σ_k^+ by

$$\Lambda_A = \{s = (s_0 s_1 \cdots) \in \Sigma_k^+ \mid a_{s_i s_{i+1}} = 1, \forall i \geq 0\}.$$

Show that Λ_A is a nonempty closed invariant set of σ^+. That is, Λ_A is nonempty, closed and

$$\sigma^+(\Lambda_A) \subset \Lambda_A.$$

Such matrix A is called a transition matrix. □

Exercise 6.39 Consider a dynamical system (Λ_A, σ_A^+), where σ_A^+ is the restriction of σ^+ to λ_A and A is defined by Exercise 6.38. Prove that σ_A^+ is topologically transitive if and only if A is irreducible. □

Exercise 6.40 Show that the following conditions are equivalence.

(i) σ_A^+ is topologically mixing;
(ii) σ_A^+ is topological weakly mixing;
(iii) A is aperiodic. □

6.5 Topological Conjugacy and Semi-conjugacy

The concept of conjugacy arises in many subject of mathematics. It also is an important concept in dynamical systems. All topological dynamical systems can be divided into different classes of equivalence relation by topological conjugacy. Each class has the same topological dynamics.

Let X and Y be two metric spaces and h be a map from X to Y. Recall that h is called onto if for each $y \in Y$ there is some $x \in X$ with $h(x) = y$, h is called one-to-one if $h(x_1) \neq h(x_2)$ whenever $x_1 \neq x_2$. Also, recall that a map h from X to Y is called a homeomorphism provided that h is continuous, one-to-one and onto and its inverse h^{-1} from Y to X is also continuous.

Recalling Definition 5.8, let f mapping from X to X and g mapping from Y to Y be continuous. We say that f and g are topologically conjugate, or just conjugate (denoted by $f \simeq g$), if there exists a homeomorphism h from X to Y such that the following diagram commutes:

$$X \xrightarrow{f} X$$
$$h \downarrow \quad \downarrow h$$
$$Y \xrightarrow{g} Y$$

i.e.,
$$hf = gh.$$

Such an h is called a *topological conjugacy*, from f to g. If h is merely a continuous map from X onto Y, then h is called a *semi-conjugacy*, and we say that f is semi-conjugate to g.

We give some properties of topological conjugate. We first note topological conjugate is an equivalence relation. That is

(i) $f \simeq f$: The identity $h = id$ from X to X is a topological conjugacy from f to itself;
(ii) If $f \simeq g$, then $g \simeq f$. In fact, if h is a topological conjugacy from f to g, then its inverse h^{-1} is a topological conjugacy from g to f;
(iii) If $f \simeq g$ and $g \simeq q$, where q is a continuous map from a metric space Z to itself, then $f \simeq q$. In fact, let h_1 and h_2 are the topological conjugacy from f to g and from g to q, respectively. Then $h = h_2 h_1$ is a topological conjugacy from f to q.

Next, from $hf = gh$ it follows that

$$hf^2 = (hf)f = (gh)f = g(hf) = g(gh) = g^2 h,$$

and f^2 and g^2 are conjugate by h. Continuing by induction

$$hf^n = (hf^{n-1})f = (g^{n-1}h)f$$
$$= g^{n-1}(hf) = g^{n-1}(gh) = g^n h,$$

and f^n and g^n are conjugate by h for any $n > 0$.

Let $x_0 \in X$ and $x_n = f^n(x_0)$ be the point on the orbit generated by x_0 under f. Let $y_0 = h(x_0)$. Then

$$y_n = g^n(y_0)$$
$$= hf^n h^{-1}(y_0)$$
$$= hf^n(h^{-1}(y_0))$$
$$= hf^n(x_0) = h(x_n).$$

Thus the orbit of x_0 under f is mapped to the orbit of $h(x_0)$ under g, i.e.,

$$h(orb(x, f)) = orb(h(x), g).$$

6.5 Topological Conjugacy and Semi-conjugacy

Lemma 6.41 *Let f and g are conjugate by h. Then for every $x \in X$, we have*

$$h(\omega(x, f)) = \omega(h(x), g).$$

Proof Let $x_0 \in \omega(x, f)$. Then by the definition, there exists a increasing integer sequence $n_1 < n_2 < \cdots$ s such that

$$\lim_{i \to \infty} f^{n_i}(x) = x_0.$$

Thus

$$\lim_{i \to \infty} g^{n_i}(h(x)) = \lim_{i \to \infty} h f^{n_i}(x) = h(x_0),$$

and $h(x_0) \in \omega(h(x), g)$. So we have

$$h(\omega(x, f)) \subset \omega(h(x), g). \tag{6.22}$$

From (6.22), we have

$$\omega(x, f) \subset h^{-1}(\omega(h(x), g)). \tag{6.23}$$

On the other hand, since f and g are conjugate by h, g and f are conjugate by h^{-1}. It follows from (6.23) that

$$\omega(y, g) \subset h(\omega(h^{-1}(y), f)), \qquad \forall y \in Y,$$

and

$$\omega(h(x), g) \subset h(\omega(x, f)), \qquad \forall x \in X,$$

by taking $x = h^{-1}(y)$. Thus

$$h(\omega(x, f)) = \omega(h(x), g). \tag{6.24}$$

□

Theorem 6.42 *Let f and g are conjugate by h. Then*

(i) *$p \in X$ is a periodic point of f with period m if and only if $h(p) \in Y$ is a periodic point of g with period m;*
(ii) *$h(Per(f)) = Per(g)$;*
(iii) *$Per(f)$ is dense in x if and only if $Per(g)$ is dense in Y;*
(iv) *f is topological transitive if and only if g is topological transitive.*

Proof For (i), let $p \in X$ be a periodic point of f with period m, i.e.,

$$f^i(p) \neq p, \quad i = 1.2, \ldots m - 1, \qquad f^m(p) = p.$$

Then

$$g^i(h(p)) = h(f^i(x)) \neq h(p), \quad i = 1, 2, \ldots m-1,$$

since h is one-to-one. And

$$g^m(h(p)) = h(f^m(p)) = h(p).$$

That is $h(p)$ is a periodic point of g with period m.

The converse follows from the fact that g and f are conjugate by h^{-1}.

(ii) follows from (i).

For (iii), assume that $Per(f)$ is dense in X. Let V be a nonempty open set in Y. Then $h^{-1}(V)$ is a nonempty open set in X. By the denseness of $Per(f)$, there exists $p \in Per(f)$ such that $p \in h^{-1}(V)$. So $h(p) \in Per(f)$ and $h(p) \in V$. That is, $Per(g)$ is dense in Y. The converse follows from the fact that g and f are conjugate by h^{-1}.

For (iv), assume that f is topological transitive, i.e., there exists a point $x_0 \in X$ such that $orb(x_0, f)$ is dense in X. Let V be any nonempty open set in Y. Then $h^{-1}(V)$ is a nonempty open set in X. Thus there exists n_0 such that

$$f^{n_0}(x_0) \in h^{-1}(V),$$

and

$$g^{n_0}(h(x_0)) = h(f^{n_0}(x_0)) \in V.$$

Thus the orbit $orb(h(x_0), g)$ is dense in Y.

The converse follows from the same way by noticing that g and f are conjugate by h^{-1}. □

From Theorems 6.42 and 6.26, we have the following conclusion.

Corollary 6.43 *Let f and g be conjugate. Then f is chaotic in the sense of Devaney if and only if g is either.* □

Theorem 6.44 *Let f and g be conjugate by h. Then f has Li-Yorke's chaos if and only if g has.*

Proof Assume that f is Li-Yorke's chaos. From (i) in Theorem 6.42, it suffices to prove that g has property (2) in Definition 6.29.

To do this end, assume that $C \subset X - Per(f)$ is a uncountable set with the property (2) in Definition 6.29. We claim that $h(C) \subset Y - Per(g)$ is a a uncountable set with the property (2) with respect to g.

First, since C is uncountable and h is one-to-one, $h(C)$ is uncountable. And by (2) in Theorem 6.42 $h(C) \subset Y - Per(g)$.

Next, for any $h(x), h(y) \in h(C)$, $x, y \in C$ with $h(x) \neq h(y)$, we have $x \neq y$, since h is one-to-one. Thus

6.5 Topological Conjugacy and Semi-conjugacy

$$\limsup_{n \to \infty} d_X(f^n(x), f^n(y)) > 0,$$

where d_X is the distance function on X. Since h is homeomorphism, we have

$$\limsup_{n \to \infty} d_Y(g^n(h(x)), g^n(h(y))) = \limsup_{n \to \infty} d_X(h(f^n(x)), h(f^n(y))) > 0.$$

By the same reason, for any $x, y \in C$ we have

$$\liminf_{n \to \infty} d_Y(g^n(h(x)), g^n(h(y))) = \liminf_{n \to \infty} d_X(h(f^n(x)), h(f^n(y))) = 0.$$

Finally, for any $h(x) \in h(C)$ and any $h(p) \in Per(g) = h(Per(f))$, we have

$$\limsup_{n \to \infty} d_Y(g^n(h(x)), g^n(h(p))) = \limsup_{n \to \infty} d_X(h(f^n(x)), h(f^n(p))) > 0.$$

So g satisfies (2) in Definition 6.29. □

Exercise 6.45 Definition: Let (X, f) and (Y, g) be compact dynamical systems and

$$f(X) = X, \qquad g(Y) = Y.$$

We call that f and g are topological semi-conjugate if the h in Definition 5.8 is only continuous and onto. In this case, f is said to be an extension of g, g is a factor of f, and h is said to be a topological semi-conjugacy.

Assume that $h : X \to Y$ is a topological conjugacy from f to g. Prove that if f is topologically transitive (weakly mixing or mixing), so is g. □

Exercise 6.46 Let $h : X \to Y$ be a topological semi-conjugacy. Prove that there is a subset $X_h \subset X$ with the following properties:

(i) X_h is closed;
(ii) X_h is invariant set of f;
(iii) $h(X_h) = Y$;
(iv) There is no proper subset in X_h with the above three conditions.

Such X_h is called a h-minimal cover of Y.

Let $h : X \to Y$ be a topological semi-conjugacy and X_h be a h-minimal cover of Y. Prove that if g is topologically transitive, so is $f|_{X_h}$. □

6.6 Shift Invariant Sets

From the preceding sections, the dynamics of symbolic systems is now rather well understood. Especially, we know that σ and σ^+ have complex dynamical behaviors, such as manifesting chaos in the sense of both Devaney and Li-Yorke. Also, we have learned that two topologically conjugate systems have the same topological properties. In the rest of this chapter, we consider a class of topological systems which are conjugate to either the symbolic dynamical system Σ_k^+ or Σ_k.

6.7 Construction Of Shift Invariant Sets

Definition 6.47 Let X be a metric space and f be continuous from X to X. Let $\Lambda \subset X$ be a closed invariant set, i.e., $f(\Lambda) \subset \Lambda$. If the subsystem

$$f|_\Lambda : \Lambda \to \Lambda$$

is topological conjugate to σ^+ or σ. That is, there exists a homeomorphism h from Λ to Σ_k^+ or Σ_k such that the following diagram commutes

$$\begin{array}{ccc} \Lambda & \xrightarrow{f} & \Lambda \\ h \downarrow & & \downarrow h \\ \Sigma_k^+ & \xrightarrow{\sigma^+} & \Sigma_k^+ \end{array}$$

i.e.

$$hf|_\Lambda = \sigma^+ h$$

then Λ is called a shift invariant set of f of order k.

If h above is only continuous and onto, we call that Λ is a quasi-shift invariant set of order k. □

Thus, if f has a shift invariant set, then it has complex behavior. In particular, it has chaos both in the sense of Devaney and Li-Yorke. In the following, we give necessary and sufficient conditions for a topological system having a shift invariant set with respect to one-side shift. In the next chapter, we shall find the conditions that f has a shift invariant set with respect to two-side shift σ.

We need the following lemma.

Lemma 6.48 *Let X, Y be two sets, f be a map from X to Y and $A \subset X$, $B \subset Y$. We have*

$$f(A \cap f^{-1}(B)) = f(A) \cap B.$$

6.7 Construction Of Shift Invariant Sets

Proof Straightforward verification. □

Example 6.49 First we give an interval map that has a shift invariant set with respect to one-side shift. Consider the interval map $f : \mathbf{R} \to \mathbf{R}$:

$$f(x) = -3x^2 + \frac{4}{3}.$$

Let $J = [-1, 1]$. Then $f^{-1}(J)$ is composed of two subintervals U_0 and U_1 in J, which has empty intersection:

$$f^{-1}(J) = U_0 \cup U_1,$$

$$U_0 = \left[-\frac{\sqrt{7}}{3}, -\frac{1}{3}\right], \quad U_1 = \left[\frac{1}{3}, \frac{\sqrt{7}}{3}\right],$$

and

$$|U_0| < \frac{1}{2}|J| = 1, \quad |U_1| < \frac{1}{2}|J| = 1.$$

See Fig. 6.3. Here $|J|$ is the length of the interval J. Thus we have

$$f(U_0) = f(U_1) = J \supset U_0 \cup U_1. \tag{6.25}$$

From (6.25), there are two subinterval U_{00}, U_{01} and U_{10}, U_{11} in U_0 and U_1, respectively, with empty intersection such that

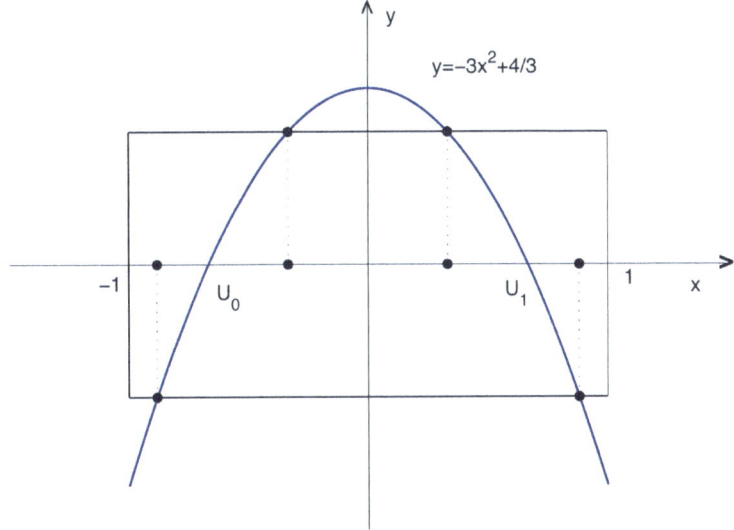

Fig. 6.3 The map f and sets U_0, U_1 for Example 6.49

$$f(U_{00}) = U_0, \quad f(U_{01}) = U_1, \tag{6.26}$$
$$f(U_{10}) = U_0, \quad f(U_{11}) = U_1, \tag{6.27}$$

and

$$|U_{ij}| < \frac{1}{2}|U_j| < \frac{1}{2}, \quad i,j = 1,2. \tag{6.28}$$

Continuing by induction, for any positive integer k, we define

$$U_{s_0 s_1 \cdots s_k} = U_{s_0} \cap f^{-1}(U_{s_1}) \cap \cdots \cup f^{-k}(U_{s_k}), \tag{6.29}$$

where $s_0, s_1, \ldots, s_k \in \{0,1\}$.

We continue the study of f throughout the following, up to Eq. (6.33). $\square$

Lemma 6.50 *Let $U_{s_0 s_1 \cdots s_k}$ be defined as (6.26). We have*

(i) $f(U_{s_0 s_1 \cdots s_k}) = U_{s_1 \cdots s_k}$;
(ii) $|U_{s_0 s_1 \cdots s_k}| < \frac{1}{2^k}$.

Proof Property (i) is true for $k = 1$ by (6.26)–(6.28).

Now we prove (i) and (ii) for the general case $k > 1$.

Using Lemma 6.48 and $f(U_{s_0}) = J$, we have

$$\begin{aligned} f(U_{s_0 s_1 \cdots s_k}) &= f(U_{s_0} \cap f^{-1}(U_{s_1 \cdots s_k})) \\ &= f(U_{s_0}) \cap U_{s_1 \cdots s_k} \\ &= U_{s_1 \cdots s_k}. \end{aligned}$$

For (ii), since f expands with an amplification factor >2, we have

$$|U_{s_1 \cdots s_k}| = |f(U_{s_0 s_1 \cdots s_k})| > 2|U_{s_0 s_1 \cdots s_k}|.$$

Thus

$$|U_{s_0 s_1 \cdots s_k}| < \frac{1}{2}|U_{s_1 \cdots s_k}| < \frac{1}{2} \cdot \frac{1}{2^{k-1}} = \frac{1}{2^k}. \tag{6.30}$$

$\square$

For $s = (s_0, s_1, s_2, \ldots) \in \Sigma_2^+$, define

$$U(s) = \cap_{j=0}^{\infty} f^{-j}(U_{s_j}) = \cap_{k=0}^{\infty} U_{s_0 s_1 \cdots s_k}. \tag{6.31}$$

Lemma 6.51 *The $U(s)$ defined by (6.31) satisfies*

(i) $f(U(s)) = U(\sigma^+(s))$;

6.7 Construction Of Shift Invariant Sets

(ii) $\sharp(U(s)) = 1$, i.e., $U(s)$ is a set with single point.

Proof (i) From $f(U_{s_0}) = J$, it follows that

$$\begin{aligned} f(U(s)) &= f(\cap_{j=0}^{\infty} f^{-j}(U_{s_j})) \\ &= f(U_{s_0}) \cap_{j=1}^{\infty} f^{1-j}(U_{s_j}) \\ &= \cap_{k=1}^{\infty} U_{s_1 \cdots s_k} \\ &= U(\sigma^+(s)). \end{aligned}$$

(ii) It follows directly from (ii) in Lemma 6.50. □

Let

$$\Lambda = \cap_{j=0}^{\infty} f^{-j}(J) = \cup_{s \in \Sigma_2^+} U(s). \tag{6.32}$$

Theorem 6.52 *The set Λ defined by (6.32) is a compact invariant set of f and there exists a topological conjugacy h from σ^+ to $f|_\Lambda$. Thus, Λ is a compact shift invariant set of f.*

Proof It is easy to see from (6.32) and the compactness of Σ_2^+ that Λ is a compact invariant set of f.

We now construct a topological conjugacy from σ^+ to $f|_\Lambda$. since, for every $s \in \Sigma_2^+$, the set $U(s)$ contains only a single point by (ii) in Lemma 6.51, we may identify it as the point contained in it. Define $h : \Sigma_2^+ \to \Lambda$ by

$$h(s) = U(s).$$

It suffices to prove that h is a conjugacy from σ^+ to $f|_\Lambda$.

We first claim that h is a homeomorphism. In fact, for $s, t \in \Sigma_2^+$ with

$$d(s, t) < \frac{1}{2^n},$$

we have that s and t must agree with each other in the first n bits by (6.16). Thus

$$h(s), h(t) \in U_{s_0 s_1 \cdots s_n} = U_{t_0 t_1 \cdots t_n},$$

and

$$|h(s) - h(t)| < \frac{1}{2^n}.$$

So h is continuous.

Let $s, t \in \Sigma_2^+$, $s \neq t$. Then there exists $k > 0$ such that $s_k \neq t_k$. On the other hand, by the definition of $U(s)$, we have

$$h(s) \in U_{s_0 s_1 \cdots s_k}, \qquad h(t) \in U_{t_0 t_1 \cdots t_k},$$

and
$$f^k(h(s)) \in U_{s_k}, \qquad f^k(h(t)) \in U_{t_k}.$$

But
$$U_{s_k} \cap U_{t_k} = \phi.$$

Therefore
$$f^k(h(s)) \neq f^k(h(t)),$$

and so
$$h(s) \neq h(t).$$

That is, h is one-to-one.

It is clear from the definition that h is onto Λ.

Since Σ_2^+ is compact space, Λ is Hausdorff space and $h : \Sigma_2^+ \to \Lambda$ is one-to-one, onto, and continuous, we have that h is a homeomorphism. This proves our claim.

Finally, it follows from (i) in Lemma 6.51 that

$$f|_\Lambda h = h\sigma^+. \tag{6.33}$$

That means that h is a topological conjugacy from σ^+ to $f|_\Lambda$. $\square$

From the above example, we see that the condition (6.25) is the key in proving the existence of shift invariant set. In general, we have the theorem as follow.

Theorem 6.53 *Let X be a compact metric space and f be continuous from X to X. Then f has a shift invariant set of order k with respect to one-side shift if and only if there exist compact subsets $A_0, A_1, \ldots, A_{k-1} \subset X$ with empty intersection with each other such that*

(i) $f(A_i) \supset \cup_{j=0}^{k-1} A_j$, $i = 0, 1, \ldots, k-1$;
(ii) $\sharp \left(\cap_{j=0}^\infty f^{-j}(A_{s_j}) \right) \leq 1$, $\forall (s_0 s_1 \cdots) \in \Sigma_k^+$.

Proof We follow Zhang [6]. First, prove the necessity condition. Let Λ be a shift invariant set of f of order k, i.e., there exists a homeomorphism $h : \Lambda \to \Sigma_k^+$ such that

$$hf|_\Lambda = \sigma^+ h.$$

Denote
$$B_i = \{s = (s_0 s_1 \cdots) \in \Sigma_k^+ \mid s_0 = i\},$$
$$A_i = h^{-1}(B_i), \quad i = 0, 1, \ldots, k-1.$$

It is easy to see that $A_0, A_1, \ldots, A_{k-1}$ are closed subsets in Λ with empty intersection each other, and

6.7 Construction Of Shift Invariant Sets

(i)
$$\begin{aligned}f(A_i) &= fh^{-1}(B_i) \\ &= h^{-1}\sigma^+(B_i) \\ &= h^{-1}(\Sigma_k^+) \\ &= \Lambda \supset \cup_{j=0}^{k-1} A_j, \qquad i = 0, 1, \cdots, k-1.\end{aligned}$$

(ii)
$$\begin{aligned}\cap_{j=0}^\infty f^{-j}(A_{s_j}) &= \cap_{j=0}^\infty f^{-j}h^{-1}(B_{s_j}) \\ &= \cap_{j=0}^\infty h^{-1}(\sigma^+)^{-j}(B_{s_j}) \\ &= h^{-1}\cap_{j=0}^\infty (\sigma^+)^{-j}(B_{s_j}) \\ &= h^{-1}((s_0 s_1 \cdots)).\end{aligned}$$

Since h is a homeomorphism, $h^{-1}((s_0 s_1 \cdots))$ is a set with a single point. Thus we have

$$\sharp\left(\cap_{j=0}^\infty f^{-j}(A_{s_j})\right) = 1,$$

and (i) and (ii) in the Theorem hold.

Then we show sufficiency. We first claim that for every $l > 0$ and any block $(s_0 s_1 \cdots s_l)$ in which each bit is taken in $\{0, 1, \ldots, k-1\}$, we have

$$f(A_{s_0 s_1 \cdots s_l}) = A_{s_1 \cdots s_l} \qquad (6.34)$$
$$f^l(A_{s_0 s_1 \cdots s_l}) = A_l, \qquad (6.35)$$

where
$$A_{s_0 s_1 \cdots s_l} = \cap_{j=0}^l f^{-j}(A_{i_j}).$$

In fact, the proof of (6.34) is the same as that in Lemma 6.50. For (6.35). It is true for $l = 1$, since
$$f(A_{s_0} \cap f^{-1}(A_{s_1})) = f(A_{s_0}) \cap A_{s_1} = A_{s_1},$$
by Lemma 6.48 and the Assumption (i).

Assume that it is also true for l. Again from Lemma 6.48, we have

$$f^{l+1}(A_{s_0s_1\cdots s_{l+1}}) = f^{l+1}\left(\cap_{j=0}^{l+1} f^{-j}(A_{s_j})\right)$$
$$= f \circ f^l \left(\cap_{j=0}^{l} f^{-j}(A_{s_j}) \cap f^{-l} \circ f^{-1}(A_{s_{(l+1)}})\right)$$
$$= f \left(f^l \left(\cap_{j=0}^{l} f^{-j}(A_{s_j})\right) \cap f^{-1}(A_{s_{(l+1)}})\right)$$
$$= f \left(A_{s_l} \cap A_{s_{l+1}}\right)$$
$$= A_{s_{l+1}}.$$

We have that (6.35) is true for every $l > 0$ by induction. This proves our claim.
From the compactness of X and (6.35), we have

$$\cap_{j=0}^{\infty} f^{-j}(A_{s_j}) \neq \phi, \quad \forall (s_0 s_1 \cdots) \in \Sigma_k^+,$$

by Cantor intersection property. Thus the assumption (ii) implies that for every $s = (s_0 s_1 \cdots) \in \Sigma_k^+$

$$U(s) \triangleq \cap_{j=0}^{\infty} f^{-j}(A_{s_j})$$

is a set with a single point. In the following, we identify it as the single point contained in it.
Similar to (i) in Lemma 6.51, we have

$$f(U(s)) = U(\sigma^+(s)). \tag{6.36}$$

Next, denote

$$C = A_0 \cup \cdots \cup A_{k-1},$$

and

$$\Lambda = \cap_{j=0}^{\infty} f^{-j}(C) = \cup_{s \in \Sigma_k^+} U(s).$$

Since C is compact and so is $f^{-j}(C)$ by the continuity of f, Λ is compact by Tychonov's Theorem. And

$$f(\Lambda) \subset \Lambda.$$

Define $h : \Sigma_k^+ \to \Lambda$ by

$$h(s) = U(s).$$

Similar to the proof of Theorem 6.52, we can show that h is one-to-one and onto Λ. To show that h is a homeomorphism, it suffices to prove that either h or h^{-1} is continuous. Here we prove the latter for the self-contained, since we have proved the continuity of h in Theorem 6.52.
For any $x \in \Lambda$, let

$$h^{-1}(x) = s = (s_0 s_1 \cdots).$$

6.8 Snap-Back Repeller as a Shift Invariant Set

For any $\varepsilon > 0$, by (6.16) there exists a $n > 0$ such that

$$[s_0 s_1 \cdots s_{n-1}]_0 \triangleq \{t = (t_0 t_1 \cdots) \in \Sigma_k^+ \mid t_0 = s_0, \ldots, t_{n-1} = s_{n-1}\}$$
$$\subset \{t \in \Sigma_k^+ \mid d(s, t) < \varepsilon\}.$$

Since f is continuous, $A_0, \ldots, A_{k-1}$ has empty intersection with each other and

$$f^j(x) \in A_{s_j}, \qquad j = 0, 1, \ldots, n-1,$$

there exists $\delta > 0$ such that when $y \in O_\delta(x) \cap \Lambda$, we have

$$f^j(y) \in A_{s_j}, \qquad j = 0, 1, \ldots, n-1.$$

By noticing that $y \in \Lambda$, we have

$$y = A_{s_0} \cap f^{-1}(A_{s_1}) \cap \cdots \cap f^{-(n-1)}(A_{s_{n-1}}) \cap \bigcap_{j=n}^{\infty} f^{-j}(A_{i_j}),$$

for some $(i_n i_{n+1} \cdots) \in \Sigma_k^+$. Thus

$$h^{-1}(y) \in [s_0 \cdots s_{n-1}]_0,$$

by the definition of h^{-1}. This prove the continuity of h^{-1}.

Finally, it follows from (6.36) that

$$f|_\Lambda h = h\sigma^+.$$

Thus h is a topological conjugacy from f to σ^+. □

From the proof of Theorem 6.53, it is easy to see the following.

Corollary 6.54 *Let X be a compact metric space and f be continuous from X to X. Then f has a quasi-shift invariant set of order k with respect to one-side shift if and only if the condition (i) in Theorem 6.53 holds.*

6.8 Snap-Back Repeller as a Shift Invariant Set

In this section, we present a dynamical system on $\mathbb{R}^N$ that has a shift invariant set.

Let $x \in \mathbb{R}^N$ and f be a differentiable map from an open set $\mathcal{O} \subseteq \mathbb{R}^N$ into $\mathbb{R}^N$. Thus, $f = (f_1, f_2, \ldots, f_N)$, where $f_1, f_2, \ldots, f_N$ are the components of f. We denote

$$Df(x) = \begin{bmatrix} \frac{\partial f_1}{\partial x_1} & \frac{\partial f_2}{\partial x_2} & \cdots & \frac{\partial f_N}{\partial x_1} \\ \frac{\partial f_1}{\partial x_2} & \frac{\partial f_2}{\partial x_2} & \cdots & \frac{\partial f_N}{\partial x_2} \\ \vdots & & & \\ \frac{\partial f_1}{\partial x_N} & \frac{\partial f_2}{\partial x_N} & \cdots & \frac{\partial f_N}{\partial x_n} \end{bmatrix}(x).$$

Definition 6.55 Let $f : \mathbb{R}^N \longrightarrow \mathbb{R}^N$ be C^1. Let p be a fixed point of f. We say that f is a *repelling fixed point* if all of the eigenvalues of $Df(p)$ have absolute value larger than 1. A repelling fixed point p is called a *snap-back repeller* if, for any neighborhood V of p, there exist $q \in V$, $q \neq p$ and an integer m, such that $f^m(p) = q$ and $\det Df^m(p) \neq 0$. Such a point q is called a snap-back point. □

This definition was first introduced by Marotto and he established that f has chaos if f has a snap-back repeller. See [7]. Here, we shall show that there exists a positive integer r such that f^r has a shift invariant set with respect to one-side shift if f has a snap-back repeller.

Lemma 6.56 Let A be an $N \times N$ constant matrix such that all of its eigenvalues have absolute values larger than μ. Then exists a norm $|\cdot|$ in $\mathbb{R}^N$ such that the associated operator-norm of A^{-1}, $\|A^{-1}\|$, satisfies $\|A^{-1}\| \leq \mu$. Consequently, A satisfies $|Ax| \geq \mu|x|$ for all $x \in \mathbb{R}^N$.

Proof This is left as an exercise. □

For $x \in \mathbb{R}^N$, let $\mathcal{N}(x)$ denote the set of all closed neighborhood V of x, where

$$V = \{y \in \mathbb{R}^N \mid |y - x| \leq r\}, \quad r > 0.$$

Lemma 6.57 Let $f : \mathbb{R}^N \longrightarrow \mathbb{R}^N$ be C^1. If f has a snap-back repeller p, then there exists $U \in \mathcal{N}(p)$ such that

(i) $U \subset f(U)$;
(ii) $\bigcap_{k=0}^{\infty} f^{-k}(U) = \{p\}$;
(iii) let $q \in \text{int } U$ be the associated snap-back point. Then there exists $V \in \mathcal{N}(q)$ such that $f^m : V \to f^m(V)$ is diffeomorphism.

Proof Since p is a snap-back repeller, all of the eigenvalues of $Df(p)$ have absolute values larger than 1. By Lemma 6.56, we can define a norm $|\cdot|$ on $\mathbb{R}^N$ such that

$$|Df(p)x| \geq \mu|x|, \quad \forall x \in \mathbb{R}^N, \qquad (6.37)$$

for some $\mu > 1$.

6.8 Snap-Back Repeller as a Shift Invariant Set

Fix μ' with $1 < \mu' < \mu$. By continuity, we can find $U \in \mathcal{N}(p)$ such that

$$\|Df(x) - Df(p)\| \leq \mu - \mu', \quad \forall x \in U. \tag{6.38}$$

Let $\omega = (f - Df(p))|_U$. By (6.38), we have

$$|\omega(x') - \omega(x)| \leq \left(\int_0^t \|D\omega(x + t(x' - x))\| dt\right) |x' - x|$$

$$\leq (\mu - \mu')|x' - x|, \quad \forall x', x \in U.$$

Thus, for any $x', x \in U$, we have

$$|f(x') - f(x)| = |(f(x') - Df(p)x') - (f(x) - Df(p)x) + Df(p)(x' - x)|$$
$$\geq |Df(p)(x' - x)| - |\omega(x') - \omega(x)|$$
$$\geq \mu|x' - x| - (\mu - \mu')|x' - x| = \mu'|x' - x|. \tag{6.39}$$

Therefore f expands U and (i) holds.

For (ii), since $f^{-1}(U) \subset U$ by (i), we have

$$\cdots \subset f^{-k}(U) \subset f^{-(k-1)}(U) \subset \cdots \subset f^{-1}(U) \subset U.$$

And from (6.39) it follows that

$$|f^{-1}(x') - f^{-1}(x)| \leq \frac{1}{\mu'}|x' - x|, \quad \forall x', x \in U.$$

Thus

$$|f^{-k}(x') - f^{-k}(x)| \leq \frac{1}{\mu'}|f^{-(k-1)}(x') - f^{-(k-1)}(x)|$$
$$\leq \cdots\cdots$$
$$\leq \frac{1}{\mu'^k}|x' - x| \to 0, \quad \text{as } k \to \infty,$$

for all $x', x \in U$. We have (ii).

Property (iii) follows from $\det Df^m(p) \neq 0$ and the inverse function theorem. $\square$

Theorem 6.58 *Let $f : \mathbb{R}^N \longrightarrow \mathbb{R}^N$ be C^1. If f has a snap-back repeller p, then there exists a positive integer r such that f^r has a shift invariant set of order 2 with respect to one-side shift.*

Proof Let U and V be defined by Lemma 6.57. By Theorem 6.53 it suffices to prove that there exist two nonempty compact sets $A_1, A_2 \subset U$ with empty intersection such that (i) and (ii) in Theorem 6.53 hold with f instead by f^r.

Let $q \in U$ be the snap-back point. Since $q \neq p$, there exist two nonempty neighborhoods $V_0 \in \mathcal{N}(p)$ and $W_0 \in \mathcal{N}(q)$ with empty intersection, which are contained in U and $U \cap V$, respectively. Let $W = f^{-m}(V_0) \cap W_0$. Since f^m is continuous and $W \neq \phi$ (in fact $q \in W$), it follows that $W \in \mathcal{N}(q)$. Let $V_1 = f^m(W)$. Then $V_1 \in \mathcal{N}(p)$ since f^m is a diffeomorphism from V to $f^m(V)$. By Lemma 6.48, $V_1 = f^m(f^{-m}(V_0) \cap W_0) = V_0 \cap f^m(W_0)$. So $V_1 \cap W \neq \phi$.

On the other hand, since f^{-1} is strict contraction on U, there exists an integer $l \geq 0$ such that
$$\cap_{j=0}^{l} f^{-j}(U) \subset V_1.$$

Let $r = m + l$, $A_1 = \cap_{j=0}^{r-1} f^{-j}(U)$ and $A_2 = W$. Then A_1 and A_2 are nonempty compact and $A_1 \cap A_2 \neq \phi$. It remains to prove that they satisfy (i) and (ii) in Theorem 6.53.

By applying Lemma 6.48 r times, we have
$$f^r(A_1) = f(U) \supset U \supset A_1 \cup A_2.$$

$$f^r(A_2) = f^l(f^m(W)) = f^l(V_1) \supset f^l\left(\cap_{j=0}^{l} f^{-j}(U)\right) \supset U \supset A_1 \cup A_2.$$

Thus (i) in Theorem 6.53 holds.

The condition (ii) in Theorem 6.53 follows from f^{-r} is strict contraction and that
$$\cap_{j=0}^{\infty} f^{-rj}(A_{s_j}) \subset \cap_{j=0}^{\infty} f^{-rj}(U) = \{p\},$$

for any $(s_0 s_1 \cdots) \in \Sigma_2^+$. □

Exercise 6.59 Let $f \in C^1(I)$. If $|f'(x)| \leq 1, \forall x \in I$, prove that f has no shift invariant sets. □

Exercise 6.60 Let $f(x) = 4x(1-x)$. Show that there exist two nonempty closed subintervals K_0 and K_1 of $[0, 1]$ with empty intersection such that
$$f^2(K_0) \cap f^2(K_1) \supset [0, 1].$$
□

Exercise 6.61 Let $f \in C^0(I)$. Prove that if g has a periodic point whose period is not a power of 2, then there exist two nonempty closed subintervals K_0 and K_1 of I with empty intersection and a positive integer m such that
$$f^m(K_0) \cap f^m(K_1) \supset K_0 \cup K_1.$$
□

Exercise 6.62 Prove Corollary 6.54. □

Exercise 6.63 Construct a piecewise continuous linear map f on I such that f has a quasi-shift invariant set of order k, for any positive integer k. □

Notes for Chapter 6

Symbolic dynamics originated as a method and representation to study general dynamical systems. The original ideas may be traced back to the works of several mathematicians: J. Hadamard, M. Morse, E. Artin, P.J. Myrberg, P. Koebe, J. Nielsen, and G.A.Hedlund. But the first formal treatment was made by Morse and Hedlund [8] in 1938.

A salient feature of symbolic dynamics is that time is measured in discrete time, and the orbits are represented by a string of symbols. The role of the system dynamics or time evolution becomes the shift operator. Important concepts such as periodicity, denseness of orbits, topological mixing, topological transitivity, and sensitive dependence on initial data all have particularly elegant discussions or representations in symbolic dynamics.

C. Shannon used symbolic sequences and shifts of finite type in his seminal paper [9] to study information theory in 1948. Today, symbolic dynamics finds many applications in computer science such as data storage, transmission and manipulation, and other areas.

Our treatment in Sect. 6.4 is mostly based on Zhou [4], in particular, Lemmas 6.30–6.34 and Theorem 6.35. Theorems 6.52–6.53 comes from Zhang [6, 10]. Theorem 6.58 comes from [11].

References

1. R.L. Devaney, *An Introduction to Chaotic Dynamical Systems*, 2nd ed., Addison-Wesley, New York, 1989. Cited on page(s) 19, 36, 75, 86, 209, 211
2. J. Banks, J. Brooks, G. Cairns and P. Stacey, On Devaney's definition of chaos, *Amer. Math. Monthly* **99** (1992), 332–334. https://doi.org/10.2307/2324899
3. R.C. Robinson, *An Introduction to Dynamical Systems: Continuous and Discrete*, Pearson Education Asia Limited and China Machine Press, 2005.
4. Z.L. Zhou, *Symbolic Dynamics*, Shanghai Scientific and Technological Education Publishing House, Shanghai, China, 1997 (in Chinese).
5. T.Y. Li and J.A. Yorke, Period three implies chaos, *Amer. Math. Monthly* **82** (1975), 985–992. https://doi.org/10.2307/2318254
6. Z.S. Zhang, Shift-invariant sets of endomorphisms, *Acta Math. Sinica* **27** (1984), 564–576 (in Chinese).
7. F.R. Marotto, Snap-back repellers imply chaos in R^n, *J. Math. Anal. Appl.* **63** (1978), 199–223. https://doi.org/10.1016/0022-247X(78)90115-4
8. M. Morse and G.A. Hedlund, Symbolic dynamics, *Amer. J. Math.* **60** (1938), 815–866. https://doi.org/10.2307/2371264
9. C.E. Shannon, A mathematical theory of communication, *Bell System Technical J.* **27** (1948), 379–423. https://doi.org/10.1145/584091.584093

10. Z.S. Zhang, *Principles of Differential Dynamical Systems*, Science Publishing House of China, Beijing, China, 1997.
11. G. Chen, S.B. Hsu and J. Zhou, Chaotic vibrations of the one-dimensional wave equation due to a self-excitation boundary condition. Part III, natural hysteresis memory effects, *Int. J. Bifur. Chaos* **8** (1998), 447–470. https://doi.org/10.1142/S0218127498001236

The Smale Horseshoe

The Smale horseshoe offers a model for pervasive high-dimensional nonlinear phenomena, as well as a powerful technique for proving chaos. Here in this chapter, we present the famous Smale horseshoe and show that it has a shift invariant set with respect to the two-sided shift. We first introduce the standard Smale horseshoe and then discuss the general case.

7.1 The Standard Smale Horseshoe

Let $\varepsilon > 0$ be small. Consider the two square regions P and Q in $\mathbb{R}^2$:

$$P = (-1-\varepsilon, 1+\varepsilon) \times (-1-\varepsilon, 1+\varepsilon), \tag{7.1}$$

and

$$Q = [-1, 1] \times [-1, 1]. \tag{7.2}$$

The horseshoe map is described geometrically on P as the composition of two maps. One map is linear which expands in vertical direction by λ ($\lambda > 2$) and contracts in horizontal direction by $1/\lambda$. The other is a nonlinear smooth map which bends the image of the linear map into a horseshoe-shaped object and places the image on P (see Fig. 7.1).

By the above construction, we have defined a horseshoe map

$$\varphi \colon P \to \mathbb{R}^2,$$

which is a diffeomorphism from P to $\varphi(P)$. We now show that φ has a shift invariant set of order 2 with respect to a two-sided shift.

First, we observe that

$$V = \varphi(Q) \cap Q$$

Fig. 7.1 The standard Smale horseshoe. The square with dotted lines is P, while the square with the solid lines is Q

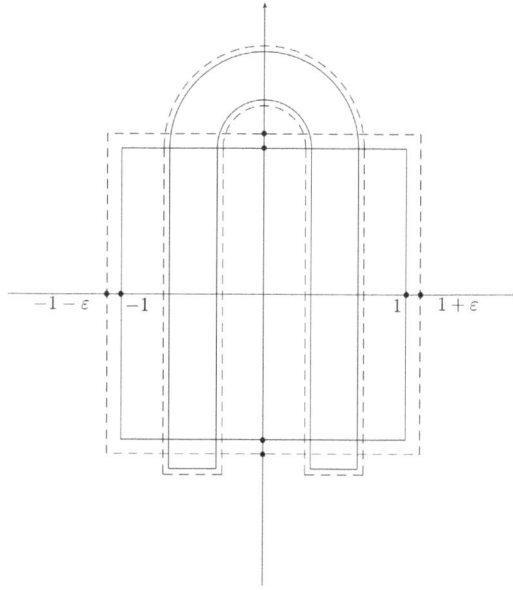

is composed of two vertical strips V_0 and V_1 with empty intersection, i.e.,

$$V = V_0 \cup V_1.$$

The width of each strip is less than the width of Q:

$$\theta(V_0) < 1, \qquad \theta(V_1) < 1.$$

Here $\theta(V_i)$ denotes twice the width of the vertical strip V_i, $i = 0, 1$.

Next, we observe that $U = \varphi^{-1}(V)$ is composed of two horizontal strips $U_0 = \varphi^{-1}(V_0)$ and $U_1 = \varphi^{-1}(V_1)$, i.e.,

$$U = U_0 \cup U_1, \quad \text{and} \quad U_0 \cap U_1 = \emptyset.$$

The thickness of each horizontal strip is less than half the thickness of Q:

$$\theta(U_0) < 1, \qquad \theta(U_1) < 1.$$

Here $\theta(U_i)$ denotes twice the thickness of the horizontal strip U_i (see Fig. 7.2).

In the following, denote

$$U_{ij} = \varphi^{-1}(V_i \cap U_j) = U_i \cap \varphi^{-1}(U_j), \qquad i, j = 0, 1;$$
$$V_{ij} = \varphi(U_i \cap V_j) = V_i \cap \varphi(V_j), \qquad i, j = 0, 1.$$

7.1 The Standard Smale Horseshoe

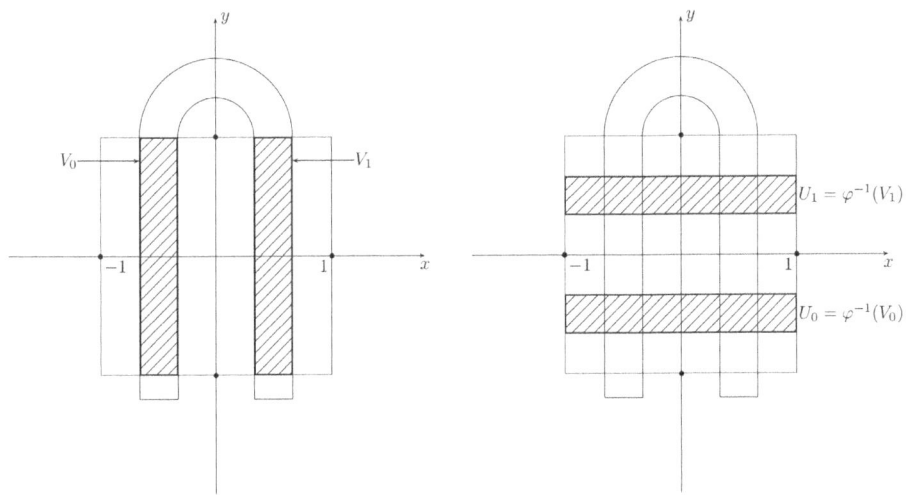

Fig. 7.2 Illustrations of the sets V_0, V_1, and U_0, U_1

It is easy to see that U_{ij} is a horizontal strip contained in U_i. Its thickness is less than half the thickness of U_i:

$$\theta(U_{ij}) < \frac{1}{2}\theta(U_i) < \frac{1}{2}.$$

Likewise, V_{ij} is a vertical strip contained in V_i. Its width is less than half the width of V_i:

$$\theta(V_{ij}) < \frac{1}{2}\theta(V_i) < \frac{1}{2}.$$

See Fig. 7.3.

In general, for

$$s_{-k}, \ldots, s_{-1}, s_0, s_1, \ldots, s_k \in \{0, 1\} = S(2), \quad (\text{cf. } (6.11)),$$

and $k > 1$, we define $U_{s_0 \ldots s_k}$ and $V_{s_{-k} \ldots s_{-2} s_{-1}}$ inductively as follows:

$$U_{s_0 \ldots s_k} = \varphi^{-1}(V_{s_0} \cap U_{s_1 \ldots s_k}) = U_{s_0} \cap \varphi^{-1}(U_{s_1 \ldots s_k})$$
$$= U_{s_0} \cap \varphi^{-1}(U_{s_1}) \cap \cdots \cap \varphi^{-k}(U_{s_k});$$
$$V_{s_{-k} \ldots s_{-2} s_{-1}} = \varphi(U_{s_{-1}} \cap V_{s_{-k} \ldots s_{-2}}) = V_{s_{-1}} \cap \varphi(V_{s_{-k} \ldots s_{-2}})$$
$$= V_{s_{-1}} \cap \varphi(V_{s_{-2}}) \cap \cdots \cap \varphi^{k-1}(V_{s_{-k}}).$$

We can see that $U_{s_0 \ldots s_k}$ is a horizontal strip contained in $U_{s_0 \ldots s_{k-1}}$ and $V_{s_{-k} \ldots s_{-2} s_{-1}}$ is a vertical strip contained in $V_{s_{-k+1} \ldots s_{-2} s_{-1}}$.

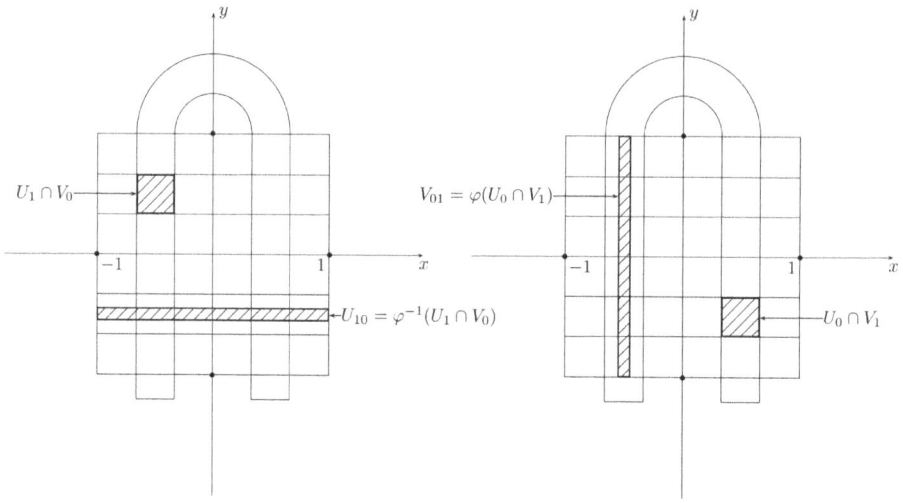

Fig. 7.3 Illustration of the sets U_{10} and V_{01}

Lemma 7.1 *We have*

(i) $\varphi(U_{s_0...s_k}) = V_{s_0} \cap U_{s_1...s_k}$, $\quad \varphi(V_{s_{-k}...s_{-2}s_{-1}}) \cap V_{s_0} = V_{s_{-k}...s_{-1}s_0}$;
(ii) $\theta(U_{s_0...s_k}) < \frac{1}{2}\theta(U_{s_1...s_k}) < \frac{1}{2^k}$, $\theta(V_{s_{-k}...s_{-2}s_{-1}}) < \frac{1}{2}\theta(V_{s_{-k}...s_{-2}}) < \frac{1}{2^k}$.

Proof (i) Follows directly from the definition.

For (ii), by definition,

$$U_{s_0...s_k} = \varphi^{-1}(V_{s_0} \cap U_{s_1...s_k}).$$

Since φ^{-1} is a strict contraction in the vertical direction with a contracting rate $<1/2$, we have

$$\theta(U_{s_0...s_k}) < \frac{1}{2}\theta(U_{s_1...s_k}) < \frac{1}{2^k}.$$

Similarly, we have

$$\theta(V_{s_{-k}...s_{-2}s_{-1}}) < \frac{1}{2}\theta(V_{s_{-k}...s_{-2}}) < \frac{1}{2^k}.$$

$\square$

For $s = (\ldots, s_{-2}, s_{-1}, s_0, s_1, s_2, \ldots) \in \Sigma_2$, we introduce

$$U(s) = \bigcap_{j=0}^{\infty} \varphi^{-j}(U_{s_j})$$

$$= \bigcap_{k=0}^{\infty} U_{s_0 s_1...s_k} \left(= \bigcap_{k=1}^{\infty} U_{s_0 s_1...s_k} \right),$$

7.1 The Standard Smale Horseshoe

$$V(s) = \bigcap_{j=1}^{\infty} \varphi^{j-1}(V_{s_{-j}})$$

$$= \bigcap_{k=1}^{\infty} V_{s_{-k}...s_{-1}} \left(= \bigcap_{k=2}^{\infty} V_{s_{-k}...s_{-1}} \right).$$

Lemma 7.2 *We have*

(i) $\varphi(V(s) \cap U(s)) = V(\sigma(s)) \cap U(\sigma(s))$;
(ii) $\sharp(V(s) \cap U(s)) = 1$.

Proof (i) By part (i) in Lemma 7.1, it follows that

$$\varphi(U(s)) = \bigcap_{k=1}^{\infty} \varphi(U_{s_0 s_1 ... s_k})$$

$$= V_{s_0} \cap \bigcap_{k=1}^{\infty} U_{s_1 ... s_k}$$

$$= V_{s_0} \cap U(\sigma(s)),$$

$$\varphi(V(s)) \cap V_{s_0} = \bigcap_{k=1}^{\infty} \varphi(V_{s_{-k}...s_{-1}}) \cap V_{s_0}$$

$$= \bigcap_{k=1}^{\infty} V_{s_{-k}...s_{-1} s_0}$$

$$= V(\sigma(s)).$$

Therefore

$$\varphi(V(s) \cap U(s)) = \varphi(V(s)) \cap \varphi(U(s))$$
$$= \varphi(V(s)) \cap V_{s_0} \cap U(\sigma(s))$$
$$= V(\sigma(s)) \cap U(\sigma(s)).$$

We have proved (i).
(ii) For any $k \in \mathbb{N}$, we have

$$V(s) \cap U(s) \subset V_{s_{-k}...s_{-1}} \cap U_{s_0...s_k}.$$

But by part (ii) in Lemma 7.1, we have

$$\theta(V_{s_{-k}...s_{-1}}) < \frac{1}{2^{k-1}}, \quad \theta(U_{s_0...s_k}) < \frac{1}{2^k}.$$

So we have $\sharp(V(s) \cap U(s)) = 1$. □

Denote
$$\Lambda = \bigcup_{s \in \Sigma_2} (V(s) \cap U(s)).$$

We identify the singleton set $V(s) \cap U(s)$ with the singleton itself contained therein and define a map $h: \Sigma_2 \to \Lambda$ by

$$h(s) = V(s) \cap U(s), \quad \forall s \in \Sigma_2.$$

Theorem 7.3 (Smale [1, 2]) *The set Λ is a compact invariant set of φ, and $\varphi|_\Lambda$ is topologically conjugate to the two-sided shift $\sigma: \Sigma_2 \to \Sigma_2$ with conjugacy h. Therefore, the horseshoe map φ has a shift invariant set of order 2 with respect to the two-sided shift.*

Proof First, we claim that $h: \Sigma_2 \to \Lambda$ is continuous. In fact, if $s, t \in \Sigma_2$ satisfies

$$d(s, t) < \frac{1}{2^k}.$$

Then $h(s)$ and $h(t)$ must belong to a rectangle which has its width $<1/2^{k-1}$ and its thickness $<1/2^k$:

$$h(s), h(s) \in V_{s_{-k}...s_{-1}} \cap U_{s_0 s_1...s_k}.$$

So we have proved our claim.

Next, we prove that h is one-to-one. Let $s, t \in \Sigma_2$. If there exists a $k \in \mathbb{Z}$, $k \leq 0$, such that $s_k \neq t_k$, then $U_{s_k} \cap U_{t_k} = \emptyset$ and

$$\varphi^k(h(s)) \in U_{s_k}, \quad \varphi^k(h(t)) \in U_{s_t}.$$

Thus $h(s) \neq h(t)$.

If there exists an $\ell \in \mathbb{Z}$, $\ell > 0$, such that $s_{-\ell} \neq t_{-\ell}$, then $V_{s_{-\ell}} \cap V_{t_{-\ell}} = \emptyset$, and

$$\varphi^{-(\ell-1)}(h(s)) \in V_{s_{-\ell}}, \quad \varphi^{-(\ell-1)}(h(t)) \in V_{t_{-\ell}}.$$

Thus, we also have $h(s) \neq h(t)$.

The proof of the remainders is the same as that of Theorem 6.52. □

7.2 The General Horseshoe

Throughout, we use Q to denote the unit square $[-1, 1] \times [-1, 1]$ as in (7.1). The Smale horseshoe map discussed in Sect. 7.1 appears quite artificial and restrictive. In this section, we introduce the Conley–Moser condition which can also generate a "horseshoe" type shift invariant set, but is much more general.

7.2 The General Horseshoe

Let $u: [-1, 1] \to [-1, 1]$ be a continuous function. We say that the curve $y = u(x)$ is a μ_h-horizontal curve if the function satisfies

$$|u(x_1) - u(x_2)| \le \mu_h |x_1 - x_2|, \qquad \forall x_1, x_2 \in [-1, 1].$$

That is, u is a Lipschitz function with Lipschitz constant μ_h. Similarly, for $v : [-1, 1] \to [-1, 1]$, we say that $x = v(y)$ is a μ_v-vertical curve if

$$|v(y_1) - v(y_2)| \le \mu_v |y_1 - y_2|, \qquad \forall y_1, y_2 \in [-1, 1].$$

Definition 7.4 (i) Let $y = u_1(x)$ and $y = u_2(x)$ be two non-intersecting μ_h-horizontal curves such that

$$-1 \le u_1(x) < u_2(x) \le 1.$$

We call

$$U = \{(x, y) \in \mathbb{R}^2 \mid -1 \le x \le 1, u_1(x) \le y \le u_2(x)\}$$

a μ_h-horizontal strip.

(ii) Let $x = v_1(y)$ and $x = v_2(y)$ be two non-intersecting μ_v-vertical curves such that

$$-1 \le v_1(y) < v_2(y) \le 1.$$

We call

$$V = \{(x, y) \in \mathbb{R}^2 \mid -1 \le y \le 1, v_1(y) \le x \le v_2(y)\}$$

a μ_v-vertical strip. □

See Fig. 7.4 for an illustration.
We define the thickness of U and the width of V, respectively, by

$$\theta(U) = \max_{-1 \le x \le 1} \{u_2(x) - u_1(x)\}, \quad \theta(V) = \max_{-1 \le y \le 1} \{v_2(y) - v_1(y)\}.$$

Lemma 7.5 *Let $U^1 \supset U^2 \supset \cdots \supset U^k \supset U^{k+1} \supset \cdots$ be a nested sequence of μ_h-horizontal strips such that*

$$\lim_{k \to +\infty} \theta(U^k) = 0.$$

Then $U^\infty \equiv \bigcap_{k=1}^\infty U^k$ is a μ_h-horizontal curve.
Similarly, for a nested sequence of μ_v-vertical strips $V^1 \supset V^2 \supset \cdots \supset V^k \supset V^{k+1} \supset \cdots$ satisfying $\lim_{k \to +\infty} \theta(U^k) = 0$, then $V^\infty \equiv \bigcap_{k=1}^\infty V^k$ is a μ_v-vertical curve.

Proof Let

$$U^k = \{(x, y) \in \mathbb{R}^2 \mid -1 \le x \le 1, u_1^{(k)}(x) \le y \le u_2^{(k)}(x)\}.$$

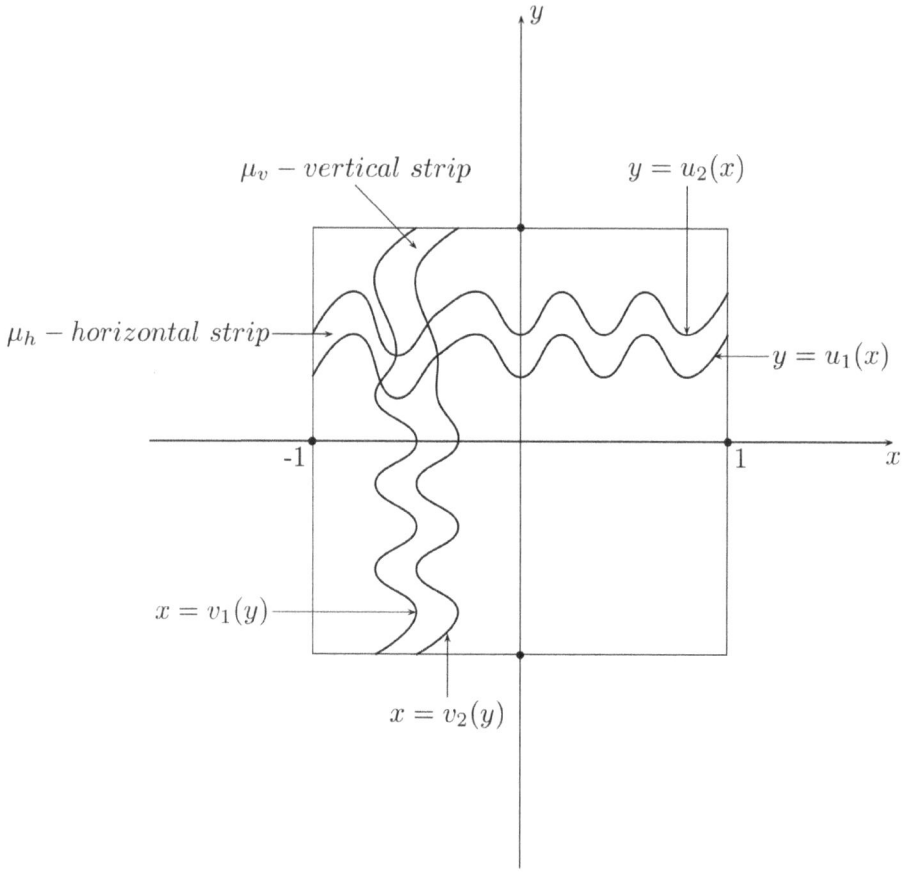

Fig. 7.4 Illustration of a μ_v-vertical strip and a μ_h-horizontal strip

By the given assumptions, for fixed $x \in [-1, 1]$, $\{u_1^{(k)}(x)\}$ is increasing and $\{u_2^{(k)}(x)\}$ is decreasing and $u_1^{(k)}(x) < u_2^{(k)}(x)$. There exists the squeezed limit

$$\lim_{k \to +\infty} u_1^{(k)}(x) = \lim_{k \to +\infty} u_2^{(k)}(x) \equiv u(x).$$

We thus obtain a function $u \colon [-1, 1] \to [-1, 1]$ with

$$|u(x_1) - u(x_2)| \leq \mu_h |x_1 - x_2|, \qquad \forall x_1, x_2 \in [-1, 1].$$

Thus $y = u(x)$ is the intersection of the horizontal strips U^k, $k = 1, 2, \ldots$. □

7.2 The General Horseshoe

Lemma 7.6 *Assume that $0 \le \mu_h \mu_v < 1$. Then each μ_h-horizontal curve and each μ_v-vertical curve intersect at a unique point.*

Proof It suffices to prove that the equation

$$\begin{cases} y = u(x) \\ x = v(y) \end{cases}$$

has a unique solution.

Substituting the first equation into the second, we have

$$x = v(u(x)).$$

Since

$$|v(u(x_1)) - v(u(x_2))| \le \mu_v |u(x_1) - u(x_2)|$$
$$\le \mu_v \mu_h |x_1 - x_2|, \quad \forall x_1, x_2 \in [-1, 1],$$

and $0 \le \mu_h \mu_v < 1$, the function $v \circ u \colon [-1, 1] \to [-1, 1]$ has a unique fixed point which is the unique intersection point of the two curves. $\square$

In the following, assume that $\mu_h = \mu_v = \mu < 1$. By Lemma 7.6, a μ-horizontal curve $y = u(x)$ and a μ-vertical curve $x = v(y)$ has a unique intersecting point, denoted by $z = (x, y)$. Let

$$\|u\| = \max_{-1 \le x \le 1} \{|u(x)|\},$$
$$\|v\| = \max_{-1 \le x \le 1} \{|v(x)|\},$$
$$|z| = \max\{|x|, |y|\}.$$

Lemma 7.7 *Let $z_j = (x_j, y_j)$ be the intersecting point of a μ-horizontal curve u_j and a μ-vertical curve v_j, $j = 1, 2$. Then we have*

$$|z_1 - z_2| \le \frac{1}{1-\mu} \max\{\|u_1 - u_2\|, \|v_1 - v_2\|\}.$$

Proof We have

$$|x_1 - x_2| = |v_1(y_1) - v_2(y_2)|$$
$$= |v_1(y_1) - v_1(y_2)| + |v_1(y_2) - v_2(y_2)|$$
$$\leq \mu |y_1 - y_2| + \|v_1 - v_2\|$$
$$\leq \mu |z_1 - z_2| + \|v_1 - v_2\|,$$
$$|y_1 - y_2| = |u_1(x_1) - u_2(x_2)|$$
$$= |u_1(x_1) - u_1(x_2)| + |u_1(x_2) - x_2(x_2)|$$
$$\leq \mu |x_1 - x_2| + \|u_1 - u_2\|$$
$$\leq \mu |z_1 - z_2| + \|u_1 - u_2\|.$$

Thus

$$|z_1 - z_2| = \mu |z_1 - z_2| + \max\{\|u_1 - u_2\|, \|v_1 - v_2\|\}$$
$$|z_1 - z_2| \leq \frac{1}{1-\mu} \max\{\|u_1 - u_2\|, \|v_1 - v_2\|\}.$$

□

Corollary 7.8 *Let V be a μ-vertical strip composed of two μ-vertical curves v_1 and v_2. A μ-horizontal curve u intersects with v_1 and v_2 at z_1 and z_2, respectively. Then*

$$|z_1 - z_2| \leq \frac{1}{1-\mu}\theta(V).$$

Similarly, if U is a μ-horizontal strip composed of two μ-horizontal curves u_1 and u_2. A μ-vertical curve v intersects with u_1 and u_2 at z_1 and z_2, respectively. Then

$$|z_1 - z_2| \leq \frac{1}{1-\mu}\theta(U).$$

Proof It follows from Lemma 7.7 by taking $u_1 = u_2$ and $v_1 = v_2$, respectively. □

Let $\psi \colon Q \to \mathbb{R}^2$ be a homeomorphism. Let $U_0, \ldots, U_{N-1}$ be a collection of disjoint μ-horizontal strips; $V_0, \ldots, V_{N-1}$ be a collection of disjoint μ-vertical strips.

We make the following assumptions on ψ:

(A1) For $j = 0, \ldots, N-1$,
$$\psi(U_j) = V_j,$$
and the horizontal boundaries of U_j are mapped onto the horizontal boundaries of V_j, the vertical boundaries of U_j are mapped onto the vertical boundaries of V_j.

(A2) Let U be a μ-horizontal strip contained in $\bigcup_{j=0}^{N-1} U_j$. Then
$$\widetilde{U}_k = \psi^{-1}(V_k \cap U)$$

7.2 The General Horseshoe

is a μ-horizontal strip, which is contained in U_k. Furthermore, there exists ν: $0 < \nu < 1$ such that
$$\theta(\tilde{U}_k) \leq \nu\theta(U_k).$$
The same holds true about a μ-vertical strip V in the vertical direction.

Under assumptions (A1) and (A2), we can prove that ψ has a shift invariant set of order N with respect to the two-sided shift.

Similar to the approach in Sect. 7.1, for
$$s_{-k}, \ldots, s_{-1}; s_0, s_1, \ldots, s_k \in \{0, 1, \ldots, N-1\} = S(N),$$
we define $U_{s_0 \ldots s_k}$ and $V_{s_{-k} \ldots s_{-2} s_{-1}}$ inductively as follows
$$U_{s_0 \ldots s_k} = \psi^{-1}(V_{s_0} \cap U_{s_1 \ldots s_k}) = U_{s_0} \cap \psi^{-1}(U_{s_1 \ldots s_k})$$
$$= U_{s_0} \cap \psi^{-1}(U_{s_1}) \cap \cdots \cap \psi^{-k}(U_{s_k});$$
$$V_{s_{-k} \ldots s_{-2} s_{-1}} = \psi(U_{s_{-1}} \cap V_{s_{-k} \ldots s_{-2}}) = V_{s_{-1}} \cap \psi(V_{s_{-k} \ldots s_{-2}})$$
$$= V_{s_{-1}} \cap \psi(V_{s_{-2}}) \cap \cdots \cap \psi^{k-1}(V_{s_{-k}}).$$

From assumptions (A1) and (A2), it is easy to show that $U_{s_0 \ldots s_k}$ is a μ-horizontal strip contained in $U_{s_0 \ldots s_{k-1}}$, and that $V_{s_{-k} \ldots s_{-1}}$ is a μ-vertical strip contained in $V_{s_{-k+1} \ldots s_{-1}}$.

Similar to Lemma 7.1, we have the following.

Lemma 7.9 *Under assumptions (A1) and (A2), we have*

(i) $\psi(U_{s_0 \ldots s_k}) = V_{s_0} \cap U_{s_1 \ldots s_k}$, $\psi(V_{s_{-k} \ldots s_{-2} s_{-1}}) \cap V_{s_0} = V_{s_{-k} \ldots s_{-1} s_0}$;
(ii) $\theta(U_{s_0 \ldots s_k}) < \nu\theta(U_{s_1 \ldots s_k}) < \nu^k$, $\theta(V_{s_{-k} \ldots s_{-2} s_{-1}}) < \nu\theta(V_{s_{-k} \ldots s_{-2}}) < \nu^{k-1}$. $\square$

Next, For $s = (\ldots, s_{-2}, s_{-1}; s_0, s_1, s_2, \ldots) \in \Sigma_N$, we define
$$U(s) = \bigcap_{j=0}^{\infty} \psi^{-j}(U_{s_j})$$
$$= \bigcap_{k=0}^{\infty} U_{s_0 s_1 \ldots s_k} \left(= \bigcap_{k=1}^{\infty} U_{s_0 s_1 \ldots s_k}\right),$$
$$V(s) = \bigcap_{j=1}^{\infty} \psi^{j-1}(V_{s_{-j}})$$
$$= \bigcap_{k=1}^{\infty} V_{s_{-k} \ldots s_{-1}} \left(= \bigcap_{k=2}^{\infty} V_{s_{-k} \ldots s_{-1}}\right).$$

Lemma 7.10 *We have*

(i) $\psi(V(s) \cap U(s)) = V(\sigma(s)) \cap U(\sigma(s))$;
(ii) $\sharp(V(s) \cap U(s)) = 1$.

Proof Part (i) is similar to the proof for part (i) in Lemma 7.2. We leave it to the readers as an exercise.

Part (ii) follows from the fact that the μ-horizontal curve $U(s)$ and the μ-vertical curve $V(s)$ have a unique intersecting point by Lemma 7.6. □

Finally, we define
$$\Delta = \bigcap_{s \in \Sigma_N} (V(s) \cap U(s)). \tag{7.3}$$

Theorem 7.11 *The set Δ in (7.3) is a compact invariant set of ψ, and $\psi|_\Delta$ is topologically conjugate to the two-sided shift $\sigma: \Sigma_N \to \Sigma_N$. Therefore, ψ has a shift invariant set of order N with respect to the two-sided shift.*

Proof By identifying a singleton set with the singleton itself therein, we define the map $h: \Sigma_N \to \Delta$ as follows:
$$h(s) = V(s) \cap U(s).$$

It remains to show that h is a topological conjugacy of ψ and σ. Here we just check the continuity of h. The proof of the remainders is similar to that of Theorem 6.52.

For the continuity, let $s, t \in \Sigma_N$ with
$$d(s, t) < \frac{1}{2^k}.$$

Then
$$h(s), h(t) \in V_{s_{-k}...s_{-1}} \cap U_{s_0...s_k}.$$

It follows from Lemma 7.7 that
$$|h(s) - h(t)| \leq \frac{1}{\mu} \max\{\theta(V_{s_{-k}...s_{-1}}), \theta(U_{s_0...s_k})\}$$
$$\leq \frac{\nu^{k-1}}{1 - \mu}.$$

□

In the following exercises, we offer some alternative and easily checkable conditions which substitute the condition (A2).

7.2 The General Horseshoe

Exercise 7.12 Consider the Henon map $F: \mathbb{R}^2 \to \mathbb{R}^2$:

$$F(x, y) = (1 + y - ax^2, bx).$$

Fix $a = 1.4$ and $b = 0.3$. Draw a rectangle in the (x, y) plane such that its image under the Henon map intersects the rectangle two times. □

Exercise 7.13 Consider the Henon map as in Exercise 7.12. Show that it has a Smale horseshoe. □

Exercise 7.14 Construct a smooth function $\beta: \mathbb{R} \to \mathbb{R}$ with the following properties:

(i) $\beta(t) > 0$, if $|t| < 1$, and $\beta(t) = 0$, if $|t| \geq 1$;
(ii) $\beta(-t) = \beta(t)$, $\forall t \in \mathbb{R}$;
(iii) $\int_{-\infty}^{+\infty} \beta(t)dt = 1$. □

Exercise 7.15 Let $v: [-1, 1] \to \mathbb{R}$ satisfy

$$|v(t_1) - v(t_2)| \leq \mu |t_1 - t_2|, \quad \forall t_1, t_2 \in [-1, 1],$$
$$|v(t)| \leq 1, \quad \forall t \in [-1, 1].$$

Prove that for any $\varepsilon > 0$, there exists a *smooth* function $\tilde{v}: \mathbb{R} \to \mathbb{R}$ such that

$$|\tilde{v}(t_1) - \tilde{v}(t_2)| \leq \mu |t_1 - t_2|, \quad \forall t_1, t_2 \in \mathbb{R},$$
$$|\tilde{v}(t)| \leq 1, \quad \forall t \in \mathbb{R},$$
$$|\tilde{v}(t) - v(t)| < \varepsilon, \quad \forall t \in [-1, 1].$$

□

In addition to condition (A1), we introduce the following condition: Assume that $\psi: P \to \psi(P)(\subset \mathbb{R}^2)$ is a diffeomorphism that satisfies
(A3) For any $p \in \bigcup_{j=0}^{N-1} U_j$, the map

$$\begin{pmatrix} \xi_1 \\ \eta_1 \end{pmatrix} = D\psi_p \begin{pmatrix} \xi_0 \\ \eta_0 \end{pmatrix}, \quad \forall \begin{pmatrix} \xi_0 \\ \eta_0 \end{pmatrix} \in \mathbb{R}^2,$$

satisfies the property that if $|\xi_0| \leq \mu |\eta_0|$, then

$$|\xi_1| \leq \mu |\eta_1|, \quad |\eta_1| \geq \mu^{-1} |\eta_0|.$$

Likewise, for any $q \in \bigcup_{j=0}^{N-1} V_j$, the map

$$\begin{pmatrix} \xi_0 \\ \eta_0 \end{pmatrix} = D\psi_q^{-1} \begin{pmatrix} \xi_1 \\ \eta_1 \end{pmatrix}, \quad \forall \begin{pmatrix} \xi_1 \\ \eta_1 \end{pmatrix} \in \mathbb{R}^2,$$

satisfies the property that if $|\eta_1| \leq \mu|\xi_1|$, then

$$|\eta_0| \leq \mu|\xi_0|, \quad |\xi_0| \geq \mu^{-1}|\xi_1|. \qquad \Box$$

Exercise 7.16 Let $\psi: P \to \mathbb{R}^2$ be a diffeomorphism from P to $\psi(P)$ that satisfies assumptions (A1) and (A3). Let $\gamma \subset \bigcup_{j=0}^{N-1} V_j$ be a μ-vertical curve and $\delta \subset \bigcup_{j=0}^{N-1} U_j$ be a μ-horizontal curve. Prove that the image $\psi(\hat{\gamma})$ of $\hat{\gamma} = U_k \cap \gamma$ is a μ-vertical curve, and the image $\psi(\check{\delta})$ of $\check{\delta} = V_l \cap \delta$ is a μ-horizontal curve. $\qquad \Box$

Exercise 7.17 Prove that under the assumptions of Exercise 7.16 with $0 < \mu < \frac{1}{2}$, assumption (A2) holds. $\qquad \Box$

Exercise 7.18 (*Baker map*) Consider a square set $S = [0, 1] \times [0, 1]$. Moreover, set

$$S_0 = [0, 1] \times \left[0, \frac{1}{2}\right], \quad S_1 = [0, 1] \times \left(\frac{1}{2}, 1\right].$$

It's clear that $S_0 \cup S_1 = S$. Define a map $f : S \to \mathbb{R}^2$ as follow

$$f(x, y) = \begin{cases} (\frac{1}{2}x, 2y) & (x, y) \in S_0 \\ (1 - \frac{1}{2}x, 2 - 2y) & (x, y) \in S_1. \end{cases}$$

1. Prove that f preserves the 2-dimensional Lebesgue measure, i.e., preserves the area.
2. Let $p \in S$. Compute the lyapunov exponent $\lambda(p)$ of f at p, which is given by

$$\lambda(p) = \liminf_{n \to +\infty} \frac{1}{n} \ln \|Df^n(p)\|,$$

where D is the differential operator. (In fact, the Baker map looks exactly like the tent map in the y direction, in which f is expanding and independent of x. Therefore, the Lyapunov exponent should be equivalent to that of the tent map and is $\ln 2$.)
3. Prove that f has a Smale horseshoe. $\qquad \Box$

Notes for Chapter 7

The Smale horseshoe, due to Smale [1] in 1963, is the earliest, most prominent example of higher-dimensional chaotic map. It is crucial in the understanding of how and why certain dynamics becomes chaotic, and then in developing the analysis to rigorously prove the occurrence of chaos. The horseshoe map is a geometrical and global concept. Its invariant set is a Cantor set, with infinitely many periodic points and uncountably many non-periodic orbits and yet it is "structurally stable."

Our treatment in this chapter is based mostly on that in Wiggins [3, Chapter 4], Zhang [4] and Zhou [5]. Section 7.1 studies the standard Smale horseshoe by symbolic dynamics by showing that the horseshoe diffeomorphism is topologically conjugate to a full shift on two-symbols on an invariant Cantor set. Section 7.2 does the general horseshoe with the Conley-Moser condition [6], leading to a topological conjugacy to a full shift on N-symbols. Further generalizations to dimensions higher than two can be found in Wiggins [7].

Theorem 7.11 in Sect. 7.2 is adopted from Zhang [4].

References

1. S. Smale, Diffeomorphisms with many periodic points, in *Differential and Combinatorial Topology*, S.S. Cairns (ed.), Princeton University Press, Princeton, New Jersey, 1963, pp. 63–80.
2. S. Smale, *The Mathematics of Time: Essays on Dynamical Systems, Economic Processes and Related Topics*, Springer, New York, 1980.
3. S. Wiggins, *Introduction to Applied Nonlinear Dynamical Systems and Chaos*, 2nd ed., Springer, New York, 2003.
4. Z.S. Zhang, *Principles of Differential Dynamical Systems*, Science Publishing House of China, Beijing, China, 1997.
5. Z.L. Zhou, *Symbolic Dynamics*, Shanghai Scientific and Technological Education Publishing House, Shanghai, China, 1997 (in Chinese).
6. J. Moser, *Stable and Random Motions in Dynamical Systems*, Princeton University Press, Princeton, New Jersey, 1973.
7. S. Wiggins, *Global Bifurcations and Chaos: Analytical Methods*, Springer Verlag, New York-Heidelberg-Berlin, 1988.

8 Fractal

We begin this chapter by giving some simple constructions.

8.1 Examples of Fractals

Example 8.1 (The classical Cantor set) The Cantor ternary set constructed in Example 5.12 is a standard example of a fractal, the subject of this chapter. □

Example 8.2 (The Sierpinski gasket) Let S_0 be a triangle with sides of unit-length. Connecting the middle point of each side, we obtain four triangles, each side has equal-length $\frac{1}{2}$. Deleting the interior of the middle one, the remainder part is denoted by S_1, which is composed of three triangles. Repeating the procedure for each triangle in S_1, we have S_2. Continuing this procedure, we obtain $S_0, S_1, S_2, \ldots$ The nonempty set

$$\mathcal{G} = \bigcap_{i=0}^{\infty} S_i$$

is the *Sierpinski gasket*. See Fig. 8.1. It is also called the Sierpinski triangle or the Sierpinski Sieve. The fractal is named after the Polish mathematician Wactlaw Sierpiński, who described it in 1915. Cf. also [1]. □

Example 8.3 (The Koch Curve) The Koch curve, due to the Swedish mathematician Helge von Koch, is constructed by first drawing an equilateral triangle in $\mathbb{R}^2$, then recursively alter each line segment as follows:

Fig. 8.1 The Sierpinski gasket (Example 8.2)

Fig. 8.2 The Koch curve

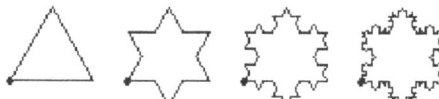

1. divide each side of the triangle into three segments of equal length;
2. draw an outward pointing equilateral triangle that has the middle segment from step (1) as its base;
3. remove the side that is the base of the triangle from step (2).

Performing just one iteration of this process, one obtains the result in the shape of the Star of David.

By iterating indefinitely, the limit is the Koch curve, which is also called the Koch snowflake, the Koch star or the Koch island. See Fig. 8.2, and [2]. □

8.2 Hausdorff Dimension and the Hausdorff Measure

Let A be a set in $\mathbb{R}^N$. The diameter of A is defined as

$$|A| = \sup\{|x - y|, \ x, \ y \in A\}.$$

Let $E \subset \mathbb{R}^N$. We call $\alpha = \{U_i, \ i > 0\}$ a (countable) cover of E if $E \subset \cup_{i>0} U_i$. Let $\delta > 0$. The collection α is called a δ-cover of E if

$$E \subset \cup_{i>0} U_i, \quad 0 < |U_i| \le \delta, \quad \forall i > 0.$$

8.2 Hausdorff Dimension and the Hausdorff Measure

Let $s \geq 0$, $\delta > 0$, define

$$\mathcal{H}^s_\delta(E) = \inf \sum_{i=1}^{\infty} |U_i|^s. \tag{8.1}$$

Here the infimum is taken over all the δ-covers of E. It can be checked that $\mathcal{H}^s_\delta$ is a metric outer measure on $\mathbb{R}^N$.

Define

$$\mathcal{H}^s(E) = \lim_{\delta \to 0} \mathcal{H}^s_\delta(E) = \sup_{\delta \to 0} \mathcal{H}^s_\delta(E).$$

The limit exists, but may be infinite, since $\mathcal{H}^s$ is decreasing as the function of δ. $\mathcal{H}^s$ is also a metric outer measure. The restriction of $\mathcal{H}^s$ to the σ-field of $\mathcal{H}^s$-measurable sets is called *the Hausdorff s-dimensional measure* of the set E.

Note that an equivalent definition of Hausdorff measure is obtained if the infimum in (8.1) is taken over δ-covers of E by convex sets rather than by arbitrary sets. If the infimum is taken over δ-covers of open (closed) sets, a different value of $\mathcal{H}^s_\delta$ may be obtained, but the value of the limit $\mathcal{H}^s$ is the same.

Exercise 8.4 For any E it is clear that $\mathcal{H}^s(E)$ is non-increasing as s increases from 0 to ∞. Furthermore, if $s < t$, then

$$\mathcal{H}^s_\delta(E) \geq \delta^{s-t} \mathcal{H}^t_\delta(E),$$

which implies that if $\mathcal{H}^t(E)$ is positive, then $\mathcal{H}^s(E)$ is infinite. □

Thus, from Exercise 8.4, there is a unique value, denoted by $\dim_\mathcal{H}(E)$, called the Hausdorff dimension of E, such that

$$\mathcal{H}^s(E) = \infty, \quad \text{if } 0 \leq s < \dim_\mathcal{H}(E)$$
$$\mathcal{H}^s(E) = 0, \quad \text{if } \dim_\mathcal{H}(E) < s < \infty.$$

Alternatively, we also have

$$\dim_\mathcal{H}(E) = \inf\{s : \mathcal{H}^s(E) = 0\} = \sup\{s : \mathcal{H}^s(E) = \infty\}.$$

In the following, denote $s = \dim_\mathcal{H}(E)$. In general, for a set E, the Hausdorff dimension of E, $s = \dim_\mathcal{H}(E)$ may not be an integer or even a fraction.

An measurable set $E \subset \mathbb{R}^N$ is said to be a s-set if $0 < \mathcal{H}^s(E) < \infty$. It is obvious that $\dim_\mathcal{H}(E) = s$ provided that E is a s-set. In this chapter, we will see many examples of s-sets.

Some elementary properties about Hausdorff dimension and Hausdorff measure are listed in the following.

Lemma 8.5 *(i) Monotonicity: Let E, $F \subset \mathbb{R}^N$. If $E \subset F$, then $\mathcal{H}^s(E) \leq \mathcal{H}^s(E)$ and $\dim_{\mathcal{H}}(E) \leq \dim_{\mathcal{H}}(F)$.*

(ii) Countable stability: Let $\{F_i\}_{i=1}^{\infty}$ be a countable sequence of sets in $\mathbb{R}^N$. Then

$$\dim_{\mathcal{H}}(\cup_{i>0} F_i) = \sup_{i>0}\{\dim_{\mathcal{H}}(F_i)\}.$$

(iii) If F is a countable set in $\mathbb{R}^N$, then $\dim_{\mathcal{H}}(F) = 0$. If F is a finite set, then $\mathcal{H}^0(F) = \text{card}\{F\}$.

(iv) Let $E \subset \mathbb{R}^N$. If $f : F \to \mathbb{R}^N$ is Hölder map, that is, there exist constants $c > 0$ and $\alpha > 0$ such that

$$|f(x) - f(y)| \leq c|x - y|^{\alpha}, \quad \forall x, y \in F,$$

then

$$\mathcal{H}^{s/\alpha}(f(F)) \leq c^{s/\alpha} \mathcal{H}^s(F), \quad \dim_{\mathcal{H}}(f(F)) \leq \frac{1}{\alpha} \dim_{\mathcal{H}}(F).$$

(v) In particular, if f is Lipschitz with Lipschitz constant c, then

$$\mathcal{H}^s(f(F)) \leq c^s \mathcal{H}^s((F)), \quad \dim_{\mathcal{H}}(f(F)) \leq \dim_{\mathcal{H}}(F).$$

Furthermore, if f is bi-Lipschitz, i.e., there exist $0 < c_1 < c$ such that

$$c_1|x - y| \leq |f(x) - f(y)| \leq c|x - y|, \quad \forall x, y \in F,$$

then

$$c_1^s \mathcal{H}^s((F)) \leq \mathcal{H}^s(f(F)) \leq c^s \mathcal{H}^s((F)), \quad \dim_{\mathcal{H}}(f(F)) = \dim_{\mathcal{H}}(F).$$

(vi) Let and $E \subset \mathbb{R}^N$ and $\lambda > 0$. Denote $\lambda E = \{\lambda x \mid x \in E\}$. Then

$$\mathcal{H}^s(\lambda E) = \lambda^s \mathcal{H}^s(E).$$

Proof (i) This follows directly from the definition.

(ii) By (i), we have

$$\dim_{\mathcal{H}}(\cup_{i>0} F_i) \geq \dim_{\mathcal{H}}(F_i), \quad \forall i > 0.$$

On the other hand, if $s > \dim_{\mathcal{H}}(F_i)$, $\forall i > 0$. then $\mathcal{H}^s(F_i) = 0$, $\forall i > 0$ by definition. It follows that $\mathcal{H}^s(\cup_{i>0} F_i) = 0$ by the additivity of measure, Thus we have

$$\dim_{\mathcal{H}}(\cup_{i>0} F_i) \leq \sup_{i>0}\{\dim_{\mathcal{H}}(F_i)\}.$$

(iii) Obviously, $\mathcal{H}^0(F) = 1$ when F only contains a single point. The first part follows from the countable stability (ii). The second part can be proved directly from the definition.

(iv) Let $\{U_i\}$ be a δ-cover of F. By assumption,
$$|f(F \cap U_i)| \leq c|U_i|^\alpha,$$
which implies that $\{f(F \cap U_i)\}$ is a $c\delta^\alpha$-cover of $f(E)$. Therefore
$$\sum_{i>0} |f(F \cap U_i)|^{s/\alpha} \leq c^{s/\alpha} \sum_{i>0} |U_i|^s,$$
$$\mathcal{H}^{s/\alpha}_{c\delta^\alpha}(f(F)) \leq c^{s/\alpha} \mathcal{H}^s_\delta(F).$$
Letting $\delta \to 0$, (so $c\delta^\alpha \to 0$), we get
$$\mathcal{H}^{s/\alpha}(f(F)) \leq c^{s/\alpha} \mathcal{H}^s(F), \quad \dim_\mathcal{H}(f(F)) \leq \frac{1}{\alpha} \dim_\mathcal{H}(F).$$

(v) Is a particular case that $\alpha = 1$.

(vi) Let $\{U_i\}$ be a δ-cover of E. Then $\{\lambda U_i\}$ be a $\lambda\delta$-cover of λE. So
$$\mathcal{H}^s_{\lambda\delta}(\lambda E) \leq \sum_i |\lambda U_i|^s = \lambda^s \sum_i |U_i|^s \leq \lambda^s \mathcal{H}^s_\delta(E).$$
Letting $\delta \to 0$, we have
$$\mathcal{H}^s(\lambda E) \leq \lambda^s \mathcal{H}^s(E).$$
The converse inequality follows by taking $\frac{1}{\lambda}$ instead of λ. □

There are many other properties about Hausdorff dimension and Hausdorff measure. For interested readers, please see [3–6].

Finally in this subsection, we state the relationship between N-dimensional Lebesgue measure $\mathcal{L}^N$ and N-dimensional Hausdorff measure. For the proof, see [7, Theorem 1.12, p. 13]

Lemma 8.6 *If $E \subset \mathbb{R}^N$, then*
$$\mathcal{L}^N(E) = c_N \mathcal{H}^N(E),$$
where $c_N = \pi^{\frac{1}{2}N}/2^N(\frac{1}{2}N)!$ is the volume of the ball in $\mathbb{R}^N$ with diameter 1. □

8.3 Iterated Function Systems (IFS)

This section presents an easy way to generate complicated sets, the so-called self-similar sets, which are generated by iterated function systems.

A map $S : \mathbb{R}^N \to \mathbb{R}^N$ is called a *contraction* if $|S(x) - S(y)| \leq c|x - y|$ for all $x, y \in \mathbb{R}^N$, for some $c : 0 < c < 1$. We call the infimum of such c the ratio of contraction. Furthermore, if $|S(x) - S(y)| = c|x - y|$ for all $x, y \in \mathbb{R}^N$ for some $c < 1$, then S is called a *similitude*. Geometrically, a similitude maps every subset of $\mathbb{R}^N$ to a similar set. Thus, it is a composition of a dilation with a rotation, a translation and perhaps a reflection as well. Such a map can be written as

$$S(x) = cQx + r, \quad x, r \in \mathbb{R}^N,$$

where Q is an orthogonal matrix.

An iterated function system on $\mathbb{R}^N$ is a finite collection of contractions $\{S_0, S_1, \ldots, S_{m-1}\}$ with $m > 1$. We refer to such a system as an IFS. A nonempty compact set E in $\mathbb{R}^N$ is said to be an invariant set of the IFS if

$$E = \bigcup_{i=0}^{m-1} S_i(E).$$

In the following, we show that an IFS has a unique invariant set. Let $\mathcal{C}(\mathbb{R}^N)$ denote the class of all nonempty compact sets of $\mathbb{R}^N$. For any $E, F \in \mathcal{C}(\mathbb{R}^N)$, define the distance between E and F:

$$\rho_H(E, F) = \max\left\{\max_{x \in E} d(x, F), \max_{y \in F} d(y, E)\right\}, \tag{8.2}$$

where for a nonempty closed set S of $\mathbb{R}^N$, $d(x, S)$ is the distance from a point x to the set S:

$$d(x, S) = \min_{y \in S}\{|x - y|\}.$$

Because the set is closed, the minimum is attained. We can prove that ρ_H is a metric on $\mathcal{C}(\mathbb{R}^N)$ (Exercise), which is called *the Hausdorff distance* and that $(\mathcal{C}(\mathbb{R}^N), \rho_H)$ become a complete metric space.

Since

$$\max_{x \in E}\{d(x, F)\} = \min\{\delta \geq 0 : E \subset F(\delta)\}$$

$$\max_{y \in F}\{d(y, E)\} = \min\{\delta \geq 0 : F \subset E(\delta)\}$$

an equivalent definition of ρ_H is

$$\rho_H(E, F) = \inf\{\delta > 0 : E \subset F(\delta), \text{ and } F \subset E(\delta)\} \tag{8.3}$$

Here

$$E(\delta) = \left\{x \in \mathbb{R}^N : d(x, E) \leq \delta\right\}, \quad F(\delta) = \left\{x \in \mathbb{R}^N : d(x, F) \leq \delta\right\}$$

are δ-closed neighborhoods of E and F, respectively.

Lemma 8.7 *Assume that S is a single contraction on $\mathbb{R}^N$ with contraction ratio $c < 1$. Then S induces a contraction on $\mathcal{C}(\mathbb{R}^N)$ with the same contraction ratio c.*

8.3 Iterated Function Systems (IFS)

Proof Let E and F be two sets in $\mathcal{C}(\mathbb{R}^N)$. For any $\rho > \rho_H(E, F)$, by (8.3), it follows that $E \subset F(\rho)$ and $F \subset E(\rho)$. That is,

$$d(x, F) < \rho, \quad \forall x \in E,$$

which implies

$$d(S(x), S(F)) \leq cd(x, F) < c\rho, \quad \forall x \in E.$$

Thus $S(E) \subset S(F)(c\rho)$. By the same way, we have $S(F) \subset S(E)(c\rho)$. Again by (8.3), we have

$$\rho_H(S(E), S(F)) \leq c\rho.$$

By the arbitrary of $\rho > \rho_H(E, F)$, we get

$$\rho_H(S(E), S(F)) \leq c\rho_H(E, F).$$

Thus, a single contraction induces a contraction on $\mathcal{C}(\mathbb{R}^N)$. □

Theorem 8.8 *Given an IFS $\{S_0, S_1, \ldots S_{m-1}\}$ with contraction ratios $0 < c_i < 1$ for $i = 0, 1, \ldots, m-1$, there exists a unique invariant set.*

Proof The induce map $\mathcal{S}$ on the complete metric space $(\mathcal{C}(\mathbb{R}^N), \rho_H)$ by the IFS is defined as

$$\mathcal{S}(E) = \cup_{i=0}^{m-1} S_i(E), \quad \forall E \in \mathcal{C}(\mathbb{R}^N).$$

Any fixed point of $\mathcal{S}$ is an invariant set of the IFS. To prove the theorem, it suffices to prove that the map $\mathcal{S}$ on $\mathcal{C}(\mathbb{R}^N)$ has a unique fixed point. This is done if we can show that $\mathcal{S}$ a contraction, since the space $\mathcal{C}(\mathbb{R}^N)$ is complete.

For any $E, F \in \mathcal{C}(\mathbb{R}^N)$, we claim

$$\rho_H(\mathcal{S}(E), \mathcal{S}(F)) = \rho_H\left(\bigcup_{i=0}^{m-1} S_i(E), \bigcup_{i=0}^{m-1} S_i(F)\right) \tag{8.4}$$
$$\leq \max_{0 \leq i \leq m-1} \rho_H(S_i(E), S_i(F))$$

In fact, if $\delta > 0$ such that

$$S_i(E)(\delta) \supset S_i(F), \forall i = 0, 1, \ldots, m-1,$$

then

$$\left(\cup_{i=0}^{m-1} S_i(E)\right)(\delta) \supset \cup_{i=0}^{m-1} S_i(\delta) \supset \cup_{i=0}^{m-1} S_i(F).$$

The same is also true by exchanging the positions of E and F. This proves our claims

It follows from (8.4) and Lemma 8.7 that

$$\rho_H(\mathcal{S}(E), \mathcal{S}(F)) \leq \max_{0 \leq i \leq m-1} c_i \rho_H(E, F).$$

That is, $\mathcal{S}$ is a contraction on $\mathcal{C}(\mathbb{R}^N)$ with contraction ratio $c = \max_{0 \leq i \leq m-1} c_i < 1$.

Definition 8.9 (*Self-similarity*) The unique invariant set E warranted by Theorem 8.8 is called the self-similar set generated by the family of similitudes $\{S_0, S_1, \ldots, S_{m-1}\}$. □

To obtain the invariant set, we can take the IFS as acting on the sets. In fact, by the proof of Theorem 8.8, $\mathcal{S}$ is a contraction on $\mathcal{C}(\mathbb{R}^N)$. So starting with any nonempty compact set F_0, the sequence of sets $F_n = \mathcal{S}(F_{n-1})$ converges to the fixed point in $\mathcal{C}(\mathbb{R}^N)$. It is the unique compact set in $\mathcal{C}(\mathbb{R}^N)$ that is the attractor and invariant of the IFS.

We now consider the construction of the invariant set of an IFS, we need the following.

Lemma 8.10 *Let $\{S_0, S_1, \ldots S_{m-1}\}$ be an IFS with contraction ratios $0 < c_i < 1$. Let*

$$R = \max_{0 \leq i \leq m-1} \left\{ \frac{\|S_i(0)\|}{1 - c_i} \right\}$$

$$\bar{B}(R) = \{x \in \mathbb{R}^N : \|x\| \leq R\}.$$

Then this ball $\bar{B}(R)$ is positively invariant under the IFS, i.e.

$$S_i(\bar{B}(R)) \subset \bar{B}(R).$$

Proof By the choice of R, $\|S_i(0)\| \leq R(1 - c_i)$. For $x \in \bar{B}(R)$,

$$\|S_i(x)\| \leq \|S_i(x) - S_i(0)\| + \|S_i(0)\|$$
$$\leq c_i \|x - 0\| + \|S_i(0)\|$$
$$\leq c_i R + R(1 - c_i)$$
$$= R.$$

This shows that the ball $\bar{B}(R)$ is positively invariant under all the S_i. □

Let Σ_m^+ be the symbolic space composed of $\{0, 1, \ldots, m-1\}$. Let F be a positively compact invariant set under all the S_i, $i = 0, \ldots, m-1$. For any $s = (s_0 s_1 \cdots) \in \Sigma_m^+$, we have

$$S_{s_0} \cdots S_{s_{k+1}}(F) \subset S_{s_0} \cdots S_{s_k}(F), \quad \forall k \leq 0,$$

and

$$|S_{s_0} \cdots S_{s_k}(F)| \leq c_{s_0} \cdots c_{s_k} |F| \to 0, \quad \text{as } k \to \infty.$$

Thus, the set $\cap_{k=0}^{\infty} S_{s_0} \cdots S_{s_k}(F)$ contains only a single set. Denote

8.3 Iterated Function Systems (IFS)

$$\{x_s\} = \cap_{k=0}^{\infty} S_{s_0} \cdots S_{s_k}(F),$$

and

$$E = \cup_{s \in \Sigma_m^+} \{x_s\}.$$

Theorem 8.11 *Let $\{S_0, S_1, \ldots S_{m-1}\}$ be an IFS with contraction ratios $0 < c_i < 1$ and E be defined as above.*

(i) E is the unique invariant set of the IFS.
(ii) Let (Σ_m^+, σ^+) be the one-side shift defined in Lecture 1. Define $\alpha : E \to E$ as

$$\alpha(x_s) = x_{\sigma^+(s)}, \qquad x_s \in E.$$

Then σ^+ is semiconjugate to α with a semiconjugacy $h : \Sigma_m^+ \to E$

$$h(s) = x_s, \qquad \forall s = (s_0 s_1 \cdots) \in \Sigma_m^+.$$

That is

$$h\sigma^+ = \alpha h. \tag{8.5}$$

Proof (i) By the definition of E, it is obvious that

$$S_i(E) = E, \qquad \forall i = 0, \ldots m-1,$$

which implies

$$E = \cup_{i=0}^{m-1} S_i(E).$$

This shows that E is the unique invariant set under the IFS.
(ii) From the definition, it is easy to show that

$$h\sigma^+ = \alpha h.$$

It remains to show that $h : \Sigma_m^+ \to E$ is continuous and onto. We leave it as an exercise. □

Let Σ_m^* denote the set of all the transitive points in Σ_m^+ as in Proposition 6.20. By Theorem 8.8, we have the following corollary.

Corollary 8.12 *Let $\{S_0, S_1, \ldots S_{m-1}\}$ be an IFS with contraction ratios $0 < c_i < 1$ and E be the unique attractor. We have*

(i) For any $s \in \Sigma_m^*$ and $x_0 \in E$, the orbit

$$\{S_{s_j} \cdots S_{s_0}(x_0)\}_{j=0}^{\infty}$$

is dense in E.

(ii) For any $s \in \Sigma_m^*$ and $x_0 \in \mathbb{R}^N$, the closure of the orbit

$$\{S_{s_j} \cdots S_{s_0}(x_0)\}_{j=0}^{\infty}$$

contains E. □

Intuitively, a set is *self-similar* (according to Definition 8.9) if an arbitrary small piece of it can be magnified to give the whole set. Many of the classical fractal sets are self-similar. The three classical examples are the Cantor set, the Koch curve, and the Sierpinski gasket. Self-similar sets can be generated by IFS of similitudes with the *open set condition*. An IFS $\{S_0, \ldots, S_{m-1}\}$ is called an IFS of similitudes if each contraction S_i is a similitude. Furthermore, we say that the open set condition holds for $\{S_0, \ldots, S_{m-1}\}$ if there exists a bounded set V in $\mathbb{R}^N$ such that

$$\mathcal{S}(V) = \bigcup_{i=0}^{m-1} S_i(V) \subset V$$

with disjoint union. It is easy to see that the unique attractor of an IFS of similitudes satisfying the open set condition is a self-similar set, which is also called a *self-similar* set with the open set condition.

Example 8.13 The classical Cantor set $\mathcal{C}$ is generated by S_0, S_1:

$$S_0(x) = \frac{1}{3}x, \quad S_1(x) = \frac{1}{3}x + \frac{2}{3}.$$

Let $V = (0, 1)$. Then

$$\mathcal{S}(V) = S_0(V) \cup S_1(V) \subset V, \quad S_0(V) \cap S_1(V) = \phi.$$

The open set condition holds. It is obviously that

$$\mathcal{C} = \cap_{n=0}^{\infty} \mathcal{S}^n(\overline{V}).$$

Thus the classical Cantor set is a self-similar set with open set condition. □

8.3 Iterated Function Systems (IFS)

Example 8.14 The Sierpinski gasket $\mathcal{G}$. A family of similitudes on $\mathbb{R}^2$ are defined as

$$S_0(x) = \frac{1}{2}x$$

$$S_1(x) = \frac{1}{2}x + \left(\frac{1}{2}, 0\right)$$

$$S_2(x) = \frac{1}{2}x + \left(\frac{1}{4}, \frac{\sqrt{3}}{4}\right).$$

Then $\{S_0, S_1, S_2\}$ satisfies open set condition with V a filled-in equilateral triangle whole sides are each of length 1 and that has vertices at $(0, 0)$, $(1, 0)$ and $(1/2, \sqrt{3}/2)$. And

$$\mathcal{G} = \cap_{n=0}^{\infty} S^n(\overline{V}).$$

Thus the Sierpinski gasket $\mathcal{G}$ is self-similar with open condition. □

To calculate the Hausdorff dimensions of a fractal sets is one of the main topic in fractal geometry. It is a difficult problem for a general fractal set. But there is an elegant result about the Hausdorff dimensions of self-similar sets with the open set condition.

Proposition 8.15 ([7], Theorem 8.6, p. 121) *Let $S_i : \mathbb{R}^N \to \mathbb{R}^N$ ($1 \le i \le m$) be an IFS of similitudes with open set condition. That is*

(1) S_i is a similar contraction with contraction ratio c_i ($0 < c_i < 1$), i.e.,

$$|S_i(x) - S_i(y)| = c_i |x - y|, \quad \forall x, y \in \mathbb{R}^N,$$

(2) $S_1, S_2, \ldots, S_m$ satisfy the open set condition.

Then the unique invariant set E as warranted in Theorem 8.8 (which is a self-similar set satisfying the open set condition) is an s-set, where s is determined by

$$\sum_{i=1}^{m} c_i^s = 1;$$

in particular, $0 < \mathcal{H}^s(E) < \infty$. □

From the Proposition 8.15, the Hausdorff dimension s of the classical Cantor set satisfies the equation

$$\left(\frac{1}{3}\right)^s + \left(\frac{1}{3}\right)^s = 1,$$

which implies that
$$\dim_{\mathcal{H}}(C) = s = \frac{\log 2}{\log 3}.$$

Through a tedious computation, it is possible to show that $\mathcal{H}^s(C) = 1$ [4].

By a similar argument, we have the Sierpinski gasket $\mathcal{G}$'s dimension as
$$\dim_{\mathcal{H}}(\mathcal{G}) = \frac{\log 3}{\log 2}.$$

Let $s \geq 0$ and $F \subset \mathbb{R}^N$ define
$$\mathcal{P}_\delta^s(F) = \sup \left\{ \sum_i |B_i| \right\}$$

where B_i is a ball centered in F and with radius less than or equal to $\delta (> 0)$ and $B_i \cap B_j = \emptyset$, $(i \neq j)$. Let
$$\mathcal{P}_0^s(F) = \lim_{\delta \to 0} \mathcal{P}_\delta^s(F).$$

This limit exists since $\mathcal{P}_\delta^s(F)$ is monotone as a function of δ.

Let
$$\mathcal{P}^s(F) = \inf \left\{ \sum \mathcal{P}_0^s(F_i) : | F = \cup F_i \right\}.$$

It is easy to see that $\mathcal{P}^s(\cdot)$ is a measure on $\mathbb{R}^N$ which is called packing measure with dimension s.

The *packing dimension* of F is defined as
$$\dim_{\mathcal{P}}(F) = \sup \{s : \mathcal{P}^s(F) = \infty\} = \inf \{s : \mathcal{P}^s(F) = 0\}.$$

To finish this section, we state a result on the Hausdorff dimension of product sets.

Proposition 8.16 *Let $E \subset \mathbb{R}^M$ and $F \subset \mathbb{R}^N$. Then there exists a positive constant c which only depends on s and t such that*
$$\mathcal{H}^{s+t}(E \times F) \geq c \mathcal{H}^s(E) \mathcal{H}^t(F).$$

Furthermore, if one of E and F is regular (i.e., its Hausdorff dimension is equal to its packing dimension), then
$$\dim_{\mathcal{H}}(E \times F) = \dim_{\mathcal{H}}(E) + \dim_{\mathcal{H}}(F). \qquad \square$$

For the proof, see [6, Theorem 1, p. 103 and Proposition 1, p. 106].

Exercise 8.17 Let $\rho_H(\cdot, \cdot)$ be defined as in (8.2). Prove that ρ_H is a distance on $\mathcal{C}\left(\mathbb{R}^N\right)$ and $\left(\mathcal{C}\left(\mathbb{R}^N\right), \rho_H\right)$ is a complete metric space. $\qquad \square$

8.3 Iterated Function Systems (IFS)

Exercise 8.18 Prove that the map h defined in the proof of Theorem 8.11 is continuous and onto. □

Exercise 8.19 Find the IFS of similitudes which generates the Koch curve and show that it is self-similar satisfying the open set condition. □

Exercise 8.20 Prove that the Hausdorff dimension of the Koch curve is $\log_3 4$. □

Exercise 8.21 For the classical Cantor set $\mathcal{C}$, show that

$$\mathcal{H}^s(\mathcal{C}) = 1$$

where $s = \log_3 2$. □

Exercise 8.22 Consider

$$f_\mu(x) = \begin{cases} \mu x, & 0 \leq x \leq \frac{1}{2}, \\ \mu(1-x), & \frac{1}{2} < x \leq 1, \end{cases} \quad (\mu > 0).$$

For $\mu > 2$, let

$$X = \{x \in [0, 1] \mid f_\mu^n(x) \in [0, 1], \text{ for all } n = 0, 1, 2, \ldots\}$$

be the largest invariant set of f_μ contained in $[0, 1]$. Prove that

(i) X is a self-similar set satisfying the open set condition;
(ii) The Hausdorff dimension of X is $\log_\mu 2$. □

Exercise 8.23 The Sierpinski carpet fractal is constructed as follows: The unit square is divided into nine equal boxes, and the open central box is deleted. This process is repeated for each of the remaining sub-boxes, and infimum.

(i) Find the IFS of similitudes which generates the Sierpinski carpet;
(ii) Find the Hausdorff dimension of the limit set;
(iii) Show that the Sierpinski carpet has zero area. □

Exercise 8.24 For the classical Cantor set $\mathcal{C}$, show that

(i) $\mathcal{C}$ contains no interval of positive length.
(ii) $\mathcal{C}$ has no isolated points, that is, if $x \in \mathcal{C}$, then every neighborhood of x will intersects $\mathcal{C}$ in more than one point. □

Exercise 8.25 Show that the classical Cantor set $\mathcal{C}$ contains an uncountable number of points. □

Exercise 8.26 If you instead of the classical Cantor set proceed as suggested before, and remove the 2nd and 4th fourth, i.e., in step one remove $\left(\frac{1}{4}, \frac{1}{2}\right)$, $\left(\frac{3}{4}, 1\right]$, etc., you arrive at a Cantor set $\mathcal{C}'$. Find the IFS that generates the set $\mathcal{C}'$ and determine its Hausdorff dimension. □

Exercise 8.27 The Menger sponge M is constructed starting from a cube with sides 1. In the first step, we divide it into 27 similar cubes and remove the center cube plus all cubes at the center of the faces. Then we continue to do the same to each of the 27 new cubes. Determine the Hausdorff dimension of M. □

Notes for Chapter 8

The term *fractal* was coined by the mathematician Benoit Mandelbrot in 1975, derived by him from the Latin word "fractus," which means "broken" or "fracture". A fractal is a geometric object with a self-similar property studied in this chapter. The earliest ideas of fractals can be dated back to Karl Weierstrass, Helge von Koch, Georg Cantor, and Felix Hausdorff, among others. Some of the most famous fractals are named after Cantor, Sierpinski, Peano, Koch, Harter-Heighway, Menger, Julia, etc. The major technique we note here is the IFS (iterated function systems) in Sect. 8.3 to define the fractals.

But having self-similarity alone is not sufficient for an object to be a fractal. For example, a straight line contains copies of itself at finer and finer scales. But it does not qualify as a fractal, as a straight line has the same Hausdorff dimension as the topological dimension, which is one. Section 8.2 gives a concise account of the Hausdorff measure and dimension, which will also be of major utility in Chap. 9.

References

1. The Sierpinski triangle, http://en.wikipedia.org/wiki/Sierpinski_gasket/.
2. The Koch snowflake, http://en.wikipedia.org/wiki/Koch_curve/.
3. J.-P. Eckmann, S.O. Kamphorst, D. Ruelle, and S. Ciliberto, Lyapunov exponent from time series, *Phys. Rev. A* **34** (1986), 4971–4979. http://mpej.unige.ch/~eckmann/ps_files/eckmannkamphorstruelle.pdf/. https://doi.org/10.1103/PhysRevA.34.4971.
4. K.J. Falconer, *The Geometry of Fractal Sets*, Cambridge University Press, 1985.
5. W. Hurewicz and H. Wallman, *Dimension Theory*, revised edition, Princeton University Press, Princeton, New Jersey, 1996.
6. Zhi-Ying Wen, *Mathematical Foundations of Fractal Geometry*, Advanced Series in Nonlinear Science, Shanghai Scientific and Technological Education Publishing House, Shanghai, 2000 (in Chinese).
7. K.J. Falconer, *Fractal Geometry*, John Wiley and Sons, New York, 1990.

9. Rapid Fluctuations of Chaotic Maps on $\mathbb{R}^N$

We have studied in Chap. 2 the use of total variations to characterize the chaotic behavior of interval maps. In this chapter, we will generalize such an approach to maps on multi-dimensional spaces.

First, we note that for a chaotic map on a multi-dimensional space, chaos may happen only on a lower-dimensional manifold C, on the complement of which the map's behavior can be quite orderly. The subset C generally is expected to have a fractal structure and a fractional dimensionality. Thus, to characterize chaotic properties in multi-dimensional spaces, we must rely on the use of the Hausdorff dimensions, the Hausdorff measure, and fractals developed in Chap. 8.

9.1 Total Variation for Vector-Value Maps

Let $\text{Lip}(\mathbb{R}^N)$ denote the class of all Lipschitz continuous maps from $\mathbb{R}^N$ to $\mathbb{R}^N$.

Definition 9.1 Let $f \in \text{Lip}(\mathbb{R}^N)$ and A be an s-set of $\mathbb{R}^N$. The *total variation* of f on A (with respect to the s-Hausdorff dimension) is defined by

$$\text{Var}_A^{(s)}(f) = \sup \left\{ \sum_{i=1}^n \mathcal{H}^s(f(C_i)) \mid n \in \mathbb{N} = \{1, 2, 3, \ldots\} \right.$$

$$\left. A \supset \bigcup_{i=1}^n C_i, C_i \cap C_j = \emptyset, i \neq j \right\}.$$

Since the total variation in the definition is always bounded in the case that $f \in \text{Lip}(\mathbb{R}^N)$.

The following are equivalent definitions.

Lemma 9.2 *Let $f \in Lip(\mathbb{R}^N)$ and A be an $s-$set of $\mathbb{R}^N$. Then*

$$\text{Var}_A^{(s)}(f) = \sup \left\{ \sum_{i=1}^n \mathcal{H}^s \left(f(C_i) \right) \mid \forall n \geq 1 \right. \tag{9.1}$$
$$\left. A \supset \bigcup_{i=1}^n C_i, C_i \text{ are } s\text{-sets}, C_i \cap C_j = \emptyset, i \neq j \right\}$$

$$\text{Var}_A^{(s)}(f) = \sup \left\{ \sum_{i=1}^n \mathcal{H}^s \left(f(C_i) \right) \mid \forall n \geq 1, \right. \tag{9.2}$$
$$\left. A \supset \bigcup_{i=1}^n C_i, C_i \text{ are } s\text{-sets}, \mathcal{H}^s \left(C_i \cap C_j \right) = 0, i \neq j \right\}.$$

Proof Denote by V_1 and V_2, respectively, the right-hand sides of the above two equations. It suffices to show that

$$\text{Var}_A^{(s)}(f) = V_1, \quad V_1 = V_2. \tag{9.3}$$

From (9.1), it follows $V_1 \leq \text{Var}_A^{(s)}(f)$. The converse inequality follows from the fact that a subset C of an $s-$set is not an $s-$set itself if and only if $\mathcal{H}^s(C) = 0$, which implies $\mathcal{H}^s(f(C)) = 0$ since f is Lipschitz continuous. Thus,

$$\text{Var}_A^{(s)}(f) = V_1.$$

We now prove the second equality in (9.3). It is obvious that $V_1 \leq V_2$. Conversely, Let C_i, $i = 1, \cdots, n$ be $s-$sets such that

$$A \supset \cup_{i=1}^n C_i, \ C_i \text{ are } s\text{-sets}, \quad \mathcal{H}^s(C_i \cap C_j) = 0, \ i \neq j.$$

Define C_i', $i = 1, \cdots, n$ inductively.

$$C_1' = C_1, \ C_i' = C_i - \cup_{j=1}^{i-1} C_j' = C_i - C_i \cap \cup_{j=1}^{i-1} C_j', \ i = 2, \cdots n.$$

Then C_i, $i = 1, \cdots n$ are disjoint and

$$\cup_{i=1}^n C_i' = \cup_{i=1}^n C_i.$$

Since

$$C_i \cap \cup_{j=1}^{i-1} C_j' \subset C_i \cap \cup_{j=1}^{i-1} C_j = \cup_{j=1}^{i-1}(C_i \cap C_j)$$

9.1 Total Variation for Vector-Value Maps

we have

$$\mathcal{H}^s\left(C_i \cap \cup_{j=1}^{i-1} C'_j\right) \le \sum_{j=1}^{i-1} \mathcal{H}^s\left(C_i \cap C_j\right) = 0. \tag{9.4}$$

Finally, for $i = 1, \cdots, n$, we have

$$C'_i \subset C_i = C'_i \cup \left(C_i \cap \cup_{j=1}^{i-1} C'_j\right),$$

which implies

$$f(C'_i) \subset f(C_i) = f(C'_i) \cup f\left(C_i \cap \cup_{j=1}^{i-1} C'_j\right).$$

Thus

$$\mathcal{H}^s(f(C'_i)) \le \mathcal{H}^s(f(C_i)) \le \mathcal{H}^s(f(C'_i)) + \mathcal{H}^s\left(f(C_i \cap \cup_{j=1}^{i-1} C'_j)\right) = \mathcal{H}^s(f(C'_i)),$$

since

$$\mathcal{H}^s\left(f(C_i \cap \cup_{j=1}^{i-1} C'_j)\right) = 0$$

by (9.4) and the Lipschitz property of f.

Summing up the above, we have

$$\sum_{i=1}^n \mathcal{H}^s(f(C_i)) = \sum_{i=1}^n \mathcal{H}^s(f(C'_i)).$$

Thus $V_2 \le V_1$. So $V_1 = V_2$. □

We now state some properties on the total variation of f.

Lemma 9.3 *Let $f \in Lip(\mathbb{R}^N)$ and A be an s–set of $\mathbb{R}^N$. Then*

(i) $Var_A^{(s)}(f) \le (Lip(f))^s \mathcal{H}^s(A)$.
(ii) *For each $\lambda > 0$, we have*

$$Var_A^{(s)}(\lambda f) = \lambda^s Var_A^{(s)}(f).$$

(iii) *If $f \in Lip(\mathbb{R}^1)$ and $A = [a, b]$ is a bounded interval, then*

$$Var_A^{(1)}(f) = V_{[a,b]}(f),$$

where $V_{[a,b]}(f)$ is the usual total variation of f on $[a, b]$.

Proof (i) and (ii) follows from the definition and (v) and (vi), respectively, in Lemma 8.5. For (iii), by Lemma 8.6, we have

$$\mathrm{Var}^{(1)}_{[a,b]}(f) \geq \sup\left\{\sum_{i=1}^{n} \mathcal{H}^1(f[x_{i-1}, x_i]) \mid \forall n \geq 1\ x_0 = a < x_1 < \cdots < x_n = b\right\}$$

$$= \sup\left\{\sum_{i=1}^{n} \mathcal{L}^1(f[x_{i-1}, x_i]) \mid \forall n \geq 1\ x_0 = a < x_1 < \cdots < x_n = b\right\}.$$

So
$$V_{[a,b]}(f) \leq \mathrm{Var}^{(1)}_{[a,b]}(f).$$

Conversely, for any $\varepsilon > 0$, there exist a positive integer n_0 and a partition of $[a, b]$: $x_0 = a < x_1 < \cdots < x_{n_0} = b$ such that

$$\mathrm{Var}^{(1)}_{[a,b]}(f) \leq \sum_{i=1}^{n_0} \mathcal{L}^1(f[x_{i-1}, x_i]) + \varepsilon.$$

For $i = 1, \cdots n_0$, let

$$f(\xi_i) = \max_{x_{i-1} \leq x \leq x_i}\{f(x)\},\ x_{i-1} \leq \xi_i \leq x_i,$$
$$f(\eta_i) = \min_{x_{i-1} \leq x \leq x_i}\{f(x)\},\ x_{i-1} \leq \eta_i \leq x_i.$$

Since f is continuous, f can obtains its the maximum and minimum on any bounded closed interval. Then

$$\mathcal{L}^1(f([x_{i-1}, x_i])) = |f(\xi_i) - f(\eta_i)|.$$

By adding the point ξ_i, η_i into the partition if necessary, we have a new partition: $x'_0 = a < x'_1 < \cdots x'_{n'} = b\ (n' \geq n_0)$, in which ξ_i, η_i are the partition points. Thus, by definition

$$V_{[a,b]}(f) + \varepsilon \geq \sum_{i=1}^{n'} |f(x'_i) - f(x'_{i-1})| + \varepsilon$$
$$\geq \sum_{i=1}^{n_0} |f(\xi_i) - f(\eta_i)| + \varepsilon$$
$$\geq \mathrm{Var}^{(1)}_{[a,b]}(f).$$

Hence, $\mathrm{Var}^{(1)}_{[a,b]}(f) \leq V_{[a,b]}(f)$ by the arbitrariness of ε and $\mathrm{Var}^{(1)}_{[a,b]}(f) = V_{[a,b]}(f)$. □

Property (iii) in Lemma 9.3 shows that Definition 9.1 is indeed a generalization of bounded variation of one-dimensional maps to vector-valued maps.

9.2 Rapid Fluctuations of Maps on $\mathbb{R}^N$

Similar to Definition 2.7 for one-dimensional dynamical systems, we now introduce a new notion to describe the complexity of multi-dimensional dynamical systems.

Definition 9.4 Let $D \subset \mathbb{R}^N$ and $f : D \to D$ be Lipschitz continuous. If there exists an s-set $A \subset D$ $(0 \leq s \leq N)$ such that $\mathrm{Var}_A^{(s)}(f^n)$ on A grows exponentially as $n \to \infty$, then we say that f has rapid fluctuations of dimension s.

We remark that the exact computation for the total variation $\mathrm{Var}_A^{(s)}(f^n)$ is almost impossible, since we in general can not calculate the exact value of Hausdorff measures for s-sets. Fortunately, only the estimates of the lower bound of $\mathrm{Var}_A^{(s)}(f^n)$ are needed if we want to show f has rapid fluctuations.

For a Lipschitz continuous map f, f may have rapid fluctuations of different dimensions. The supremum of such dimensionality numbers is called *the dimension of rapid fluctuations*.

Example 9.5 Consider the tent map $f_\mu : [0, 1] \to \mathbb{R}$ defined by

$$f_\mu(x) = \begin{cases} \mu x, & 0 \leq x \leq 1/2, \\ \mu(1-x), & 1/2 < x \leq 1, \end{cases} \quad (\mu > 0)$$

cf. the special case in Example 4.4 and Fig. 5.1. For $\mu > 2$, let

$$X = \left\{ x \in [0, 1] \mid f_\mu^n(x) \in [0, 1], \text{ for all } n = 0, 1, 2, \ldots \right\}.$$

X is the largest invariant set of f_μ contained in $[0,1]$. It is an extant result that f_μ is chaotic on X. In fact, X is also a self-similar set generated by the IFS of similitudes

$$S_0(x) = \frac{1}{\mu}x, \quad S_1(x) = \frac{1}{\mu}x + 1 - \frac{1}{\mu}.$$

Since $\mu > 2$, the open set condition holds by taking

$$V = (0, 1).$$

The set X is self-similar and has a Hausdorff dimension of $\ln 2 / \ln \mu$ by Proposition 8.15. Therefore, we conclude that for $\mu > 2$, f_μ has rapid fluctuations on an s-set X with $s = \ln 2 / \ln \mu$.

In fact, for any positive integer n, there exists 2^n subintervals $J_i^n = [a_i^n, a_{i+1}^n]$, $i = 0, 1, \ldots, 2^n - 1$ with

$$a_0^n = 0, \quad a_i^n < a_{i+1}^n, \quad a_{2^n}^n = 1,$$

such that f_μ^n is strictly monotone on J_i^n and $f_\mu^n(J_i^n) = [0, 1]$, $i = 0, 1, \ldots, 2^n - 1$. Therefore,

$$\text{Var}_X^{(s)}(f_\mu^n) \geq \sum_{i=0}^{2^n-1} \mathcal{H}^s\left(f_\mu^n(X \cap J_i^n)\right) = 2^n \mathcal{H}^s(X).$$

$\square$

A more interesting example is the Standard Smale horseshoe map.

Example 9.6 Consider the standard Smale horseshoe map φ introduced in Lecture 4 with $\lambda = 3$. We know that from Theorem 6.52 that φ is chaotic on the invariant set Λ. We now consider the rapid fluctuation property of φ.

First, since φ is homeomorphism, it has no rapid fluctuation of dimension 2. But it has rapid fluctuation of dimension 1, since each vertical line in the square $Q = [0, 1] \times [0, 1]$ is split into to two vertical lines under φ.

Next, we shall show that for each $s \in (0, \log 6/\log 3 - 1)$, φ has rapid fluctuation of dimension $1 + s$, which illustrates that φ has *higher dimensional chaos*! In fact, for any $s \in (0, \log 6/\log 3 - 1)$, taking an arbitrary s-set X in $[-1, 1]$ and letting $A = X \times [-1, 1]$, we have $A \subset Q$ is an s-set and

$$\dim_\mathcal{H}(A) = \dim_\mathcal{H}(X) + \dim_\mathcal{H}([-1, 1]) = s + 1,$$

by Proposition 8.16. To prove our claim, it suffices to show that $\text{Var}_A^{(s+1)}(f^n)$ grows exponentially as $n \to \infty$.

By definition of the maps φ, the image of A under φ contains 6 smaller "copies" of A under similitude mappings with contractive ratio $\frac{1}{3}$. Therefore

$$\mathcal{H}^{1+s}(f(A)) \geq 6\left(\frac{1}{3}\right)^{1+s}\mathcal{H}^{1+s}(A) = (1 + \delta)\mathcal{H}^{1+s}(A)$$

where

$$\delta = 6\left(\frac{1}{3}\right)^{1+s} - 1$$

is a positive constant which depends only on s. Inductively, we have

$$\mathcal{H}^{1+s}(f^n(A)) \geq (1 + \delta)^n \mathcal{H}^{1+s}(A).$$

Thus $\text{Var}_A^{(s+1)}(f^n) \geq (1 + \delta)^n \mathcal{H}^{1+s}(A)$ grows exponentially as $n \to \infty$. $\square$

We show next that rapid fluctuations are unchanged under Lipschitz conjugacy.

9.3 Rapid Fluctuations of Systems with Quasi-Shift Invariant Sets

Definition 9.7 Let $D_1 \subset \mathbb{R}^{N_1}$, $D_2 \subset \mathbb{R}^{N_2}$, $f : D_1 \to D_1$ and $g : D_2 \to D_2$ be Lipschitz continuous maps. We say that f and g are *Lipschitz conjugate* if there exists a bi-Lipschitz map h from D_1 to D_2 such that
$$h \circ f = g \circ h. \tag{9.5}$$
□

Lemma 9.8 *Let f and g have Lipschitz conjugacy. Then*

(i) *f has rapid fluctuations of dimension s if and only if g has rapid fluctuations of dimension s.*
(ii) *Let A be a s−set in D_1 for some $s \in (0, N]$. Then $\operatorname{Var}_A^{(s)}(f^n)$ grows unbounded if and only if $\operatorname{Var}_{h(A)}^{(s)}(g^n)$ grows unbounded as $n \to \infty$, where the bi-Lipschitz map $h : D_1 \to D_2$ is given by (9.5).*

Proof (i) Let h be bi-Lipschitz continuous such that $h \circ f = g \circ h$. Then $f = h^{-1} \circ g \circ h$. Let A be an s−set. Then $h(A)$ is also an s−set by (v) in Lemma 8.5. For any $k > 1$, subsets $C_1, C_2, \cdots, C_k$ satisfy $C_i \cap C_j = \phi$, $i \neq j$ and $A \supset \bigcup_{i=1}^{k} C_i$ if and only if compact subsets $h(C_1), h(C_2), \cdots, h(C_k)$ satisfy the very same property with $h(A) \supset \bigcup_{i=1}^{k} h(C_i)$. Thus, by the definition of conjugacy and by part (v) of Lemma 8.5, we have
$$\operatorname{Var}_A^{(s)}(f^n) \leq (\operatorname{Lip}(h^{-1}))^s \operatorname{Var}_{h(A)}^{(s)}(g^n).$$
Therefore, if there is an s−set A such that $\operatorname{Var}_A(f^n)$ grows exponentially as $n \to \infty$, then so does $\operatorname{Var}_{h(A)}(g^n)$, and vice versa.

The proof of part (ii) is similar, so we omit it. □

9.3 Rapid Fluctuations of Systems with Quasi-Shift Invariant Sets

From Sect. 6.6, it is known that a compact dynamical system has complex behavior if it has a quasi-shift invariant set. In the section, we will show that f has rapid fluctuation provided that f has a quasi shift invariant set.

Theorem 9.9 *Let (X, f) be a compact dynamical system and f be Lipschitz continuous. Assume the assumption (i) in Theorem 6.53 holds. That is, there exist k subsets $A_0, A_1, \cdots, A_{k-1}$ of X with $k \geq 2$ which are mutually disjoint such that*
$$f(A_i) \supset \cup_{j=0}^{k-1} A_j, \quad i = 0, 1, \cdots, k-1. \tag{9.6}$$

If there exists an $i_0 \in \{0, 1, \cdots k-1\}$ such that A_{i_0} is an s-set, then A_i is also an s-set and $\mathrm{Var}_{A_i}^{(s)}(f^n)$ grows exponentially for every $i \in \{0, 1, \cdots k-1\}$. Consequently, f has rapid fluctuation of dimension s.

It follows from Corollary 6.54 that the condition (9.6) implies that f has a quasi-shift invariant set.

Proof Denote by L the Lipschitz constant of f. By assumption, for any $i = 0, 1, \cdots k-1$, by Lemma 8.5, we have

$$0 < \mathcal{H}^s(A_{i_0}) \le \mathcal{H}^s(\bigcup_{j=0}^{k-1} A_j) \le \mathcal{H}^s(f(A_i)) \le L^s \mathcal{H}^s(A_i)$$

$$\le L^s \mathcal{H}^s(\bigcup_{j=0}^{k-1} A_j) \le L^s \mathcal{H}^s(f(A_{i_0})) \le L^{2s} \mathcal{H}^s(f(A_{i_0})) < \infty.$$

Since A_{i_0} is a s-set, so is A_i.

Now we prove the second part of the theorem. Denote

$$\delta = \min\{\mathcal{H}^s(A_i), \quad i = 0, 1, \cdots k-1\}.$$

Then $\delta > 0$ since A_i ($i = 0, 1, \cdots, k-1$) is an s-set. For a given i, let

$$J_{ij} = f^{-1}(A_j) \cap A_i, \quad j = 0, 1, \cdots, k-1.$$

Then J_{ij} ($j = 0, 1, \cdots, k-1$) is mutually disjoint and

$$f(J_{ij}) = A_j. \tag{9.7}$$

Thus, by definition, we have

$$\mathrm{Var}_{A_i}^{(s)}(f) \ge \sum_{j=0}^{k-1} \mathcal{H}^s(f(J_{ij})) = \sum_{j=0}^{k-1} \mathcal{H}^s(A_j) \ge k\delta.$$

On the other hand, from (9.7) and the assumptions, it follows that

$$f^2(J_{ij}) \supset \bigcup_{l=1}^{k-1} A_l, \quad j = 0, 1, \cdots, k-1.$$

By the same way, we can find subsets of J_{ijl} of J_{ij} which are mutually disjoint such that

$$f^2(J_{ijl}) = A_l, \quad j, l = 0, 1, \cdots, k-1.$$

9.3 Rapid Fluctuations of Systems with Quasi-Shift Invariant Sets

Thus

$$\mathrm{Var}^{(s)}_{A_i}(f^2) \geq \sum_{j,l=0}^{k-1} \mathcal{H}^s(f^2(J_{ijl}))$$

$$= \sum_{j,l=0}^{k-1} \mathcal{H}^s(A_l) \geq k^2 \delta.$$

Repeating the above procedure, we can prove by induction that

$$Var(A_i, f^n) \geq k^n \delta.$$

So $\mathrm{Var}^{(s)}_{A_i}(f^n)$ grows exponentially. This completes the proof. □

We know from Theorem 6.58 that for a C^1 map $f : \mathbb{R}^N \longrightarrow \mathbb{R}^N$ with a snap-back repeller p, there exists a positive integer r such that f^r has a shift invariant set of order 2 with respect to one-side shift. Thus we have

Corollary 9.10 *Let $f : \mathbb{R}^N \longrightarrow \mathbb{R}^N$ be C^1. If f has a snap-back repeller p, then f has rapid fluctuation of dimension N.*

Proof From the proof in Theorem 6.58, there exist two closed neighborhoods A_1, A_2 of p and a positive integer r such that

$$A_1 \cap A_2 = \phi, \quad f^r(A_1) \cap f^r(A_2) \supset A_1 \cup A_2.$$

Thus A_i $(i = 1, 2)$ are N-sets. By Theorem 9.9, $\mathrm{Var}^{(N)}_{A_i}((f^r)^n)$ grow exponentially as $n \to \infty$.

On the other hand, since f is C^1, f is locally Lipschitz. Denote by L the Lipschitz constant on the neighborhood of p which contains both A_1 and A_2. For any positive integer $m > r$, let $m = rn - l$ with $0 \leq l < r$. Then we have

$$\mathrm{Var}^{(N)}_{A_i}((f^r)^n) \leq L^{lN} \mathrm{Var}^{(N)}_{A_i}(f^m).$$

Thus $\mathrm{Var}^{(N)}_{A_i}(f^m)$ grow exponentially as $n \to \infty$. That is, f has rapid fluctuation of dimension N. □

9.4 Rapid Fluctuations of Systems Containing Topological Horseshoes

In Chap. 7, we have discussed that the standard Smale horseshoe and its general case have shift invariant sets with respect to the two-sided shift σ. So they have complex dynamical behavior. In this section, we discuss the more general dynamical systems of topological horseshoes which include the Smale horseshoe as a special case. We show that such a map with topological horseshoe has rapid fluctuations.

Definition 9.11 Assume that X is a separable metric space and consider the dynamical system (X, f). If there is a locally connected and compact set Q of X such that

(i) The map $f : Q \to X$ is continuous;
(ii) There are two disjoint and compact sets Q_0 and Q_1 of Q such that each component of Q intersects both Q_0 and Q_1;
(iii) Q has crossing number $M \geq 2$(see below);

then (X, f) is said to have topological horseshoes. □

From the above, we define a connection Γ between Q_0 and Q_1 as a compact connected subset of Q that intersects both Q_0 and Q_1. A preconnection γ is defined as a compact connected subset of Q for which $f(\gamma)$ is a connection. We define the crossing number M to be the largest number such that every connection contains at least M mutually disjoint preconnections.

Example 9.12 Let φ be the standard Smale horseshoe map defined in Sect. 4.1. Then φ satisfies the topological horseshoe assumption on $V = Q \cap \varphi(Q)$ with crossing number $M = 2$ and
$$Q_0 = V \cap \{y = -1\}, \quad Q_1 = V \cap \{y = 1\},$$
where $Q = [-1, 1] \times [-1, 1]$. □

Example 9.13 Let ψ be the general Smale horseshoe map defined in Sect. 7.2. That is, $\psi : Q \to \mathbb{R}^2$ satisfies the assumptions (A1) and (A2) in Sect. 7.2, where $Q = [-1, 1] \times [-1, 1]$. Let
$$V = \cup_{j=0}^{N-1} V_j.$$
Then ψ satisfies the topological horseshoe assumption on V with the crossing number $M = N$. The proof of this example is left as an exercise. □

Kennedy and Yorke [1] studied the chaotic properties of dynamical systems with topological horseshoes. We state one of the main theorems therein, below.

9.4 Rapid Fluctuations of Systems Containing Topological Horseshoes

Lemma 9.14 ([1] (Theorem 1)) *Assume that f has topological horseshoes. Then there is a closed invariant set $Q_I \subset Q$ for which $f|_{Q_I} : Q_I \to Q_I$ is semiconjugate to a one side M-shift.* □

We now discuss the rapid fluctuation of dynamical systems with topological horseshoe. We shall prove that such kind of systems has rapid fluctuation of dimension at least 1. To this end, we first need the following.

Lemma 9.15 *If Γ is a connection of Q in the sense of Definition 9.11, then the Hausdorff dimension of Γ is at least 1.*

Proof Let $x_0 \in \Gamma$ fixed. Define $h : \Gamma \to \mathbb{R}^+$ by

$$h(x) = d(x_0, x),$$

where $d(\cdot, \cdot)$ is the distance in X.

Let $a = \max_{x \in \Gamma} h(x)$. Then $h : \Gamma \to [0, a]$ is a non-expanding and onto map by the connectedness of Γ. Thus

$$0 < \mathcal{H}^1([0, a]) = \mathcal{H}^1(h(\Gamma)) \le \mathcal{H}^1(\Gamma).$$

This implies that $\dim_{\mathcal{H}}(\Gamma) \ge 1$. □

Assume that f has topological horseshoes. Denote by $\mathcal{S}$ the set of all connections of Q and

$$s = \inf\{dim_{\mathcal{H}}(\Gamma), \Gamma \in \mathcal{S}\}. \tag{9.8}$$

We have $s \ge 1$ from Lemma 9.15.

Theorem 9.16 *Assume that f has topological horseshoes and is Lipschitz continuous. Let s be defined as (9.8). If there exists a $\Gamma_0 \in \mathcal{S}$ such that $\dim_{\mathcal{H}}(\Gamma_0) = s$ and*

$$\mathcal{H}^s(\Gamma_0) = \inf\left\{\mathcal{H}^s(\Gamma), \mid \Gamma \in \mathcal{S} \text{ and } \dim_{\mathcal{H}}(\Gamma) = s\right\} > 0, \tag{9.9}$$

then $Var_{\Gamma_0}(f^n)$ grows exponentially as $n \to \infty$. Thus f has rapid fluctuation of dimension s.

Proof Since $\Gamma_0 \in \mathcal{S}$, by assumptions Γ_0 has at least M mutually disjoint preconnections, which are denoted by $\Gamma_{01}, \Gamma_{02} \cdots, \Gamma_{0M}$. That is $\Gamma_{0i} \subset \Gamma_0$ and $f(\Gamma_{0i}) \in \mathcal{S}$, for $i = 1, 2, \cdots M$. Hence, we have

$$\mathcal{H}^s(f(\Gamma_{0i})) \le L\mathcal{H}^s(\Gamma_{0i}) \le L\mathcal{H}^s(\Gamma_0),$$

where L is the Lipschitz constant. It follows from (9.9) that $\dim_{\mathcal{H}}(f(\Gamma_{0i})) = s$ and

$$Var_{\Gamma_0}(f) \geq \sum_{i=0}^{M} \mathcal{H}^s(f(\Gamma_{0i})) \geq M\mathcal{H}^s(\Gamma_0).$$

For each i ($1 \leq i \leq M$), since $f(\Gamma_{0i}) \in \mathcal{S}$, again $f(\Gamma_{0i})$ has at least M mutually disjoint preconnections, denoted by $\gamma_{0i}^1, \cdots, \gamma_{0i}^M$. So

$$f(\gamma_{0i}^l) \in \mathcal{S}, \quad \bigcup_{l=1}^{M} \gamma_{0i}^l \subset f(\Gamma_{0i}).$$

The latter implies that there exist M mutually disjoint subsets $\Gamma_{0i}^1, \cdots, \Gamma_{0i}^M$ of Γ_{0i} such that $\gamma_{0i}^l = f(\Gamma_{0i}^l)$. We thus obtain M^2 mutually disjoint subsets Γ_{0i}^l, $i, l = 1, \cdots, M$ of Γ_0. Therefore

$$Var_{\Gamma_0}(f^2) \geq \sum_{i,l=1}^{M} \mathcal{H}^s(f^2(\Gamma_{0i}^l)) = \sum_{i,l=1}^{M} \mathcal{H}^s(f(\gamma_{0i}^l)) \geq M^2 \mathcal{H}^s(\Gamma_0).$$

Repeating the above procedure, we can prove by induction that

$$Var_{\Gamma_0}(f^n) \geq M^n \mathcal{H}^s(\Gamma_0).$$

That is, $Var_{\Gamma_0}(f^n)$ grows exponentially as $n \to \infty$. Thus, f has rapid fluctuation of dimension s. □

9.5 Examples of Applications of Rapid Fluctuations

In this section, we give two examples to illustrate rapid fluctuations: one from nonlinear economic dynamics, and the other from a predator-prey model.

Example 9.17 Benhabib and Day in [2, 3] used two types of one-dimensional systems as models for dynamic consumer behavior. One is the well-known logistic type. The second is an exponential type. Then Dohtani [4] presented two classes of examples of the Benhabib-Day model in multi-dimensional space, which are governed by the Lotka-Volterra type

$$x_i(n+1) = f_i(x(n)), \quad f_i(x_1, \cdots, x_N) = x_i \left(a - \sum_{j=1}^{N} b_{ij} x_j \right), \tag{9.10}$$

or by the exponential type

9.5 Examples of Applications of Rapid Fluctuations

$$x_i(n+1) = g_i(x(n)), \quad g_i(x_1, \cdots, x_N) = x_i \exp\left(a - \sum_{j=1}^{N} b_{ij} x_j\right), \tag{9.11}$$

where $i \in H = \{1, 2, \cdots, N\}$, x_i ($i \in H$) is the amount of the ith good consumed within a given period, and the constants $a > 0$, b_{ij} are the parameters with respect to the economic environment. See [4] for details. The chaotic behavior of (9.10) and (9.11) will be proved in Theorem 9.18.

□

Theorem 9.18 *Let*

$$Q = (1, \cdots, 1)^T \in \mathbb{R}^N, \quad B = [b_{ij}] \in \mathbb{R}^{N \times N}.$$

Suppose that the matrix B is nonsingular and each entry in $B^{-1}Q$ is positive.

(1) If $1 < a < 4$ and the logistic map

$$\alpha(x) = ax(1-x),$$

from $[0, 1]$ to itself, has a periodic orbit whose period is not a power of 2, then the Lotka-Volterra system (9.10) has rapid fluctuations of dimension 1.

(2) If the exponential map

$$\beta(x) = x \exp(a - x), \tag{9.12}$$

from $\mathbb{R}_+$ to itself, has a periodic orbit whose period is not a power of 2, then the exponential system (9.11) has rapid fluctuations of dimension 1.

□

Diamond [4] introduced the following definition.

Definition 9.19 (*Chaos in the sense of P. Diamond*) Let $D \subset \mathbb{R}^N$ and $f : D \to D$ be continuous. We said that f is chaotic in the sense of Diamond if

(i) for every $n = 1, 2, 3, \ldots$, there is an n-periodic orbit in D;
(ii) there is an uncountable set S of D, which contains no periodic points and satisfies

(a) $f(S) \subset S$;
(b)
$$\limsup_{n \to \infty} |f^n(p) - f^n(q)| > 0, \quad \forall p, q \in S, \quad p \neq q,$$
$$\liminf_{n \to \infty} |f^n(p) - f^n(q)| = 0, \quad \forall p, q \in S;$$

(c) for every p in S and every periodic point q in D;

$$\limsup_{n \to \infty} |f^n(p) - f^n(q)| > 0.$$

□

Remark 9.20 It is known that the logistic map $\alpha(x)$ in Theorem 9.18 has a periodic point whose period is not a power of 2 if $a > a^* = 3.59 \cdots$. Since the exponential map $\beta(x)$ is strictly increasing in $[0, 1]$ and strictly decreasing in $[1, \infty]$, respectively, and $\beta(x) \to 0$ as $x \to +\infty$, $\beta(x)$ has a period-three point if $\beta^3(1) < 1$. A sufficient condition for this is $a \geq 3.13$ ([4]). If $\beta^3(1) \leq a$, then $\beta(x)$ has a period-6 point. This is the case when $a \geq 2.888$.

Dohtani [4] has established that the Lotka-Volterra system (9.10) generates chaos in the sense of Diamond if $a > 3.84$ and the exponential type system (9.11) is chaotic in the same sense if $a > 3.13$.

□

Proof of Theorem 9.18. Let $W = aB^{-1}Q = (w_1, w_2, \cdots, w_N)^T$. Then $W > 0$ by the assumptions. Denote

$$\Omega = \{rW \mid 0 \leq r \leq 1\}.$$

Then Ω is the line segment in $\mathbb{R}^N$ that connects the origin with W, so it is a 1−set. For any $x \in \Omega$, there exists a positive constant $r \in [0, 1]$ such that $x = rW$. Then, by (9.10), we have

$$f(x) \equiv (f_1(x), f_2(x), \cdots, f_N(x))^T$$
$$= r\,\text{diag}(w_1, w_2, \cdots, w_N)(aQ - rBW) = r(1-r)aW, \quad (9.13)$$

where $\text{diag}(\cdots)$ is a diagonal matrix with the indicated diagonal entries. Noting that $r(1-r) \leq \frac{1}{4}$ for any $0 \leq r \leq 1$, we have $0 \leq r(1-r)a \leq 1$ for $1 < a < 4$. It follows from (9.13) that $f(x) \in \Omega$. This means that Ω is an invariant set under f.

If the logistic map $\alpha(x)$ has a periodic point whose period is not a power of 2, then there exists a positive integer k such that $\alpha^k(\cdot)$ is strictly turbulent. That is, there are compact intervals $J, K \subset [0, 1]$ with $J \cap K = \phi$ and

$$\alpha^k(J) \cap \alpha^k(K) \supset J \cup K.$$

Let

$$J' = JW \equiv \{rW \mid r \in J\}, \qquad K' = KW \equiv \{rW \mid r \in K\}.$$

Then J' and K' are two compact 1−sets in $\mathbb{R}^N$ with empty intersection and

$$f^k(J') \cap f^k(K') \supset J' \cup K'.$$

Thus, system (9.10) has rapid fluctuations of dimension 1 by Theorem 9.16.

9.5 Examples of Applications of Rapid Fluctuations

The corresponding property for the exponential type system (9.11) can be proved similarly. □

Example 9.21 We consider the following predator-prey model

$$\begin{cases} \frac{dx}{dt} = x(t)[\mu_1 - \mu_1 x(t) - s_1 y(t)] \\ \frac{dy}{dt} = y(t)[-\mu_2 + s_2 y(t)]. \end{cases} \quad (9.14)$$

Let $t_n = nh$ where h is a step size. Applying the variation of parameter formula to each equation in (9.14), one obtains

$$\begin{aligned} x(t_{n+1}) &= x(t_n)\exp(\int_{t_n}^{t_{n+1}}[\mu_1 - \mu_1 x(\xi) - s_1 y(\xi)]d\xi) \\ &\approx x(t_n)\exp(\mu_1 - \mu_1 x(t_n) - s_1 y(t_n)]h) \\ y(t_{n+1}) &= y(t_n)\exp(\int_{t_n}^{t_{n+1}}[-\mu_2 + s_2 y(\xi)]d\xi) \\ &\approx y(t_n)\exp([-\mu_2 + s_2 y(t_n)]h). \end{aligned} \quad (9.15)$$

Thus, the following difference equation system gives a numeric scheme for (9.15):

$$\begin{aligned} x(t_{n+1}) &= x(t_n)\exp(\mu_1 - \mu_1 x(t_n) - s_1 y(t_n)]h) \\ y(t_{n+1}) &= y(t_n)\exp([-\mu_2 + s_2 y(t_n)]h). \end{aligned} \quad (9.16)$$

Denoting $x(t_n) = x_n$ and $y(t_n) = y_n$, re-scaling $hs_1 y_n \to y_n$ and also re-scaling parameters $h\mu_1 \to \mu_1$ and $hs_2 \to h$, we rewrite (9.16) as

$$\begin{aligned} x_{n+1} &= x_n \exp[\mu_1 - \mu_1 x_n - y_n] \\ y_{n+1} &= y_n \exp[-\mu_2 + sx_n], \end{aligned} \quad (9.17)$$

where $s = s_2$. We let $F = F_{\mu_1,\mu_2,s} : \mathbb{R}^2 \to \mathbb{R}^2$ be the map satisfying the right-hand side of (9.17), i.e.,

$$F(x, y) = (x\exp[\mu_1 - \mu_1 x - y], y\exp[-\mu_2 + sx]).$$

Now we consider the dynamical system $(\mathbb{R}^2, F)$. We show that F has rapid fluctuation of dimension 1.
Denote

$$\begin{aligned} h(x) &= xe^{\mu_1 - \mu_1 x}, \\ g_1(x, y) &= \mu_1 h(x)(1 - e^{-y}) - y(1 + e^{-\mu_2 + sx}), \\ g_2(x, y) &= -2\mu_2 + s(x + h(x)e^{-y}). \end{aligned} \quad (9.18)$$

Then by a direct calculation F^2 can be written as

$$F^2(x, y) = (h^2(x)e^{g_1(x,y)}, ye^{g_2(x,y)}).$$

The map $h(x)$ of exponential type is unimodel (cf. Example 1.3 in Chap.1), which has been studied extensively in the discrete dynamical system literature. The function $h(x)$ has

two fixed points 0 and 1. It is strict increasing in $[0, \frac{1}{\mu_1}]$ and strict decreasing in $[\frac{1}{\mu_1}, +\infty)$ and
$$\lim_{x \to +\infty} h(x) = 0.$$

Thus $h(x)$ has global maximum value at $x = \frac{1}{\mu_1}$ and
$$M \triangleq h(\frac{1}{\mu_1}) = \frac{1}{\mu_1} e^{\mu_1 - 1}.$$

Therefore, when $\mu_1 > 1$ there exist r_1 and r_2 with $0 < r_1 < 1 < r_2$ such that
$$h^2(r_1) = h^2(r_2) = h(\frac{1}{\mu_1}) = M. \tag{9.19}$$

See the graphics of h and h^2 in Fig. 9.1.

In Theorem 9.24 below, we show that F^2 has a horseshoe. Therefore, the map F is chaotic. □

Lemma 9.22 *There exists a constant μ_1^0 such that when $\mu_1 > \mu_1^0$, we have*
$$h^3(\frac{1}{\mu_1}) < \frac{1}{\mu_1}. \tag{9.20}$$

Thus if $\mu_1 > \mu_1^0$, then
$$h^2(\frac{1}{\mu_1}) < r_1.$$

Proof A routine calculation shows that
$$h^3(\frac{1}{\mu_1}) = \frac{1}{\mu_1} \exp(3\mu_1 - 1 - e^{\mu_1 - 1} - e^{2\mu_1 - 1 - e^{\mu_1 - 1}}).$$

(9.20) holds if and only if
$$3\mu_1 - 1 - e^{\mu_1 - 1} - e^{2\mu_1 - 1 - e^{\mu_1 - 1}} < 0.$$

A sufficient condition for the above inequality is
$$3\mu_1 - 1 - e^{\mu_1 - 1} < 0.$$

This is obviously true if μ_1 is large enough. □

9.5 Examples of Applications of Rapid Fluctuations

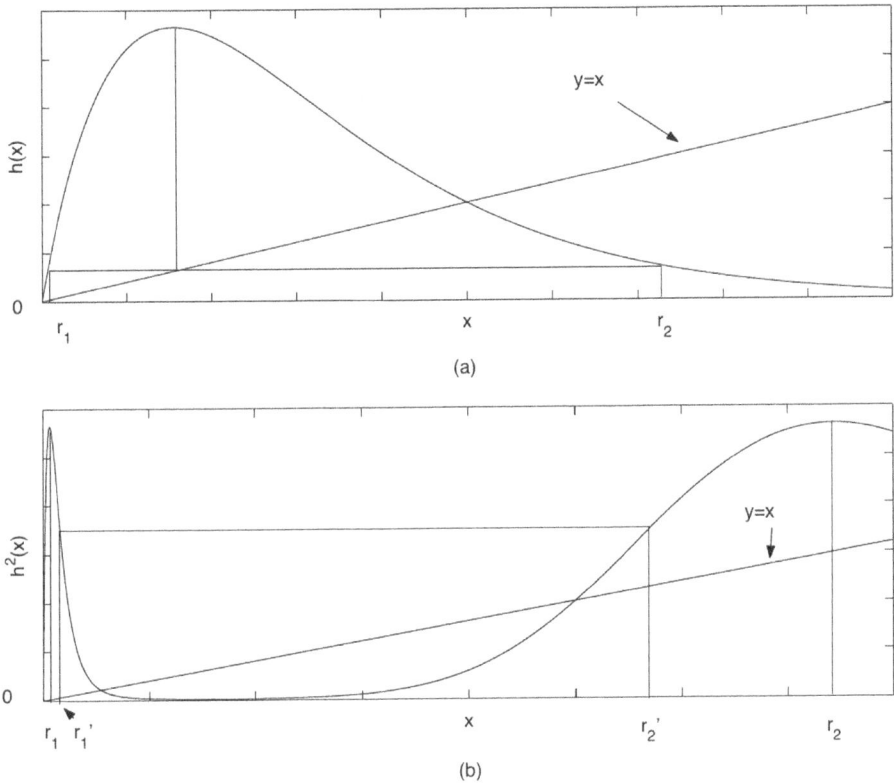

Fig. 9.1 The graphics of the exponential type $h(x)$ and $h^2(x)$ with $\mu_1 = 3.2$, cf. h defined in (9.18)

Remark 9.23 Numerical computation shows that (9.20) holds when $\mu_1 > 3.117$. □

In the following, we fix $\mu_1 > \mu_1^0$. Denote by p_2 the period two point of h in $(r_1, \frac{1}{\mu_1})$. From (9.19) and the unimodel properties of $h(x)$, it follows that for any $1 < r_2' < r_2$, there exists a unique r_1' with $r_1 < r_1' < p_2$ such that

$$h^2(r_1') = h^2(r_2') > r_2'. \tag{9.21}$$

For any $\varepsilon > 0$, let

$$Q_\varepsilon \triangleq [r_1', r_2'] \times [0, \varepsilon], \tag{9.22}$$

and

$$M_1 = \max_{r_1 < x < r_2} \{x + h(x)\}. \tag{9.23}$$

Theorem 9.24 Let $\mu_1 > \mu_1^0$, where μ_1^0 is given by Lemma 9.22. If

$$0 < s < \frac{2\mu_2}{M_1}, \tag{9.24}$$

then for any r_2' with $1 < r_2' < r_2$, there exists an $\varepsilon > 0$, such that the map $F^2 : Q_\varepsilon \to \mathbb{R}^2$ has a topological horseshoe with a cross number $m = 2$.

Proof Let r_1' and r_2' defined as above. Since $\mu_1 > \mu_1^0$, there exist r_3, r_4 such that

$$p_2 < r_3 < \frac{1}{\mu_1} < r_4 < 1, \tag{9.25}$$

$$h^2(r_3) = h^2(r_4) < r_1 < r_1'. \tag{9.26}$$

On the other hand, we have

$$g_1(x, 0) = 0,$$

for any $x > 0$.

It follows from (9.21) and (9.26) that there exists an $\varepsilon > 0$ such that

$$h^2(r_1')e^{g_1(r_1', y)} > r_2', \quad h^2(r_2')e^{g_1(r_2', y)} > r_2', \tag{9.27}$$

$$h^2(r_3)e^{g_1(r_3, y)} < r_1', \quad h^2(r_4)e^{g_1(r_4, y)} < r_1', \tag{9.28}$$

for any $y \in [0, \varepsilon]$.

On the other hand, we have

$$ye^{g_2(x, y)} < y, \tag{9.29}$$

for any $y > 0$ and $r_1 < x < r_2$.

Denote

$$Q_1 = \{(x, y) \mid x = r_1', \ 0 \le y \le \varepsilon\}, \quad Q_2 = \{(x, y) \mid x = r_2', \ 0 \le y \le \varepsilon\},$$
$$D_1 = \{(x, y) \mid r_1' \le x \le r_3, \ 0 \le y \le \varepsilon\}, \quad D_2 = \{(x, y) \mid r_4 \le x \le r_2', \ 0 \le y \le \varepsilon\}.$$

If we denote

$$(\bar{x}, \bar{y}) = F^2(x, y),$$

then by (9.27)-(9.29), for any $(\bar{x}, \bar{y}) = F^2(Q_i)$, $i = 1, 2$, we have $\bar{x} > r_2'$ and $0 < \bar{y} < \varepsilon$. Likewise, by the same argument, for any $(\bar{x}, \bar{y}) = F^2(\{r_3\} \times [0, \varepsilon])$, $i = 1, 2$, we have $\bar{x} < r_1'$ and $0 < \bar{y} < \varepsilon$. Thus If Γ is a connection, then by our discussion above, we can see that $\gamma_i \triangleq \Gamma \cap D_i$ ($i = 1, 2$) is two mutually disjoint preconnections, since the curve $F^2(\gamma_i)$ crosses Q_1 and Q_2, $i = 1, 2$. See Fig. 9.2. The proof is completed. □

From the definition of Q_ε, there exists a connection Γ_0 of Q_ε such that

$$\dim_\mathcal{H}(\Gamma_0) = 1, \quad \mathcal{H}^1(\Gamma_0) = r_2' - r_1'.$$

9.5 Examples of Applications of Rapid Fluctuations

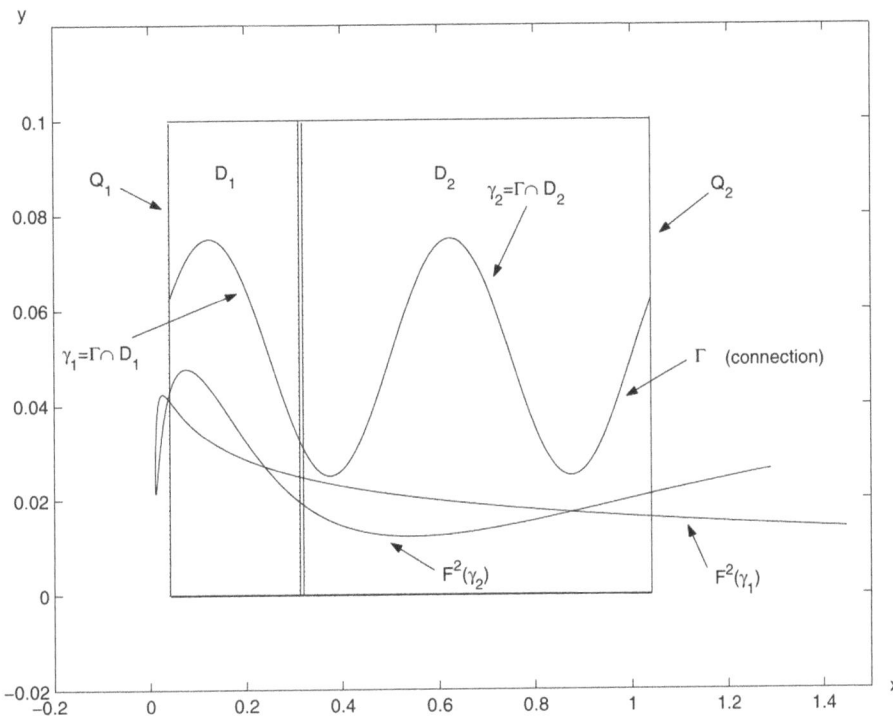

Fig. 9.2 The existence of topological horseshoe for f^2 with $\mu_1 = 3.2, \mu_2 = 1, s = 0.6$ and $\varepsilon = 0.1$

Thus, under the assumptions in Theorem 9.24, the system $(\mathbb{R}^2, f^2)$ has rapid fluctuation of dimension 1 by Theorem 9.16, and so does $(\mathbb{R}^2, f)$ by the Lipschitz property of f.

Exercise 9.25 Let $f : D \subset \mathbb{R}^N \to D$ be Lipschitz continuous with Lipschitz constant L and A be an s-set in D. Prove that for any nonnegative integer k, we have

$$\text{Var}_{f(D)}^{(s)}\left(f^{n+k-1}\right) \leq \text{Var}_D^{(s)}\left(f^{n+k}\right) \leq L^{sk} \text{Var}_D^{(s)}\left(f^n\right), \quad \forall n \geq 1.$$

□

Exercise 9.26 Let $f : D \subset \mathbb{R}^N \to D$ be Lipschitz continuous with Lipschitz constant L and A be an s-set in D. Prove that

$$\limsup_{n \to \infty} \frac{1}{n} \ln \text{Var}_D^{(s)}\left(f^n\right) = \lim_{n \to \infty} \frac{1}{n} \ln \text{Var}_D^{(s)}\left(f^n\right).$$

□

Exercise 9.27 Let $f : [0, 1] \to \mathbb{R}$ be continuously differentiable. Show that

$$V_{[0,1]}(f) = \int_0^1 |f'(x)|\, dx.$$

$\square$

Exercise 9.28 Under the assumption in Exercise 9.26, if

$$\lim_{n \to \infty} \text{Var}_D^{(s)}(f^n) = \infty$$

show that there exists $x \in D$, such that

$$\lim_{n \to \infty} \text{Var}_{D \cap U_\varepsilon(x)}^{(s)}(f^n) = \infty$$

where $U_\varepsilon(x)$ denotes a closed ε-neighborhood of x. $\square$

Exercise 9.29 Let $f : [0, 1] \to [0, 1]$ be defined as

$$f(x) = \begin{cases} 3x & 0 \leq x \leq \tfrac{1}{3} \\ 1 & \tfrac{1}{3} < x \leq \tfrac{2}{3} \\ -3x + 3 & \tfrac{2}{3} < x \leq 1 \end{cases}.$$

Show that f has rapid fluctuations of dimension s for any $s \in (0, 1]$. $\square$

Exercise 9.30 Under the assumption in Exercise 9.25, if f has rapid fluctuations on an s-set A, show that there exists at least an $x \in A$ such that

$$\lim_{n \to \infty} \frac{1}{n} \text{Var}_A^{(s)}(f^n) = \lim_{n \to \infty} \frac{1}{n} \text{Var}_{A \cap U_\varepsilon(x)}^{(s)}(f^n), \quad \forall \varepsilon > 0$$

where $U_\varepsilon(x)$ denotes a closed ε-neighborhood of x. $\square$

Exercise 9.31 Let $f : D \subset \mathbb{R}^N \to D$ be Lipschitz continuous with Lipschitz constant L and A be an s-set in D. Prove that

$$\lim_{n \to \infty} \frac{1}{n} \ln \text{Var}_A^{(s)}(f^n) = \lim_{n \to \infty} \frac{1}{n} \ln \sum_{i=1}^n \text{Var}_A^{(s)}(f^i)$$

(Here we set $0 \ln 0 = 0$.) $\square$

Exercise 9.32 Prove the assertion in Example 9.13. $\square$

Exercise 9.33 *(Baker map)* We revisit the Baker map. Consider a square set $S = [0, 1] \times [0, 1]$. Moreover, set

$$S_0 = [0, 1] \times [0, \frac{1}{2}], \quad S_1 = [0, 1] \times (\frac{1}{2}, 1].$$

It's clear that $S_0 \cup S_1 = S$. Define a map $f : S \to \mathbb{R}^2$ as follow

$$f(x, y) = \begin{cases} (\frac{1}{2}x, 2y) & (x, y) \in S_0 \\ (1 - \frac{1}{2}x, 2 - 2y) & (x, y) \in S_1. \end{cases}$$

Prove that $f(S) = S$ and has rapid fluctuations of dimension s for some $s > 0$. □

Notes for Chapter 9

This chapter is a continuation of Chap. 8. The notion of rapid fluctuations was first introduced in Huang, Chen, and Ma [5], motivated by the study of chaotic vibration of the wave equation; see Chap. 12. More results are developed in [6, 7]. Sections 9.3-9.4 follow from [7]. In particular, we have chosen Example 9.5 from [7] and Example 9.6 from [6].

In the study of chaotic dynamics of maps on one or higher dimensions, it is worth emphasizing that chaos often happens on an invariant set (the "strange attractor") with a fractal dimension. Thus, fractional Hausdorff dimensions and Hausdorff measures studied in Chap. 8 have found nice applications to the characterization of chaos.

References

1. J. Kennedy and J.A. Yorke, Topological horseshoe, *Trans. Amer. Math. Soc.* **353** (2001), 2513–2530. https://doi.org/10.1090/S0002-9947-01-02586-7
2. J. Benhabib and R.H. Day, Rational choice and erratic behaviour, *Rev. Econom. Stud.* **48** (1981), 459–471. https://doi.org/10.2307/2297158
3. J. Benhabib and R.H. Day, A characterization of erratic dynamics in the overlapping generations model, *J. Econom. Dynamics Control* **4** (1982), 37–55. https://doi.org/10.1016/0165-1889(82)90002-1
4. A. Dohtani, Occurrence of chaos in higher-dimensional discrete-time systems, *SIAM J. Appl. Math.* **52** (1992), 1707–1721. https://doi.org/10.1137/0152098
5. Y. Huang, G. Chen and D.W. Ma, Rapid fluctuations of chaotic maps on $\mathbb{R}^N$, *J. Math. Anal. Appl.* **323** (2006), 228–252. https://doi.org/10.1016/j.jmaa.2005.10.019
6. Y. Huang, X.M. Jiang and X. Zou, Dynamics in numerics: On a discrete predator-prey model, *Differential Equations Dynam. Systems* **16** (2008), 163–182. https://doi.org/10.1007/s12591-008-0010-6
7. Y. Huang and Y. Zhou, Rapid fluctuation for topological dynamical systems, *Front. Math. China* **4** (2009), 483–494. https://doi.org/10.1007/s11464-009-0030-8

10 Infinite-Dimensional Systems Induced by Continuous-Time Difference Equations

10.1 Infinite-Dimensional Discrete Dynamical System (I3DS)

Let I be a closed interval in $\mathbb{R}$, $C([0, 1], I)$ be the space of continuous functions from $[0, 1]$ to I and f be a continuous map from I into itself. That is, (I, f) is a one-dimensional discrete dynamical system. In this section, we consider the following map:

$$F: C([0, 1], I) \longrightarrow C([0, 1], I), \quad F(\varphi) = f \circ \varphi \text{ for } \varphi \in C([0, 1], I). \tag{10.1}$$

This F maps from the infinite-dimensional space $C([0, 1], I)$ into itself and will be denoted as $(C([0, 1], I), F)$. It constitutes an *infinite-dimensional discrete dynamical system (I3DS)*. The system $(C([0, 1], I), F)$ is an infinite-dimensional dynamical system generated by the one-dimensional system (I, f).

One of the motivations of studying the system (10.1) is from the continuous-time difference equation

$$x(t + 1) = f(x(t)), \quad t \in \mathbb{R}^+, \tag{10.2}$$

where f is a continuous map from I into itself. For each initial function φ from $[0, 1) \to I$, the difference equation (10.2) has a unique solution $x_\varphi: R^+ \to I$, which can be defined step by step:

$$x_\varphi(t) = (f^i \circ \varphi)(t - i) \quad \text{for } t \in [i, i + 1), \quad i = 0, 1, \ldots, \tag{10.3}$$

where f^i represents the i-th iterates of f. If only continuous solutions of (10.2) are admitted, we need *consistency conditions*

$$\varphi \in C([0, 1), I) \quad \text{and} \quad \varphi(1^-) = f(\varphi(0)). \tag{10.4}$$

If we define $\varphi(1) = \varphi(1^-)$, then $\varphi \in C([0, 1], I)$. We denote by $C^*([0, 1], I)$ the set of all such initial functions.

For $\varphi \in C^*([0, 1], I)$, there is a one-to-one correspondence between the solutions of the difference equation (10.2) and the orbits of the infinite-dimensional system (10.1).

$$x_\varphi(t + n) = F^n(\varphi)(t) \quad \text{for} \quad n \in \mathbb{Z}^+ \text{ and } t \in [0, 1], \tag{10.5}$$

where and throughout $\mathbb{Z}^+$ denotes the set of all positive integers.

The difference equation (10.2) as well as the I3DS (10.1) arise from concrete problems in applications. In particular, they are also related to nonlinear boundary value problems related to wave propagation, see [1–3]. We will give illustrative examples in Appendix II.

10.2 Rates of Growth of Total Variations of Iterates

We will characterize the asymptotic behavior and complexity of the solutions by means of the growth rates of the total variations. This approach is quite intuitive and natural for such a purpose. Let $BV([0, 1], I)$ be the space of all functions from $[0, 1]$ to I with bounded total variations and $V_{[0,1]}(\varphi)$ be the same as in (2.1). To study the asymptotic behavior of the I3DS (10.1), we will be concerned with the growth rates of $V_{[0,1]}(F^n(\varphi))$ as $n \to \infty$. The following are three distinct cases as n tends to ∞:

(1) $V_{[0,1]}(F^n(\varphi))$ remains bounded;
(2) $V_{[0,1]}(F^n(\varphi))$ grows unbounded, but may or may not be exponentially with respect to n;
(3) $V_{[0,1]}(F^n(\varphi))$ grows exponentially with respect to n.

Those properties are decided completely by the one-dimensional dynamical system (I, f). Thus, we must consider the relationship between the growth rates of the total variations of f^n as $n \to \infty$ and the complexity of (I, f).

Throughout the following, we assume that $f \in C(I, I)$ and f is *piecewise monotone* with finitely many extremal points. Let $PM(I, I)$ denote the set of all such maps.

Definition 10.1 Let $f \in PM(I, I)$ and $x \in I$.

(1) x is called a point of bounded variation of f if there exists a neighborhood J of x in I such that $V_J(f^n)$ remains bounded for all $n \in \mathbb{Z}^+$;
(2) x is called a point of unbounded variation of f if, for any neighborhood J of x, $V_J(f^n)$ grows unbounded as $n \to \infty$;
(3) x is called a chaotic or rapid fluctuation point of f if

$$\gamma(x, f) \equiv \lim_{\varepsilon \to 0} \gamma([x - \varepsilon, x + \varepsilon] \cap I, f) > 0,$$

where
$$\gamma(J, f) = \limsup_{n\to\infty} \frac{1}{n} \ln V_J(f^n), \tag{10.6}$$
for any subinterval J in I.

Denote by $B(f)$, $U(f)$, and $E(f)$ the sets of all, respectively, points of bounded variation, points of unbounded variation, and chaotic points of f. □

It is easy to see from Definition 10.1 that $B(f)$ is open in I. Thus $U(f)$ is closed in I and
$$B(f) \cap U(f) = \emptyset, \quad B(f) \cup U(f) = I, \quad E(f) \subset U(f).$$

For the orbits of the I3DS (10.1), we have the following classification.

Theorem 10.2 *For an initial function $\varphi \in C([0, 1], I)$, let $R(\varphi)$ denote the range of φ and $\dot{R}(\varphi)$ the interior of $R(\varphi)$.*

(1) *If $B(f) \neq \emptyset$, $R(\varphi) \subset B(f)$ and φ is piecewise monotone with finitely many extremal points, then $V_{[0,1]}(F^n(\varphi))$ remains bounded for all $n \in \mathbb{Z}^+$.*
(2) *If $U(f) \neq \emptyset$ and $U(f) \cap \dot{R}(\varphi) \neq \emptyset$, then $V_{[0,1]}(F^n(\varphi))$ grows unbounded as $n \to \infty$.*
(3) *If $E(f) \neq \emptyset$ and $E(f) \cap \dot{R}(\varphi) \neq \emptyset$, then $V_{[0,1]}(F^n(\varphi))$ grows exponentially with respect to n as $n \to \infty$.*

Proof For case (1), since φ is continuous and $B(f)$ is open, there exists an interval $J \subset B(f)$ such that $R(\varphi) \subset J$ by assumptions. Let $\ell(\varphi)$ denote the number of maximal closed subintegrals of $[0, 1]$ on each of which φ is monotonic. (Call this number the *lap* of φ.) Then we have
$$V_{[0,1]}(F^n(\varphi)) \leq \ell(\varphi) V_J(f^n).$$
Thus, $V_{[0,1]}(F^n(\varphi))$ remains bounded for all $n \in \mathbb{Z}^+$.

In the case of (2), it follows from the assumptions that there exist a point $x \in U(f)$ and a neighborhood J of x in I such that
$$x \in J \subset R(\varphi).$$
Thus
$$V_{[0,1]}(F^n(\varphi)) \geq V_J(f^n), \quad \forall n \in \mathbb{Z}^+.$$
This implies (2).

The proof for case (3) is the same as the one for case (2). □

In the remaining three sections of this chapter, we shall consider the properties of $B(f)$, $U(f)$, and $E(f)$, respectively.

10.3 Properties of the Set $B(f)$

We first give the following.

Theorem 10.3 *Let $f \in PM(I, I)$. Then $V_I(f^n)$ remains bounded if and only if $B(f) = I$.*

Proof Necessity follows directly from the definition.

Now we prove the sufficiency. Since $B(f) = I$, it follows that for each $x \in I$ there exists an open neighborhood U_x of x such that

$$V_{U_x}(f^n) \leq C_x$$

for some $C_x > 0$ which is independent of n.

Since $\{U_x, x \in I\}$ is an open cover of the compact interval I, there is a finite subcover such that

$$I \subset \bigcup_{i=1}^{n} U_{x_i}.$$

So

$$V_I(f^n) \leq \sum_{i=1}^{n} V_{U_{x_i}}(f^n) \leq C$$

, where

$$C = \max_{1 \leq i \leq n} \{C_{x_i}\}.$$

□

Definition 10.4 ([3]) A continuous map $g: I = [a, b] \to I$ is called an *L-map* if there exists $[x_1, x_2] \subseteq I$, $x_1 \leq x_2$ such that

$$g(x) = x, \text{ for } x \in [x_1, x_2], \tag{10.7}$$

$$x < g(x) \leq x_2, \text{ for } x < x_1, \text{ and } g(x) < x, \text{ for } x > x_2. \tag{10.8}$$

□

In [3], it is shown that every orbit of the I3DS (10.1) is compact and hence possesses a nonempty compact ω-limit set in $C([0, 1], I)$ if and only if either f or f^2 is an L-map.

Theorem 10.5 *Let $f \in C(I, I)$ be piecewise monotone with finitely many critical points. If either f or f^2 is an L-map, then $B(f) = I$.*

Proof It suffices to show that $V_I(f^n)$ remains bounded for all $n \in \mathbb{Z}^+$. Without loss of generality, we assume that $a < x_1$, where x_1 (and x_2) are defined according to Definition 10.4. Here we merely need to prove that $V_{[a,x_1]}(f^n)$ remains bounded for all $n \in \mathbb{Z}^+$ since $V_{[x_1,x_2]}(f^n)$ remains constant for all $n \in \mathbb{Z}^+$, and the boundedness of $V_{[x_2,b]}(f^n)$ can be proved similarly.

If f is an L-map, then f has only fixed points rather than periodic points. Since f has finitely many critical points and is piecewise monotone, there exists a $\delta_0 > 0$ such that $[x_1 - \delta_0, x_1] \subset I$ and f is monotonic on $[x_1 - \delta_0, x_1]$. By (10.7) and (10.8), f^n is also monotonic on $[x_1 - \delta_0, x_1]$ for every $n \in \mathbb{Z}^+$. On the other hand, $\lim_{n \to \infty} f^n(x)$ converges to a fixed point of f uniformly in $x \in I$ since f is an L-map. Thus, there exists a positive integer N such that when $n \geq N$,

$$f^n(x) > x_1 - \delta, \quad \forall x \in [a, x_1], \tag{10.9}$$

due to the fact that f has no fixed points in $[a, x_1)$.

Since $f \in PM(I, I)$, so is f^n for any $n \in \mathbb{Z}^+$. Let $\ell(f)$ denote the number of different laps on which f is strictly monotone and let

$$l_N = \max\{\ell(f^N), \ell(f^{N+1}), \ldots, \ell(f^{2N-1})\}.$$

Consider the map f^N. Let J_1 and J_2 be, respectively, the subintervals in $[a, x_1]$ such that

$$f^N(x) \leq x_1, \text{ for } x \in J_1,$$
$$(x_2 \geq) f^N(x) \geq x_1, \text{ for } x \in J_2.$$

Then $J_1 \cup J_2 = [a, x_1]$ and for any $n \in \mathbb{Z}^+$, $f^{Nn}(x) = f^N(x)$ for all $x \in J_2$, and the lap number of f^{Nn} on J_1 does not increase with n increasing. Thus,

$$V_{[a,x_1]}(f^{Nn}) = V_{J_1}(f^{Nn}) + V_{J_2}(f^{Nn})$$
$$\leq \ell(f^N)|f(x_1 - \delta) - f(x_1)| + V_{J_2}(f^N)$$
$$\leq \ell_N |f(x_1 - \delta) - x_1| + V_{[a,x_1]}(f^N). \tag{10.10}$$

Similarly, we have

$$V_{[a,x_1]}(f^{(N+i)n}) \leq \ell(f^{N+i})|f(x_1 - \delta) - x_1| + V_{[a,x_1]}(f^{N+i})$$
$$\leq \ell_N |f(x_1 - \delta) - x_1| + V_{[a,x_1]}(f^N), \tag{10.11}$$

for $i = 1, 2, \ldots, N - 1$. Inequalities (10.10) and (10.11) imply that the total variation $V_{[a,x_1]}(f^n)$ of f^n on $[a, x_1]$ remains bounded for all $n \in \mathbb{Z}^+$.

If f^2 is an L-map but f itself is not, then f has periodic points of period less than or equal to 2 only. Utilizing the same argument, we can derive that $V_I(f^{2n})$ remains bounded for all $n \in \mathbb{Z}^+$. Note that

$$V_I(f^{2n+1}) \leq \ell(f) V_I(f^{2n}),$$

hence, we have that $V_I(f^n)$ remains bounded for all $n \in \mathbb{Z}^+$. This completes the proof of the theorem. □

Remark 10.6 The converse of Theorem 10.5 need not be true. That is, there are maps in $PM(I, I)$ which are not L-maps but their total variations remain bounded for all $n \in \mathbb{Z}^+$. For instance, $f(x) = \sqrt{x}$ on $[0, 1]$. It is easy to see that $V_{[0,1]}(f^n) = 1$ for all $n \in \mathbb{Z}^+$, but neither f nor f^2 is an L-map. Thus, not every orbit of the I3DS (10.1) defined through such an f is compact. We will see from Theorem 10.20 that a necessary condition for $B(f) = I$ is that f has no periodic orbit with period great than 2. □

Theorem 10.7 Let $f \in PM(I, I)$. Then we have

(1) If $x \in \text{Fix}(f)$ (the set of all fixed points of f) and x is a local stable point of f, then $x \in B(f)$.
(2) If f has at most a periodic point in $\text{int}(I)$, then $B(f) = I$. □

Exercise 10.8 Prove Theorem 10.7. □

10.4 Properties of the Set $U(f)$

We have the following.

Theorem 10.9 Let $f \in PM(I, I)$. Then $\lim\limits_{n\to\infty} V_I(f^n) = \infty$ if and only if $U(f) \neq \emptyset$.

Proof The sufficiency immediately follows by definition. To prove the necessity, we assume that
$$\lim_{n\to\infty} V_I(f^n) = \infty.$$
Let $I = [a, b]$. Consider the two bisected subintervals $[a, (a+b)/2)]$ and $[(a+b)/2, b]$. Then the total variations of f^n grow unbounded at least on one of the two intervals as $n \to \infty$. We denote by $I_1 = [a_1, b_1]$ such an interval. Repeating the above bisecting processing, we obtain a sequence of intervals $I_k = [a_k, b_k]$ with

(i) $[a, b] \supset [a_1, b_1] \supset \cdots \supset [a_k, b_k] \supset \cdots$,
(ii) $b_k - a_k = \frac{b-a}{2^k}$,
(iii) $V_{I_k}(f^n)$ grows unbounded as $n \to \infty$ for each $k \in \mathbb{Z}^+$.

It follows from (i) and (ii) that there exists a unique $c \in I_k$ for all k such that
$$\lim_{k\to\infty} a_k = \lim_{k\to\infty} b_k = c.$$

10.4 Properties of the Set $U(f)$

For any neighborhood J of c, there exists a $k \in \mathbb{Z}^+$ such that $J \supset I_k$. Following (iii), we have that $c \in U(f)$. □

The following is the main theorem of this section.

Theorem 10.10 *Let $f \in PM(I, I)$. Then $U(f) = I$ if and only if f has sensitive dependence on initial data on I.* □

To prove the sufficiency of Theorem 10.10, we need the following lemma.

Lemma 10.11 *Assume that $f \in PM(I, I)$ and f has sensitive dependence on initial data on I. Let $J = [c, d]$ be an arbitrary subinterval of I, with $|J| \geq \delta$, where δ is the sensitive constant of f. Then there exists an A: $0 < A \leq \delta/2$, independent of J, such that*

$$|f^n(J)| \geq A, \quad \forall n \in \mathbb{Z}^+. \tag{10.12}$$

Proof Let $N = [\frac{2(b-a)}{\delta}] + 1$, where $[r]$ denotes the usual integral part of the real number r. Divide I into N equal-length subintervals I_i, $i = 1, \ldots, N$, with $I_i = [x_{i-1}, x_i]$, $x_0 = a$ and $x_N = b$, $|I_i| = (b-a)/N$. Then $|I_i| \leq \delta/2$ holds for each i.

Let $x \in \text{int}(I_i) = (x_{i-1}, x_i)$, the interior of I_i. Then from the sensitive dependence on initial data of f (cf. Definition 6.8), it follows that there is a $y \in \text{int}(I_i)$ and an $N_i \in \mathbb{Z}^+$ such that

$$|f^{N_i}(x) - f^{N_i}(y)| \geq \delta.$$

This implies

$$|f^{N_i}(I_i)| \geq \delta.$$

For $i = 1, \ldots, N$, let

$$a_i = \min\{|f^j(I_i)| \mid j = 0, 1, \ldots, N_i\}, \tag{10.13}$$

and

$$A = \min\{a_i \mid i = 1, \ldots, N\}; \quad A > 0. \tag{10.14}$$

Then $A \leq \delta/2$ because $|I_i| \leq \delta/2$ for each i. Now let $J = [c, d]$ satisfy $|J| \geq \delta$. Then there exists at least a subinterval I_{j_0}, $1 \leq j_0 \leq N$, such that $J \supset I_{j_0}$. Thus

$$f^k(J) \supset f^k(I_{j_0}), \quad \text{for } k = 0, 1, \ldots. \tag{10.15}$$

We are ready to establish (10.12). We divide the discussion into the following cases:

(i) $0 \leq n \leq N_{j_0}$. In this case, it is obvious that

$$|f^n(J)| \geq |f^n(I_{j_0})| \geq a_{j_0} \geq A,$$

by (10.13)–(10.15). So (10.12) holds.

(ii) $n > N_{j_0}$. If $0 < n - N_{j_0} \leq \min\{N_i \mid i = 1, \ldots, N\}$, then from the facts that $f^{N_{j_0}}(J) \supset f^{N_{j_0}}(I_{j_0})$ and $f^{N_{j_0}}(I_{j_0})$ having length at least δ and then further containing at least one subinterval I_{j_1}, we obtain

$$f^{n-N_{j_0}}(f^{N_{j_0}}(J)) \supset f^{n-N_{j_0}}(f^{N_{j_0}}(I_{j_0})) \supset f^{n-N_{j_0}}(I_{j_1}). \tag{10.16}$$

But $n - N_{j_0} \leq \min\{N_i \mid i = 1, \ldots, N\}$. By (10.13), (10.14) and (10.16), we have

$$|f^n(J)| = |f^{n-N_{j_0}}(f^{N_{j_0}}(J))| \geq |f^{n-N_{j_0}}(I_{j_1})| \geq A.$$

The above restriction that $0 < n - N_{j_0} \leq \min\{N_i \mid i = 1, 2, \ldots, N\}$ can actually be relaxed to $0 < n - N_{j_0} \leq N_{j_1}$, if I_{j_1} is the subinterval satisfying (10.16), by (10.13) and (10.14).

One can then extend the above arguments inductively to any $n = N_{j_0} + N_{j_1} + \cdots + N_{j_k} + R_k$, where $R_k \in \mathbb{Z}^+ \cup \{0\}$, $0 \leq R_k \leq N_{j_{k+1}}$, and where

$$f^n(J) = f^{R_k}(f^{N_{j_0}+\cdots+N_{j_k}}(J)) \supseteq f^{R_k}(f^{N_{j_k}}(I_{j_k})) \supseteq f^{R_k}(I_{j_{k+1}}) \tag{10.17}$$

is satisfied for a sequence of intervals $I_{j_0}, I_{j_1}, \ldots, I_{j_k}$ and $I_{j_{k+1}}$. From (10.13), (10.14), and (10.17), we have proved (10.12). □

Equipped with Lemma 10.11, we can now proceed to give the following.

Proof of the sufficiency of Theorem 10.10 Let $J = [c, d]$ be any subinterval of I and let $M > 0$ be sufficiently large. We want to prove the following statement:

"there exists an $N(M) \in \mathbb{Z}^+$ such that $V_J(f^n) \geq M$ for all $n \geq N(M)$". (10.18)

First, divide J into N subintervals, with $N = \left[\frac{M}{A}\right] + 1$, where A satisfies Lemma 10.11. Thus $J = J_1 \cup J_2 \cup \cdots \cup J_N$, with $J_k = [x_{k-1}, x_k]$; $x_k = c + k\left(\frac{d-c}{N}\right)$, $k = 1, 2, \ldots, N$. By the sensitive dependence of f on I, for any $k = 1, 2, \ldots, N$, there exists an N_k such that

$$|f^{N_k}(J_k)| \geq \delta. \tag{10.19}$$

Since $f^{N_k}(J_k)$ is a connected interval, by (10.19) we can apply Lemma 10.11 and obtain

$$|f^n(J_k)| = |f^{n-N_k}(f^{N_k}(J_k))| \geq A, \text{ if } n \geq N_k, \text{ for } k = 1, 2, \ldots, N.$$

Now take $N(M) = \max\{N_1, \ldots, N_N\}$. Then for $n \geq N(M)$,

$$V_J(f^n) = \sum_{k=1}^{N} V_{J_k}(f^n) \geq \sum_{k=1}^{N} |f^n(J_k)| \geq \sum_{k=1}^{N} A = NA > M.$$

10.4 Properties of the Set $U(f)$

The proof of (10.8) and, therefore, the sufficiency of Theorem 10.10 are now established. □

For the necessity of Theorem 10.10, a long sequence of proposition and lemmas is needed in order to address the case $U(f) = I$, i.e., every point $x \in I$ is a point of unbounded variation.

Proposition 10.12 *Assume $f \in PM(I, I)$ and $U(f) = I$. Then*

(i) *$f(x) \not\equiv c$ on any subinterval J of I, for any constant c;*
(ii) *$f(x) \not\equiv x$ on any subinterval J of I.*
(iii) *Let J be a subinterval of I whereupon f is monotonic. Then there exists at most one point $\bar{x} \in J$ such that $f(\bar{x}) = \bar{x}$. Consequently, f has at most finitely many fixed points on I.*
(iv) *Let J be a subinterval of I and $x_0 \in J$ satisfies $f(x_0) = x_0$. If f is increasing on J, then $f(x) > x$ for all $x > x_0$, $x \in J$, and $f(x) < x$ for all $x < x_0$, $x \in J$. This property also holds if J is an interval with x_0 either as its left or right end point.*
(v) *For any positive integer n, $f^n \in PM(I, I)$ and $U(f^n) = I$.*

Proof Part (i) is obvious. Consider part (ii). If $f(x) \equiv x$ on J, then

$$V_J(f^n) = |J| \quad \text{for every} \quad n.$$

This violates $U(f) = I$.

Now consider (iii). Let us first assume that f is monotonically decreasing on J. Define $g(x) = f(x) - x$. Then g is also decreasing on J. If there were two points $\bar{x}_1$ and $\bar{x}_2$; $\bar{x}_1, \bar{x}_2 \in J$, $\bar{x}_1 \neq \bar{x}_2$, such that $f(\bar{x}_i) = \bar{x}_i$, $i = 1, 2$, then $g(\bar{x}_1) = g(\bar{x}_2)$ and therefore $g(x) \equiv 0$ on a subinterval of J, implying $f(x) \equiv x$ on J, contradicting part (ii).

If f is monotonically increasing on J and there exist two fixed points $\bar{x}_1, \bar{x}_2 \in J, \bar{x}_1 < \bar{x}_2$, then f is monotonically increasing on $J_0 \equiv [\bar{x}_1, \bar{x}_2]$, and f^n is also increasing on J_0, such that $f^n(J_0) = J_0$ for every $n \in \mathbb{Z}^+$. Thus $V_{J_0}(f^n) = |J_0| \not\to \infty$ as $n \to \infty$, contradicting $U(f) = I$. Therefore, we have established (iii).

Further, consider (iv). If there exists an $x_1 \in J$ such that $x_1 > x_0$ and $f(x_1) \leq x_1$, then $f(x_1) < x_1$ because $f(x_1) = x_1$ is ruled out by (iii). Consider $J_1 \equiv [x_0, x_1]$, if $x_0 < x_1$. Then f is increasing on J_1, so are f^n for any $v \in \mathbb{Z}^+$, such that $f^n(J_1) \subseteq J_1$. Hence $V_{J_1}(f^n) \leq |J_1|$, violating $U(f) = I$. The case that $x_1 < x_0$ and $f(x_1) \geq x_1$ similarly also leads to a contradiction.

Finally, part (v) is obvious. Its proof is omitted. □

Remark 10.13 We see that Proposition 10.12 (iv) is actually a *hyperbolicity* result, i.e., if x_0 is a fixed point of f and $x_0 \in U(f)$, and if f is increasing and differentiable at x_0, then $|f'(x_0)| > 1$. □

Lemma 10.14 *Assume $f \in PM(I, I)$ and $U(f) = I$. Let $\bar{x}_0$ be a fixed point of f on I and U be a small open neighborhood of $\bar{x}_0$ in I. Then there exists a $\delta_0 > 0$ such that for any $x \in U \setminus \{\bar{x}_0\}$, there exists an $N_x \in \mathbb{Z}^+$, N_x depending on x, such that*

$$|f^{N_x}(x) - \bar{x}_0| > \delta_0. \tag{10.20}$$

Proof Since $f \in PM(I, I)$, we have two possibilities: (i) f is monotonic on $U = [\bar{x}_0 - \delta, \bar{x}_0 + \delta]$ for some sufficiently small $\delta > 0$; (ii) $\bar{x}_0$ is an extremal point of f.

First, consider case (i) when f is increasing on U. Since $U(f) = I$, Proposition 10.12 part (iv) gives

$$f(\bar{x}_0 - \delta) < \bar{x}_0 - \delta, \quad f(\bar{x}_0 + \delta) > \bar{x}_0 + \delta.$$

Thus, we can find $x_1 \in (\bar{x}_0 - \delta, \bar{x}_0)$, $x_2 \in (\bar{x}_0, \bar{x}_0 + \delta)$, such that

$$f(x_1) = \bar{x}_0 - \delta, \quad f(x_2) = \bar{x}_0 + \delta. \tag{10.21}$$

Define $\delta_0 = \min\{\bar{x}_0 - x_1, x_2 - \bar{x}_0\}$. We now show that (10.20) is true.

Assume the contrary that (10.20) fails for some $\hat{x} \in (\bar{x}_0 - \delta, \bar{x}_0) \cup (\bar{x}_0, \bar{x}_0 + \delta)$. Then

$$|f^n(\hat{x}) - \bar{x}_0| \leq \delta_0, \quad \text{for all} \quad n \in \mathbb{Z}^+. \tag{10.22}$$

We consider the case $\hat{x} > \bar{x}_0$. (The case $\hat{x} < \bar{x}_0$ can be similarly treated and is therefore omitted.) From (10.22), $f(\bar{x}_0) = \bar{x}_0$, and the fact that f^n is increasing on $[\bar{x}_0, \bar{x}_0 + \delta_0]$, we have $f^n([\bar{x}_0, \hat{x}]) \subseteq [\bar{x}_0, \bar{x}_0 + \delta_0]$ and thus $V_{[\bar{x}_0, \hat{x}]}(f^n) \leq \delta_0$ for any $n \in \mathbb{Z}^+$, violating $U(f) = I$.

Next, consider case (i) when f is decreasing on U. By the continuity of f and Proposition 10.12 part (i), we have $f(\bar{x}_0 - \delta) > \bar{x}_0$ and $f(\bar{x}_0 + \delta) < \bar{x}_0$. Thus we can find $x_1 \in (\bar{x}_0 - \delta, \bar{x}_0)$ and $x_2 \in (\bar{x}_0, \bar{x}_0 + \delta)$ such that

$$f(x_1) = \bar{x}_0 + \delta, \quad f(x_2) = \bar{x}_0 - \delta.$$

Let $\delta_0 = \min\{x_2 - \bar{x}_0, \bar{x}_0 - x_1\}$. If (10.20) were not true for this δ_0, then there is an $\hat{x} \in (\bar{x}_0 - \delta_0, \bar{x}_0) \cup (\bar{x}_0, \bar{x}_0 + \delta_0)$ such that

$$|f^n(\hat{x}) - \bar{x}_0| < \delta_0, \quad \text{for all} \quad n \in \mathbb{Z}^+. \tag{10.23}$$

Again, we may assume that $\hat{x} > \bar{x}_0$. (The case $\hat{x} < \bar{x}_0$ can be treated similarly.) Since f^2 is increasing on $[\bar{x}_0, \hat{x}]$ and by (10.23) and $f(\bar{x}_0) = \bar{x}_0$, we have $f^{2n}([\bar{x}_0, \hat{x}]) \subseteq [\bar{x}_0, \bar{x}_0 + \delta_0]$ for all $n \in \mathbb{Z}^+$. Therefore

$$V_{[\bar{x}_0, \hat{x}]}(f^{2n}) \leq \delta_0, \quad \text{for any} \quad n \in \mathbb{Z}^+,$$

contradicting $U(f) = I$. So case (i) implies (10.20).

10.4 Properties of the Set $U(f)$

We proceed to treat case (ii), i.e., $\bar{x}_0$, as a fixed point of f, is also an extremal point of f. Note that it is also possible that $\bar{x}_0 = a$ or $\bar{x}_0 = b$, i.e., $\bar{x}_0$ is a boundary extremal point. Let us divide the discussion into the following four subcases: (1) $\bar{x}_0 = a$; (2) $\bar{x}_0 = b$; (3) $\bar{x}_0 \in (a, b)$ is a relative maximum; and (4) $\bar{x}_0 \in (a, b)$ is a relative minimum.

Subcase (1) implies that $\bar{x}_0 = a$, as a fixed point, must be a local minimum. Let

$$\tilde{x}_1 = \min\{\tilde{x} | \tilde{x} \text{ is an extremal point}, \tilde{x} > \bar{x}_0\}.$$

Then by Proposition 10.12 part (iv), we have $f(\tilde{x}_1) > \tilde{x}_1$. Then there exists an $\hat{x}_1 \in (\bar{x}_0, \tilde{x}_1)$ such that $f(\hat{x}_1) = \tilde{x}_1$. Define $\delta_0 = \hat{x}_1 - \bar{x}_0$. Then since f is increasing on $[\bar{x}_0, \tilde{x}_1]$, the case can be treated just as in case (i) earlier.

Subcase (2) is a mirror image of subcase (1) and can be treated in the same way. So let us treat subcase (3). Let

$$\tilde{x}_1 = \max\{\tilde{x} | \tilde{x} \text{ is an extremal point}, \tilde{x} < \bar{x}_0\},$$
$$\tilde{x}_2 = \min\{\tilde{x} | \tilde{x} \text{ is an extremal point}, \tilde{x} > \bar{x}_0\}.$$

Then f is increasing on $[\tilde{x}_1, \bar{x}_0]$ and decreasing on $[\bar{x}_0, \tilde{x}_2]$. By Proposition 10.12 part (iv), we have $f(\tilde{x}_1) < \tilde{x}_1$. Therefore, there exists an $\hat{x}_1 \in (\tilde{x}_1, \bar{x}_0)$ such that $f(\hat{x}_1) = \tilde{x}_1$. If $f(\tilde{x}_2) < \tilde{x}_1$, then there is an $\hat{x}_2 = (\bar{x}_0, \tilde{x}_2)$ such that $f(\hat{x}_2) = \tilde{x}_1$. In this case, we set $\delta_0 = \min\{\bar{x}_0 - \hat{x}_1, \hat{x}_2 - \bar{x}_0\}$. If $f(\tilde{x}_2) \geq \tilde{x}_1$ then we set $\delta_0 = \bar{x}_0 - \hat{x}_1$. The remaining arguments go the same way as in (i) earlier.

Subcase (4) can be treated in the same way as Subcase (3). □

Recall $\omega(x, f)$, the ω-limit set of a point x in I under f; cf. Definition 6.32.

Lemma 10.15 *Let $f \in C(I, I)$ and $\hat{x} \in I$. If $\omega(\hat{x}, f) = \{x_0, \ldots, x_k\}$, then $f^k(x_i) = x_i$, for $i = 0, 1, \ldots, k$. That is, $\omega(\hat{x}, f)$ is a periodic orbit.* □

Exercise 10.16 Prove Lemma 10.15. □

Lemma 10.17 *Assume that $f \in PM(I, I)$ and $U(f) = I$. Let J be any subinterval of I. Then there exists an infinite sequence $\{n_j \in \mathbb{Z} | j = 1, 2, \ldots\}$, $n_j \to \infty$, such that $f^{n_j}(J)$ contains at least an extremal point of f for all n_j.*

Proof If f is not monotonic on J, take $n_1 = 0$. Then $f^{n_1}(J) = J$ contains an extremal point of f.

If f is monotonic on J, then because f cannot be constant on J, f must be either strictly increasing or strictly decreasing on J. Assume first that f is strictly increasing. Then there exists some $m_1 \geq 2$ such that f^{m_1} is not monotonic on J because otherwise

$$V_J(f^n) \leq b - a \quad \text{for all} \quad n = 1, 2, \ldots,$$

a contradiction. This implies that f is not monotonic on $f^{m_1-1}(J)$ and, therefore, $f^{m_1-1}(J)$ has an extremal point of f. We then choose $n_1 = m_1 - 1 \geq 1$ in this case. (If instead f is strictly decreasing on J, then the proof is similar.)

Since $U(f) = I$, $f^{n_1}(J)$ does not collapse to a single point by Proposition 10.12. Choose a subinterval J_1 of $f^{n_1}(J)$ where f is monotonic on J_1. Using the above arguments again, we have some $m_2 \geq 2$ such that f^{m_2} is not monotonic on J_1. Therefore $f^{m_2-1}(J_1)$ contains an extremal point of f. But $J_1 \subseteq f^{n_1}(J)$, and so $f^{m_2-1}(J_1) \subseteq f^{n_1+m_2-1}(J)$ contains an extremal point of f. Define $n_2 = n_1 + m_2 - 1$.

This process can be continued indefinitely. The proof is complete. □

Lemma 10.18 *Assume $f \in PM(I, I)$ and $U(f) = I$. Let $\tilde{x}_0$ be an extremal point of f. Then there is a $\delta > 0$ such that for any (relatively) open neighborhood U of $\tilde{x}_0$, there is an $\hat{x} \in U$ and an $\widehat{N} \in \mathbb{Z}^+$ such that $|f^{\widehat{N}}(\hat{x}) - f^{\widehat{N}}(\tilde{x}_0)| \geq \delta$.*

Proof Let $E = \{\tilde{x}_0, \tilde{x}_1, \ldots, \tilde{x}_k\}$ be the set of all extremal points of f. We may note that by Proposition 10.12 part (iv) and $f \in PM(I, I)$ that we have $a, b \in E$. Consider the orbit of $\tilde{x}_0$: $\mathcal{O}(\tilde{x}_0) = \{f^n(\tilde{x}_0) \mid n = 1, 2, \ldots\}$. There are two possibilities.

Case 1: There are n_1, n_2: $n_1 > n_2 \geq 0$, such that $f^{n_1}(\tilde{x}_0) = f^{n_2}(\tilde{x}_0)$.

Case 2: For any $n_1, n_2 \in \mathbb{Z}^+$, $f^{n_1}(\tilde{x}_0) \neq f^{n_2}(\tilde{x}_0)$ if $n_1 \neq n_2$.

Consider Case 1 first. Let $y_0 = f^{n_2}(\tilde{x}_0)$. For any interval W, $f^{n_2}(W)$ is also an interval because f^{n_2} is continuous. This interval $f^{n_2}(U)$ can never degenerate into a point by Proposition 10.12 part (i). Set $F(x) = f^{n_1-n_2}(x)$. Then $F \in PM(I, I)$ and $U(F) = I$. Pick $y_1 \in f^{n_2}(W)$ but $y_1 \neq y_0$. Let $\hat{x}_1 \in W$ satisfy $f^{n_2}(\hat{x}_1) = y_1$. Then because y_0 is a fixed point of F, by Lemma 10.14 there are a $\delta > 0$ (independent of y_1) and an $N \in \mathbb{Z}^+$ (dependent on y_1) such that

$$|F^N(y_1) - F^N(y_0)| = |F^N(y_1) - y_0| \geq \delta,$$

or

$$|f^{N(n_1-n_2)}(y_1) - f^{N(n_1-n_2)}(y_0)| \geq \delta,$$
$$|f^{N(n_1-n_2)+n_2}(\hat{x}_1) - f^{N(n_1-n_2)+n_2}(\tilde{x}_0)| \geq \delta.$$

Therefore, Lemma 10.18 holds for Case 1.

Next, consider Case 2. For the ω-limit set $\omega(\tilde{x}, f)$, there are two subcases:

Case (2.a): $\omega(\tilde{x}, f) \nsubseteq E$;

Case (2.b): $\omega(\tilde{x}, f) \subseteq E$.

Consider Case (2.a). Let $y_0 \in \omega(\tilde{x}, f)$ but $y_0 \notin E$, and let $\delta_0 = \frac{1}{2}d(y_0, E)$. By Lemma 10.17, for W there is an $N_1 \in \mathbb{Z}^+$ and a sequence $\{n_j\}$ such that $f^{n_j}(W)$ contains at least an extremal point of f for all $n_j \geq N_1$. Since $y_0 \in \omega(\tilde{x}, f)$, there is an $n_k > N_1$ such that $|f^{n_k}(\tilde{x}_0) - y_0| < \delta_0/3$. Let $\tilde{x}_j \in E$ be such that $\tilde{x}_j \in f^{n_k}(W)$, and let $\hat{x} \in W$ be such that $f^{n_k}(\hat{x}) = \tilde{x}_j$. Then

10.4 Properties of the Set $U(f)$

$$|f^{n_k}(\tilde{x}_0) - f^{n_k}(\hat{x})| \geq |y_0 - f^{n_k}(\hat{x})| - |y_0 - f^{n_k}(\tilde{x}_0)|$$
$$\geq d(y_0, E) - |y_0 - f^{n_k}(\tilde{x}_0)|$$
$$\geq \frac{1}{2}d(y_0, E) - \delta_0/3$$
$$= \delta_0/2 - \delta_0/3 = \delta_0/6.$$

Set $\delta = \delta_0/6$ and $\widehat{N} = n_k$. Then we have

$$|f^{\widehat{N}}(\tilde{x}_0) - f^{\widehat{N}}(\hat{x})| \geq \delta,$$

so Lemma 10.18 holds true.

Now consider Case (2.b). We divide this into two further subcases:

Case (2.b.i) For all $n \in \mathbb{Z}^+$, $f^n(W) \cap \omega(\tilde{x}, f) = \emptyset$;

Case (2.b.ii) There is an $n_0 \in \mathbb{Z}^+$ such that $f^{n_0}(W) \cap \omega(\tilde{x}, f) \neq \emptyset$.

Consider Case (2.b.i). Since E is finite and by Lemma 10.17, there is an $\tilde{x}_{\tilde{i}} \in E$ and a subsequence $\{n_i \in \mathbb{Z}^+ \mid i = 1, 2, \ldots\}$ such that $f^{n_i}(W)$ always contains $\tilde{x}_{\tilde{i}}$. Since $f^n(W) \cap \omega(\tilde{x}, f) = \emptyset$, $\tilde{x}_{\tilde{i}} \notin \omega(\tilde{x}, f)$. Let $\delta = \frac{1}{2}d(\tilde{x}_{\tilde{i}}, \omega(\tilde{x}, f)) > 0$. Since $\lim_{n \to \infty} d(f^n(\tilde{x}_0), \omega(\tilde{x}, f)) = 0$, there is a j_0 sufficiently large such that

$$d(f^{n_j}(\tilde{x}_0), \omega(\tilde{x}, f)) < \frac{1}{2}d(\tilde{x}_{\tilde{i}}, \omega(\tilde{x}, f)), \quad \text{for all} \quad j \geq j_0.$$

Now, choose $N = n_{j_0} > N_1$. Since $f^{n_{j_0}}(W) \ni \tilde{x}_{\tilde{i}}$, there is an $\hat{x} \in W$ such that $f^{n_{j_0}}(\hat{x}) = \tilde{x}_{\tilde{i}}$. Therefore

$$|f^{n_{j_0}}(\tilde{x}_0) - f^{n_{j_0}}(\hat{x})| = |f^{n_{j_0}}(\tilde{x}_0) - \tilde{x}_{\tilde{i}}| \geq d(\tilde{x}_{\tilde{i}}, \omega(\tilde{x}, f)) - d(f^{n_{j_0}}(\tilde{x}_0), \omega(\tilde{x}, f))$$
$$\geq \frac{1}{2}d(\tilde{x}_{\tilde{i}}, \omega(\tilde{x}, f)) = \delta.$$

Hence Lemma 10.18 holds for Case (2.b.i).

Finally, consider Case (2.b.ii). Since $f^{n_0}(W) \cap \omega(\tilde{x}, f) \neq \emptyset$, there is an $\hat{x} \in W$ such that $f^{n_0}(\hat{x}) = \tilde{x}_j$, for some $\tilde{x}_j \in \omega(\tilde{x}, f) \subseteq E$. Pick a point $y_0 \in f^{n_0}(W) \setminus \{f^{n_0}(\hat{x})\}$. Let $\hat{x} \in W$ be such that $f^{n_0}(\hat{x}) = y_0$. Since $\omega(\tilde{x}, f) \subseteq E$ and E is finite, $\omega(\tilde{x}, f)$ is finite and has, say, k_1 elements. By Lemma 10.15, we have $f^{k_1}(x^*) = x^*$ for all $x^* \in \omega(\tilde{x}, f)$. Define $F(x) = f^{k_1}(x)$. Then each $x^* \in \omega(\tilde{x}, f)$ is a fixed point of F, and F satisfies $F \in PM(I, I)$ and $U(F) = I$ as well, by Proposition 10.12 (v). By Lemma 10.15, there exists a $\delta > 0$ and N_1 (depending on $\tilde{x}_j$) such that

$$|F^{N_1}(y_0) - F^{N_1}(\tilde{x}_j)| = |F^{N_1}(y_0) - \tilde{x}_j| \geq 2\delta.$$

Let $N = N_1 k_1 + n_0$. Then

$$|f^N(\hat{x}) - f^N(\hat{x})| = |F^{N_1}(y_0) - F^{N_1}(\tilde{x}_j)| \geq 2\delta.$$

Hence, by an application of the triangle inequality, we have

$$\text{either } |f^N(\tilde{x}_0) - f^N(\hat{x})| > \delta, \quad \text{or} \quad |f^N(\tilde{x}_0) - f^N(\hat{\hat{x}})| > \delta.$$

Therefore, Lemma 10.18 holds for Case (2.b.ii).

The proof is complete. □

Proof of necessity of Theorem 10.10 Let $E = \{\tilde{x}_0, \tilde{x}_1, \ldots, \tilde{x}_k\}$ be the set of all extremal points of f. By Lemma 10.18, for any interval $W \ni \tilde{x}_i$, there is a $\delta_i > 0$ (independent of W) such that there is an $\hat{x}_i \in W \setminus \{x_i\}$ and N_i (dependent on $\hat{x}_i$) satisfying

$$|f^{N_i}(\hat{x}_i) - f^{N_i}(\tilde{x}_i)| > \delta_i, \quad i = 0, 1, 2, \ldots, k. \tag{10.24}$$

Define $2\delta \equiv \min\{\delta_i \mid i = 0, 1, \ldots, k\}$. For any $x \in I$ and any interval $W \ni x$, by Lemma 10.17, for some $N' \in \mathbb{Z}^+$, $f^{N'}(W)$ contains an extremal point, say, $\tilde{x}_{\bar{j}}$, i.e., $\tilde{x}_{\bar{j}} \in f^{N'}(W)$. Since $f^{N'}(W)$ is an interval with positive length, by (10.24) and Lemma 10.18, there is an $\hat{x} \in f^{N'}(W)$ and an $N_{\bar{j}}$ such that

$$|f^{N_{\bar{j}}}(\hat{x}) - f^{N_{\bar{j}}}(\tilde{x}_{\bar{j}})| \geq \delta_{\bar{j}} \geq 2\delta.$$

Now, let $N = N_{\bar{j}} + N'$, $y_1, y_2 \in W$ satisfy $f^{N'}(y_1) = \hat{x}$, $f^{N'}(y_2) = \tilde{x}_{\bar{j}}$. We have

$$|f^N(y_1) - f^N(y_2)| = |f^{N_{\bar{j}}}(\hat{x}) - f^{N_{\bar{j}}}(\tilde{x}_{\bar{j}})| \geq 2\delta.$$

Therefore, for any $x \in W$, by an application of the triangle inequality, we have

$$\text{either } |f^N(y_1) - f^N(x)| \geq \delta \quad \text{or} \quad |f^N(y_2) - f^N(x)| \geq \delta.$$

The sensitive dependence of f on initial data has been proven.

10.5 Properties of the Set $E(f)$

Recall from Definition 10.1 part (3) that

$$E(f) = \{x \in I \mid \gamma(x, f) > 0\}.$$

Denote

$$\gamma(f) = \limsup_{n \to \infty} \frac{1}{n} \ln V_I(f^n).$$

10.5 Properties of the Set $E(f)$

Theorem 10.19 *Let $f \in PM(I, I)$. Then*

$$\gamma(f) = \sup_{x \in I} \gamma(x, f),$$

and there exists an $x \in I$ such that $\gamma(f) = \gamma(x, f)$.

Proof We assume that $f: I \to I$ is onto. It is clear that $\gamma(x, f)$ is no greater than $\limsup_{n \to \infty} \frac{1}{n} \ln V_I(f^n)$ for each $x \in I$. Choose the two subintervals I_{11}, I_{12} of I such that their lengths satisfy $|I_{11}| = |I_{12}| = \frac{1}{2}|I|$ and $I = I_{11} \cup I_{12}$. Then,

$$\limsup_{n \to \infty} \frac{1}{n} \ln \left(V_{I_{11}}(f^n) + V_{I_{12}}(f^n) \right) = \limsup_{n \to \infty} \frac{1}{n} \ln V_I(f^n) = \gamma(f).$$

This indicates that

$$\limsup_{n \to \infty} \frac{1}{n} \ln V_{I_{1i}}(f^n) = \gamma(f) \text{ for } i = 1 \text{ or } i = 2.$$

Denote this interval by I_1. Similarly, we can find a subinterval I_2 of I_1 whose length $|I_2|$ equals $\frac{1}{2}|I_1|$ such that

$$\limsup_{n \to \infty} \frac{1}{n} \ln V_{I_2}(f^n) = \gamma(f).$$

Inductively, we can find a decreasing sequence of closed intervals $I_1 \supset I_2 \supset \cdots \supset I_n \supset \cdots$ whose lengths $|I_n| \to 0$ as $n \to \infty$. Thus, $\bigcap_{n \geq 1} I_n$ contains a single point x_0, whose fluctuations satisfy $\gamma(x_0, f) = \gamma(f)$. $\square$

We know from Theorem 10.19 that f has rapid fluctuations if and only if $E(f) \neq \emptyset$. This is a characterization of the chaotic behavior of the I3DS in (10.1) in terms of the map f.

Theorem 10.20 *Let $f \in PM(I, I)$. Then the map $\gamma(\cdot, f): I \to [0, +\infty]$ is upper semi-continuous.*

Proof For a point $x_0 \in I$ and a number $\varepsilon > 0$, since $\gamma(x_0, f) = \lim_{\delta \to 0} \gamma([x_0 - \delta, x_0 + \delta] \cap I, f)$, we can find some $\delta > 0$ such that

$$\limsup_{n \to \infty} \frac{1}{n} \ln V_{[x_0 - \delta, x_0 + \delta] \cap I}(f^n) < \gamma(x_0, f) + \varepsilon.$$

Therefore, $\gamma(x, f) < \gamma(x_0, f) + \varepsilon$ for every x in $(x_0 - \delta, x_0 + \delta) \cap I$. This is just the definition of upper semi-continuity of x_0. $\square$

From Theorem 10.20, we have the following.

10 Infinite-Dimensional Systems Induced by Continuous-Time Difference Equations

Corollary 10.21 *Let $f \in PM(I, I)$. We have that $\gamma(x, f) \le \gamma(z, f)$ for every z in the closure $\overline{\{f^n(x): n \ge 0\}}$ of the orbit $orb(x, f) = \{f^n(x): n \ge 0\}$. Thus, $\overline{\{f^n(x): n \ge 0\}} \subset E(f)$ if $x \in E(f)$.* □

The following gives the conditions under which the function $\gamma(\cdot, f)$ is almost constant.

Theorem 10.22 *Let $f \in PM(I, I)$. Then*

(1) *If f is topologically mixing, then $\gamma(x, f) = const. > 0$, for all $x \in int\ I$;*
(2) *If f is topologically transitive, then*

$$\gamma(x, f) = const. > 0, \quad \forall x \in int\ I - \{p\},$$

for some fixed point $p \in I$. □

Exercise 10.23 Prove Theorem 10.22. □

Exercise 10.24 Let $f \in C(I, I)$, piecewise monotone with finitely many critical points. Show that $B(f)$ is open in I and $B(f) \cup U(f) = I$. □

Exercise 10.25 Prove (3) in Theorem 10.2. □

Exercise 10.26 Consider the quadratic map

$$f(x) = \mu x(1 - x).$$

Show that

(i) If $1 < \mu < 3$, then $B(f) = [0, 1]$.
(ii) If $\mu > 3$, then $0 \in U(f)$.
(iii) Find a condition on μ such that $E(f) \ne \emptyset$. □

Exercise 10.27 Let $f: [0, 1] \to [0, 1]$ be defined as

$$f(x) = \begin{cases} 3x & 0 \le x \le \dfrac{1}{3} \\ 1 & \dfrac{1}{3} < x \le \dfrac{2}{3} \\ -3x + 3 & \dfrac{2}{3} < x \le 1. \end{cases}$$

Show that

(i) For any $x \in (\frac{1}{3}, \frac{2}{3})$, we have $x \in B(f)$.
(ii) Furthermore, for any $x \in [0, 1] - \mathcal{C}$, $x \in B(f)$.
(iii) $E(f) = \mathcal{C}$.

Here $\mathcal{C}$ denotes the classical Cantor set. □

Notes for Chapter 10

Chapter 10 studies maps from a more functional-analytic and function-space point of view, utilizing the notion of rapid fluctuations in Chap. 9. In this approach, we are essentially dealing with maps on an infinite-dimensional space. Such dynamical systems are called *I3DS*, which occur naturally in the study of delay equations or even PDEs.

In particular, our Sects. 10.2–10.3 are from [4] and Theorem 10.10 in Sect. 10.4 is from [5].

References

1. G. Chen, S.B. Hsu and J. Zhou, Chaotic vibrations of the one-dimensional wave equation due to a self-excitation boundary condition. Part I, controlled hysteresis, *Trans. Amer. Math. Soc.* **350** (1998), 4265–4311. https://doi.org/10.1090/S0002-9947-98-02022-4
2. G. Chen, S.B. Hsu and J. Zhou, Snapback repellers as a cause of chaotic vibration of the wave equation with a van der Pol boundary condition and energy injection at the middle of the span, *J. Math. Phys.* **39** (1998), 6459–6489. https://doi.org/10.1063/1.532670
3. A.N. Sharkovsky, Y.L. Maistrenko and E.Y. Romanenko, *Difference Equations and Their Applications* , Ser. Mathematics and its Applications, **250**, Kluwer Academic Publisher, Dordrecht, 1993.
4. Y. Huang and Z. Feng, Infinite-dimensional dynamical systems induced by interval maps, *Dyn. Contin. Discrete Impuls. Syst. Ser. A Math. Anal.* **13** (2006), no. 3–4, 509–524.
5. G. Chen, T. Huang and Y. Huang, Chaotic behavior of interval maps and total variations of iterates, *Int. J. Bifur. Chaos* **14** (2004), 2161–2186. https://doi.org/10.1142/S0218127404010242

Introduction to Continuous-Time Dynamical Systems

11

All of the preceding chapters deal with discrete maps. In this chapter, we study continuous systems of ordinary differential equations and introduce some canonical forms and maps therein. The most important map is the Poincaré first return map which oftentimes can help render the study of chaos in a continuous system into that of chaotic maps.

A nonlinear differential equation or system can be described in the form:

$$\begin{cases} \dot{x}(t) = f(x(t), t), & t_0 \le t < T, \\ x(t_0) = x_0 \in \mathbb{R}^N, \end{cases} \quad (11.1)$$

where

$$f : \mathbb{R}^N \times \mathbb{R} \to \mathbb{R}^N \quad (11.2)$$

is a sufficiently smooth function; $t_0, T \in \mathbb{R}$ and $x_0 \in \mathbb{R}^N$ are given. The state of x at time t_0, x_0, is called the initial condition. If f in (11.1) and (11.2) is independent of t, we say that (11.1) is an autonomous system. (Otherwise, the system (11.1) is called nonautonomous, time dependent, or time varying.) Assume that the solution of (11.1) exists for any $T > t_0$. One is interested in studying the asymptotic behavior of (11.1) when $T \to \infty$. This study for continuous-time systems can obviously be extremely challenging, if not more so than the discrete-time cases that we have been studying in the preceding chapters. Nevertheless, as we will see in this chapter, one can deduce important information about the asymptotic behavior of (11.1) through the use of the Poincaré sections. The main objective of this Chap. 10 is to give the readers just a little flavor of continuous-time dynamical systems, as an in-depth account would require a large tome to do the job.

© The Author(s), under exclusive license to Springer Nature Switzerland AG 2025
L. Li et al., *Chaotic Maps, Fractals, and Rapid Fluctuations*, Synthesis Lectures on Mathematics & Statistics, https://doi.org/10.1007/978-3-031-84828-5_11

11.1 The Local Behavior of Two-Dimensional Nonlinear Systems

We consider mainly autonomous systems (11.1), i.e., $f(x, t) \equiv f(x)$ for all x, t. As we have seen in Definition 11.1, the local behavior of being attracting, repelling, or neutral for fixed or periodic points is critical in the analysis. This is analyzed through linearization.

We first introduce the following.

Definition 11.1 A point $x_0 \in \mathbb{R}^N$ is called an equilibrium point of an autonomous equation $\dot{x}(t) = f(x(t))$ if $f(x_0) = 0$. □

If $x_0 \in \mathbb{R}^N$ is an equilibrium point of (11.1) and $\tilde{x}_0 \in \mathbb{R}^N$ satisfies $|x_0 - \tilde{x}_0| < \delta$ for some very small $\delta > 0$, then the solution $\tilde{x}(t)$ of

$$\begin{cases} \dot{\tilde{x}}(t) = f(\tilde{x}(t)), & t > t_0 \\ \tilde{x}(t_0) = \tilde{x}_0 \in \mathbb{R}^N \end{cases}$$

satisfies

$$|\tilde{x}(t) - x(t)| < \delta', \quad t_0 \le t \le t_1$$

for some δ' (depending on δ) for some $t_1 > t_0$. Thus,

$$\eta(t) \equiv \tilde{x}(t) - x_0$$

satisfies

$$\dot{\eta}(t) = \dot{\tilde{x}}(t) = f(\tilde{x}(t)) - f(x_0) = D_x f(x_0) \cdot \eta(t) + \mathcal{O}\left(|\eta(t)|^2\right).$$

Thus, for $\eta(t)$ small, the first-order term

$$\dot{\eta}(t) \approx D_x f(x_0) \eta(t) \tag{11.3}$$

dominates in (11.3). We call

$$\dot{y}(t) = A y(t), \quad A \equiv D_x f(x_0) \tag{11.4}$$

the linearized equation of $\dot{x}(t) = f(x(t))$ at x_0.

We may now utilize a standard procedure in linear algebra by converting the $N \times N$ constant matrix A into the Jordan canonical form by finding a (similarity) matrix S such that

$$z(t) = S y(t).$$

11.1 The Local Behavior of Two-Dimensional Nonlinear Systems

Then
$$\dot{z}(t) = S\dot{y}(t) = SAy(t) = SAS^{-1}z(t) \equiv Jz(t),$$

where
$$J = \begin{bmatrix} J_1 & 0 & \cdots & 0 \\ 0 & J_2 & \ddots & \vdots \\ \vdots & \ddots & \ddots & 0 \\ 0 & \cdots & 0 & J_k \end{bmatrix}$$

with J_ℓ, $1 \le \ell \le k$, taking one of the following two forms:

$$J_\ell = \begin{bmatrix} \lambda_\ell & 0 & \cdots & 0 \\ 0 & \lambda_\ell & \ddots & \vdots \\ \vdots & \ddots & \ddots & 0 \\ 0 & \cdots & 0 & \lambda_\ell \end{bmatrix} \quad \text{or} \quad J_\ell = \begin{bmatrix} \lambda_\ell & 1 & & 0 \\ 0 & \lambda_\ell & \ddots & \\ \vdots & \ddots & \ddots & 1 \\ 0 & \cdots & 0 & \lambda_\ell \end{bmatrix}.$$

The case $N = 2$, i.e., two-dimensional autonomous systems, is the easiest to understand and to visualize (besides the somewhat trivial case $N = 1$) as well as offers significant clues to more complicated cases for systems in higher dimensional spaces.

Thus, we consider a real 2×2 matrix A. We have the following possibilities for J:

Case (1)
$$J = \begin{bmatrix} \lambda_1 & 0 \\ 0 & \lambda_2 \end{bmatrix}, \quad \lambda_1, \lambda_2 \in \mathbb{R}; \tag{11.5}$$

Case (2)
$$J = \begin{bmatrix} \lambda & 1 \\ 0 & \lambda \end{bmatrix}, \quad \lambda \in \mathbb{R}; \tag{11.6}$$

Case (3)
$$J = \begin{bmatrix} \alpha + i\beta & 0 \\ 0 & \alpha - i\beta \end{bmatrix}, \quad \text{or, equivalently,}$$

$$J = \begin{bmatrix} \alpha & -\beta \\ \beta & \alpha \end{bmatrix}, \quad \alpha, \beta \in \mathbb{R}. \tag{11.7}$$

Case (1) may be further subdivided into the following:

(1.i) $\lambda_1 < \lambda_2 < 0$ } stable node;
(1.ii) $\lambda_2 < \lambda_1 < 0$

(1.iii) $\lambda_2 > \lambda_1 > 0$ } unstable node;
(1.iv) $\lambda_1 > \lambda_2 > 0$

(1.v) $\lambda_2 = \lambda_1 > 0$ } unstable star node; stable star node;
(1.vi) $\lambda_1 = \lambda_2 < 0$

(1.vii) $\lambda_1 = \lambda_2 = 0$
(1.viii) $\lambda_1 = 0, \lambda_2 \lessgtr 0$ } the equilibrium points may be non-isolated;
(1.ix) $\lambda_2 = 0, \lambda_1 \lessgtr 0$

(1.x) $\lambda_1 > 0 > \lambda_1$ } saddle points.
(1.xi) $\lambda_2 > 0 > \lambda_1$

We discuss them in separate groups below.

Subcase (1.I)–(1.VI)

$$\begin{bmatrix} \dot{x} \\ \dot{y} \end{bmatrix} = \begin{bmatrix} \lambda_1 & 0 \\ 0 & \lambda_2 \end{bmatrix} \begin{bmatrix} x \\ y \end{bmatrix}, \quad \begin{bmatrix} x(0) \\ y(0) \end{bmatrix} = \begin{bmatrix} x_0 \\ y_0 \end{bmatrix} \tag{11.8}$$

$$\frac{dy/dt}{dx/dt} = \frac{dy}{dx} = \frac{\lambda_2 y}{\lambda_1 x} = \gamma \left(\frac{y}{x}\right), \quad \gamma \equiv \frac{\lambda_2}{\lambda_1}$$

$$\int \frac{dy}{y} = \gamma \int \frac{dx}{x}$$

$$\ln\left|\frac{y}{y_0}\right| = \gamma \ln\left|\frac{x}{x_0}\right|^\gamma$$

$$|y| = |y_0| \left|\frac{x}{x_0}\right|^\gamma. \tag{11.9}$$

Also from (11.8),

$$x(t) = x_0 e^{\lambda_1 t}, \quad y(t) = y_0 e^{\lambda_2 t}. \tag{11.10}$$

Solutions to Subcases (1.vii)-(1.xi) can be similarly determined. The trajectories of solutions $(x(t), y(t))$ plotted on the (x, y)-plane are called phase portraits. We illustrate them through various examples and graphics in this section.

11.1 The Local Behavior of Two-Dimensional Nonlinear Systems

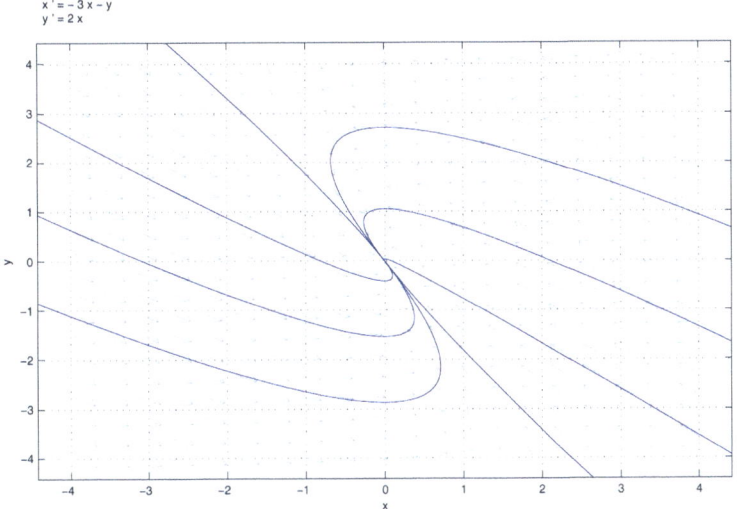

Fig. 11.1 The phase portrait for Example 11.2. This is a stable node (Subcases (1.i) and (1.ii))

Example 11.2 (A stable node (Subcases (1.i) and (1.ii))) For the differential equation

$$\frac{d}{dt}\begin{bmatrix} x(t) \\ y(t) \end{bmatrix} = \begin{bmatrix} -3 & -1 \\ 2 & 0 \end{bmatrix}\begin{bmatrix} x(t) \\ y(t) \end{bmatrix} \equiv A \begin{bmatrix} x(t) \\ y(t) \end{bmatrix},$$

we have

$$A = SJS^{-1} \equiv \begin{bmatrix} -1 & 1 \\ 2 & -1 \end{bmatrix}\begin{bmatrix} -1 & 0 \\ 0 & -2 \end{bmatrix}\begin{bmatrix} 1 & 1 \\ 2 & 1 \end{bmatrix}, \text{ with } J = \begin{bmatrix} -1 & 0 \\ 0 & -2 \end{bmatrix},$$

thus $\lambda_2 = -2 < \lambda_1 = -1 < 0$. So the equilibrium point $(x, y) = (0, 0)$ is a stable node. See Fig. 11.1 for the phase portrait. □

Example 11.3 (An unstable node (Subcases (1.iii) and (1.iv))) For the differential equation

$$\frac{d}{dt}\begin{bmatrix} x(t) \\ y(t) \end{bmatrix} = \begin{bmatrix} 0 & -1 \\ 2 & 3 \end{bmatrix}\begin{bmatrix} x(t) \\ y(t) \end{bmatrix} \equiv A \begin{bmatrix} x(t) \\ y(t) \end{bmatrix}$$

we have

$$A = SJS^{-1} = \begin{bmatrix} 1 & -1 \\ -1 & 2 \end{bmatrix} = \begin{bmatrix} 1 & 0 \\ 0 & 2 \end{bmatrix}\begin{bmatrix} 2 & 1 \\ 1 & 1 \end{bmatrix}, \text{ with } J \begin{bmatrix} 1 & 0 \\ 0 & 2 \end{bmatrix}$$

thus $\lambda_2 = 2 > \lambda_1 = 1 > 0$. So the equilibrium point $(x, y) = (0, 0)$ is an unstable node. See Fig. 11.2 for the phase portrait. □

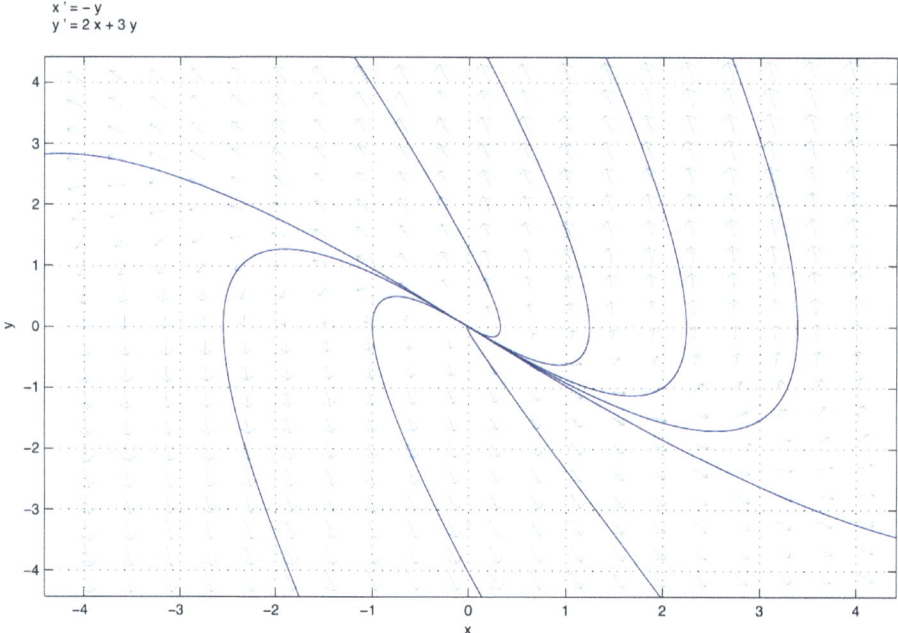

Fig. 11.2 The phase portrait for Example 11.3. This is an unstable node (Subcases (1.i) and (1.ii))

Example 11.4 (An unstable star node (Subcase (1.v))) The phase portrait of the differential equation

$$\frac{d}{dt}\begin{bmatrix} x(t) \\ y(t) \end{bmatrix} = \begin{bmatrix} 3 & 0 \\ 0 & 3 \end{bmatrix}\begin{bmatrix} x(t) \\ y(t) \end{bmatrix}$$

is illustrated in Fig. 11.3. We have $\lambda_1 = \lambda_2 = 3 > 0$, and thus an unstable star node. □

Example 11.5 (A stable star node (Subcase (1.vi))) The phase portrait of the differential equation

$$\frac{d}{dt}\begin{bmatrix} x(t) \\ y(t) \end{bmatrix} = \begin{bmatrix} -1 & 0 \\ 0 & -1 \end{bmatrix}\begin{bmatrix} x(t) \\ y(t) \end{bmatrix}$$

is illustrated in Fig. 11.4. We have $\lambda_1 = \lambda_2 = -1 < 0$, and thus a stable star node. □

11.1 The Local Behavior of Two-Dimensional Nonlinear Systems

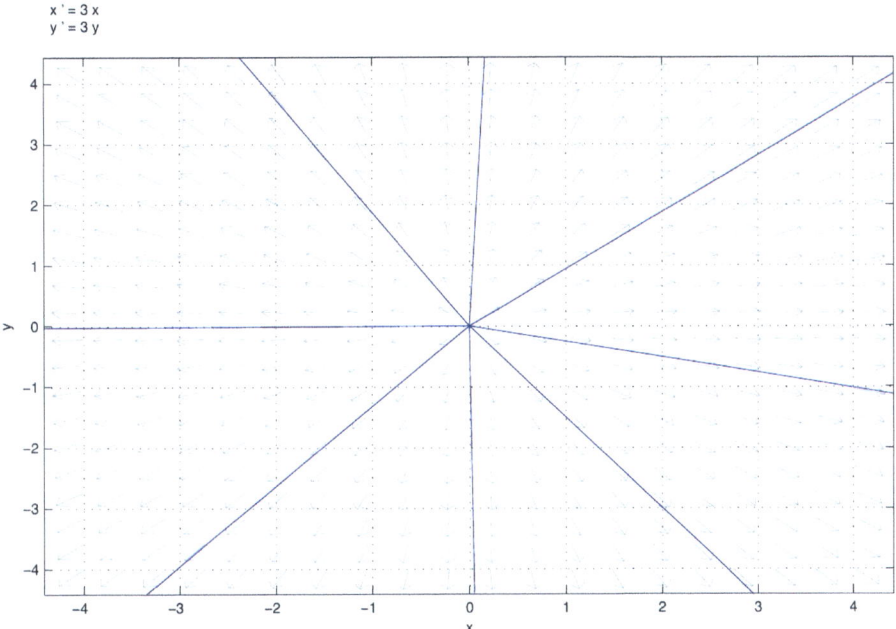

Fig. 11.3 The phase portrait for Example 11.4. This is an unstable star node (Subcase (1.v))

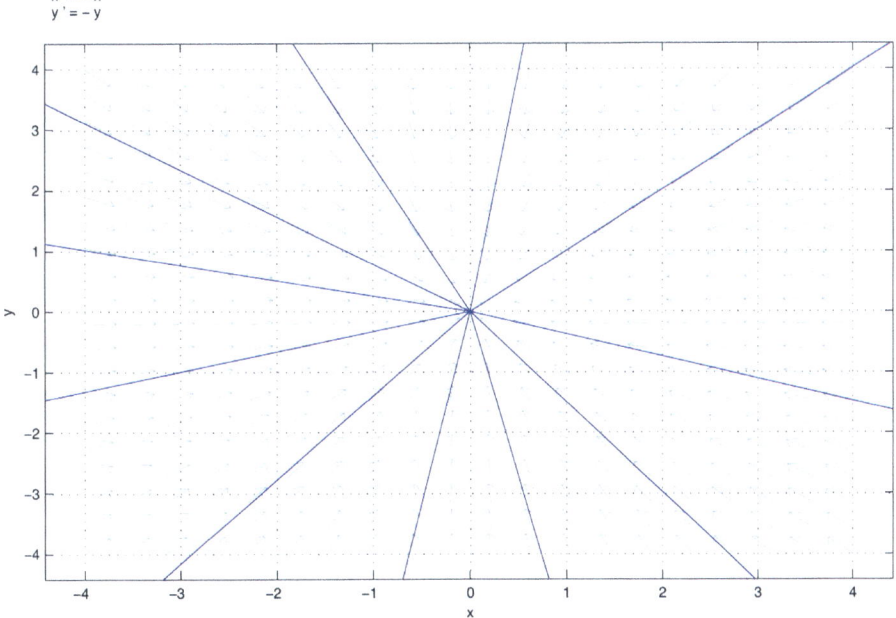

Fig. 11.4 The phase portrait for Example 11.5. This is a stable star node (Subcase (1.vi))

Example 11.6 (Non-isolated equilibrium points (Subcases (1.vii), (1.viii), (1.ix))) Consider

$$\text{(a)} \quad \frac{d}{dt}\begin{bmatrix} x(t) \\ y(t) \end{bmatrix} = \begin{bmatrix} -2 & 2 \\ -4 & 4 \end{bmatrix}\begin{bmatrix} x(t) \\ y(t) \end{bmatrix} \equiv A\begin{bmatrix} x(t) \\ y(t) \end{bmatrix}, \quad \text{and} \quad (11.11)$$

$$\text{(b)} \quad \frac{d}{dt}\begin{bmatrix} x(t) \\ y(t) \end{bmatrix} = \begin{bmatrix} 2 & -2 \\ 4 & -4 \end{bmatrix}\begin{bmatrix} x(t) \\ y(t) \end{bmatrix} = -A\begin{bmatrix} x(t) \\ y(t) \end{bmatrix}. \quad (11.12)$$

Since

$$A = SJS^{-1} = \begin{bmatrix} 1 & 1 \\ 2 & 1 \end{bmatrix}\begin{bmatrix} 2 & 0 \\ 0 & 0 \end{bmatrix}\begin{bmatrix} -1 & 1 \\ 2 & -1 \end{bmatrix}, \text{ with } J = \begin{bmatrix} 2 & 0 \\ 0 & 0 \end{bmatrix},$$

we see that (a) and (b) are examples of case (1.ix). The phase portraits of (a) and (b) are illustrated in Fig. 11.5a, b, respectively. Every trajectory consists of two semi-infinite line segments.

If

$$\frac{d}{dt}\begin{bmatrix} x(t) \\ y(t) \end{bmatrix} = \begin{bmatrix} 0 \\ 0 \end{bmatrix}, \quad \text{i.e.,} \quad A \equiv \begin{bmatrix} 0 & 0 \\ 0 & 0 \end{bmatrix}$$

this becomes case (1.vii). Then every point on the (x, y)-plane is an equilibrium point. □

Example 11.7 (A saddle point (Subcases (1.x) and (1.xi))) For the differential equation

$$\frac{d}{dt}\begin{bmatrix} x(t) \\ y(t) \end{bmatrix} = \begin{bmatrix} -4 & 3 \\ -6 & 5 \end{bmatrix}\begin{bmatrix} x(t) \\ y(t) \end{bmatrix} \equiv A\begin{bmatrix} x(t) \\ y(t) \end{bmatrix},$$

we have

$$A = SJS^{-1} = \begin{bmatrix} 1 & 1 \\ 2 & 1 \end{bmatrix}\begin{bmatrix} 2 & 0 \\ 0 & -1 \end{bmatrix}\begin{bmatrix} -1 & 1 \\ 2 & -1 \end{bmatrix}, \text{ with } J = \begin{bmatrix} 2 & 0 \\ 0 & -1 \end{bmatrix}.$$

Thus, $\lambda_2 = -1 < 0 < \lambda_1 = 2$. So the equilibrium point $(x, y) = (0, 0)$ is a saddle point. See Fig. 11.6 for the phase portrait.

Next we consider Case (2), which may be further subdivided into the following:

(2.i) $A = \begin{bmatrix} 0 & 1 \\ 0 & 0 \end{bmatrix}$, i.e., $\lambda = 0$;

(2.ii) $A = \begin{bmatrix} \lambda & 1 \\ 0 & \lambda \end{bmatrix}$, where $\lambda \neq 0$. □

Subcase (2.I)

We have $\dot{x} = y$ and $\dot{y} = 0$. Thus,

$$y(t) = y_0 = \text{constant}, \quad x(t) = y_0 t.$$

11.1 The Local Behavior of Two-Dimensional Nonlinear Systems

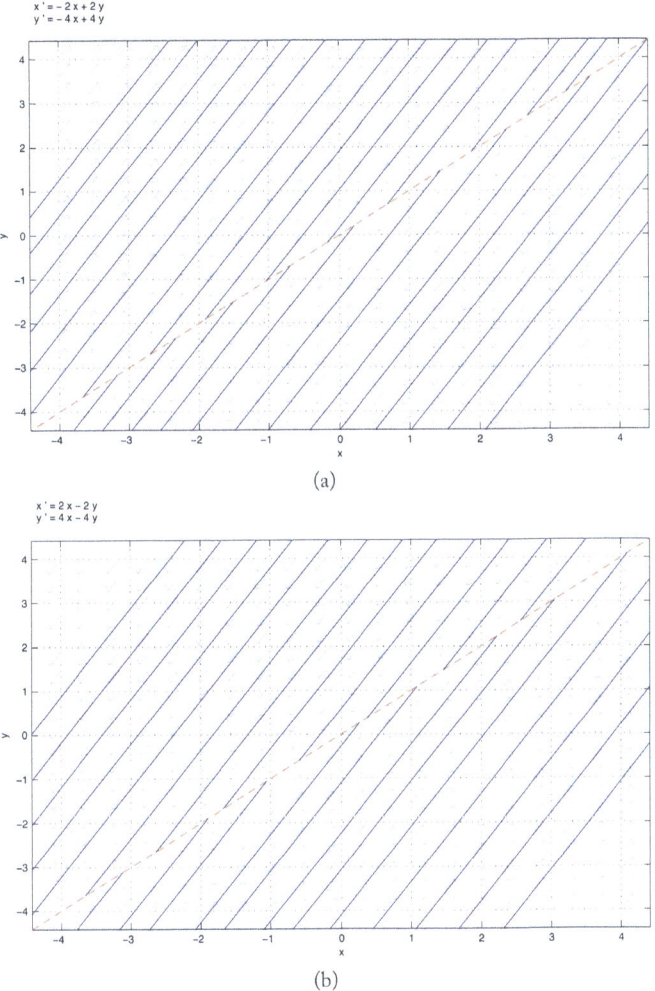

Fig. 11.5 The phase portrait for Example 11.6. (Subcases (1.viii) and (1.ix).) Here **a** and **b** correspond, respectively, to (11.11) and (11.12). Every trajectory consists of two semi-infinite line segments. The equilibrium points consist of an entire line, so they are not isolated

Example 11.8 Non-isolated equilibrium points (Subcase (2.i))) Consider the differential equation

$$\frac{d}{dt}\begin{bmatrix} x(t) \\ y(t) \end{bmatrix} = \begin{bmatrix} 2 & -1 \\ 4 & -2 \end{bmatrix}\begin{bmatrix} x(t) \\ y(t) \end{bmatrix} \equiv A\begin{bmatrix} x(t) \\ y(t) \end{bmatrix},$$

where

$$A = SJS^{-1} = \begin{bmatrix} 1 & 1 \\ 2 & 1 \end{bmatrix}\begin{bmatrix} 0 & 1 \\ 0 & 0 \end{bmatrix}\begin{bmatrix} -1 & 1 \\ 2 & -1 \end{bmatrix}, \text{ with } J = \begin{bmatrix} 0 & 1 \\ 0 & 0 \end{bmatrix}.$$

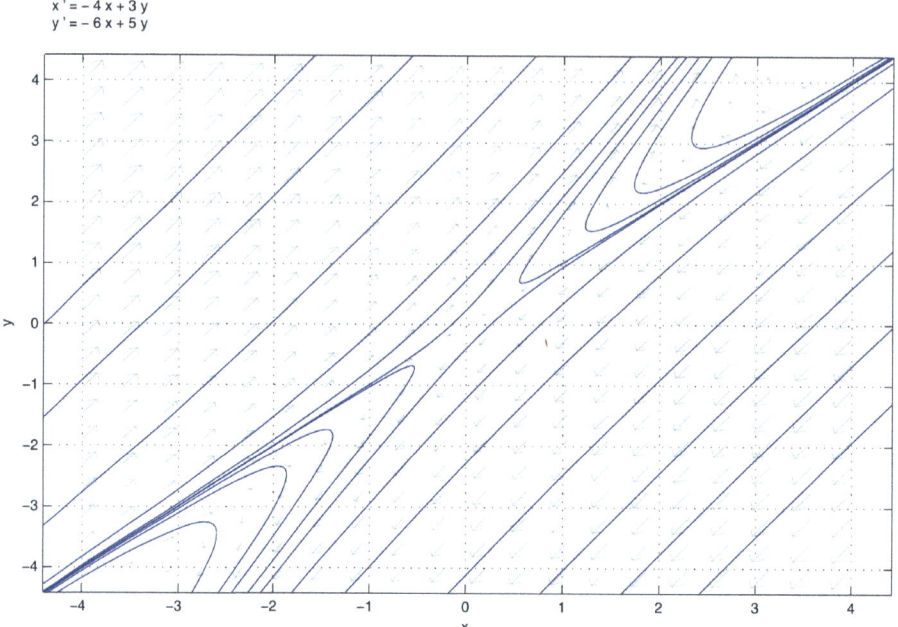

Fig. 11.6 The phase portrait for Example 11.7. This is a saddle point (Subcases (1.x) and (1.xi)). Saddle points are always unstable

Thus, this belongs to Subcase (2.i). The phase portrait is plotted in Fig. 11.7. All points on the line $2x - y = 0$ are equilibrium points. □

Subcase (2.II)

We have $\dot{x} = \lambda x + y$ and $\dot{y} = \lambda y$. Thus,

$$y(t) = y_0 e^{\lambda t} \quad \text{and} \quad x(t) = x_0 e^{\lambda t} + y_0 t e^{\lambda t}$$

$$t = \frac{1}{\lambda} \ln \left| \frac{y}{y_0} \right|,$$

$$x = \left(\frac{x_0}{y_0} + \frac{1}{\lambda} \ln \left| \frac{y}{y_0} \right| \right) y.$$

Example 11.9 (Improper stable or unstable node (Subcase (2.ii))) Consider

$$\text{(a)} \quad \frac{d}{dt} \begin{bmatrix} x(t) \\ y(t) \end{bmatrix} = \begin{bmatrix} 0 & -1 \\ 4 & -4 \end{bmatrix} \begin{bmatrix} x(t) \\ y(t) \end{bmatrix} \equiv A_1 \begin{bmatrix} x(t) \\ y(t) \end{bmatrix}, \tag{11.13}$$

11.1 The Local Behavior of Two-Dimensional Nonlinear Systems

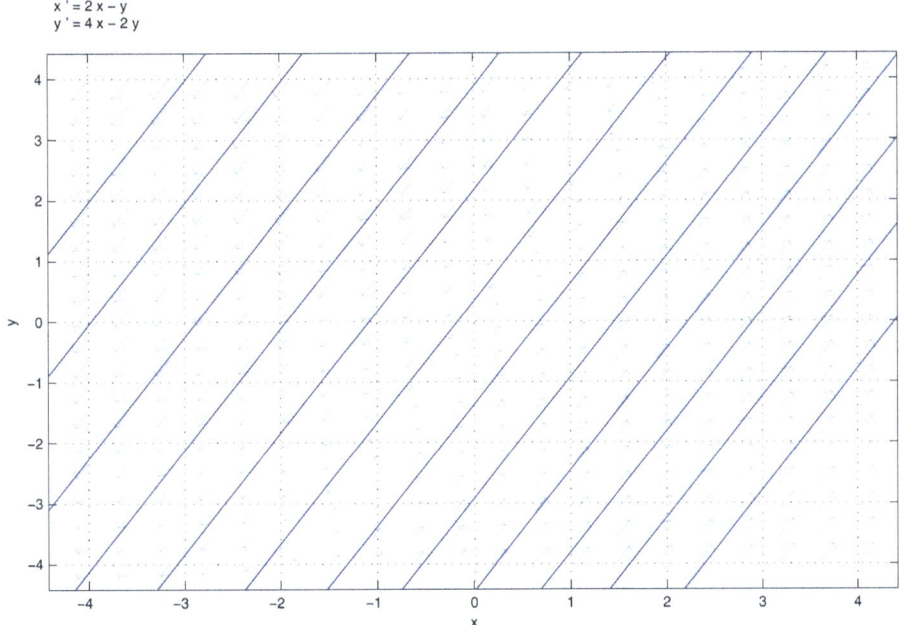

Fig. 11.7 The phase portrait for Example 11.8 where all points on the line $2x - y = 0$ are equilibrium points. The other trajectories are parallel lines (Subcase (2.i))

where

$$A_1 = SJ_1S^{-1} = \begin{bmatrix} 1 & 1 \\ 2 & 1 \end{bmatrix} \begin{bmatrix} -2 & 1 \\ 0 & -2 \end{bmatrix} \begin{bmatrix} -1 & 1 \\ 2 & -1 \end{bmatrix}, \text{ with } J_1 = \begin{bmatrix} -2 & 1 \\ 0 & -2 \end{bmatrix}$$

and

$$\text{(b)} \quad \frac{d}{dt}\begin{bmatrix} x(t) \\ y(t) \end{bmatrix} = \begin{bmatrix} 4 & -1 \\ 4 & 0 \end{bmatrix} \begin{bmatrix} x(t) \\ y(t) \end{bmatrix} \equiv A_2 \begin{bmatrix} x(t) \\ y(t) \end{bmatrix}, \tag{11.14}$$

where

$$A_2 = SJ_2S^{-1} = \begin{bmatrix} 1 & 1 \\ 2 & 1 \end{bmatrix} \begin{bmatrix} 2 & 1 \\ 0 & 2 \end{bmatrix} \begin{bmatrix} -1 & 1 \\ 2 & -1 \end{bmatrix}, \text{ with } J_2 = \begin{bmatrix} 2 & 1 \\ 0 & 2 \end{bmatrix}.$$

The phase portraits for (11.13) and (11.14) are given in Fig. 11.8. The equilibrium point $(x, y) = (0, 0)$ is, respectively, a stable and an unstable improper node. □

Finally, we treat Case (3) by writing $A = \begin{bmatrix} \alpha & -\beta \\ \beta & \alpha \end{bmatrix}$. We have

$$\begin{cases} \dot{x} = \alpha x - \beta y \\ \dot{y} = \beta x + \alpha y \end{cases} \tag{11.15}$$

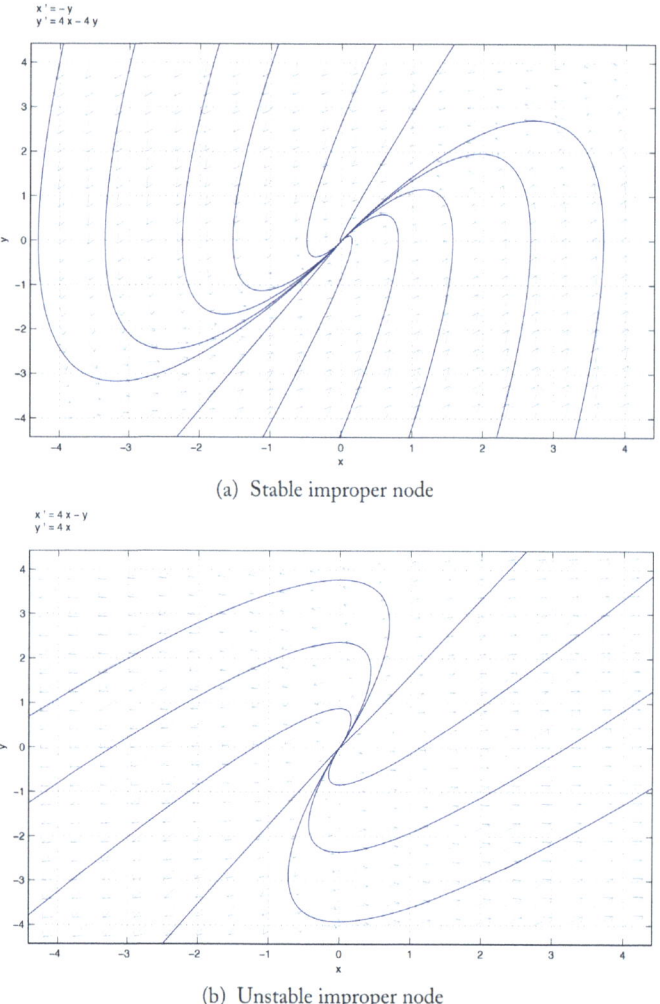

(a) Stable improper node

(b) Unstable improper node

Fig. 11.8 The phase portraits for Equations (11.13) and (11.14) are, respectively, stable and unstable improper nodes (Subcase (2.ii))

Using polar coordinates

$$x(t) = r(t)\cos(\theta(t)), \quad y(t) = r(t)\sin(\theta(t)),$$

we obtain

$$\begin{cases} \dot{r}\cos\theta - r\dot{\theta}\sin\theta = \alpha r\cos\theta - \beta r\sin\theta \\ \dot{r}\sin\theta - r\dot{\theta}\cos\theta = \beta r\cos\theta + \alpha r\sin\theta \end{cases}$$

11.1 The Local Behavior of Two-Dimensional Nonlinear Systems

Hence,

$$\begin{cases} \dot{r} = \alpha r \\ r\dot{\theta} = \beta r, \end{cases} \text{implying} \quad \begin{cases} r(t) = r_0 e^{\alpha t} \\ \theta(t) = \beta t \end{cases} \quad (11.16)$$

We see that the stability of the equilibrium point is completely determined by the sign of α.

Example 11.10 (Stable and unstable spirals (Case (3), $\alpha \neq 0$)) Consider

$$\text{(a)} \quad \frac{d}{dt}\begin{bmatrix} x(t) \\ y(t) \end{bmatrix} = \begin{bmatrix} -3 & -2 \\ 2 & -3 \end{bmatrix}\begin{bmatrix} x(t) \\ y(t) \end{bmatrix} \quad (11.17)$$

and

$$\text{(b)} \quad \frac{d}{dt}\begin{bmatrix} x(t) \\ y(t) \end{bmatrix} = \begin{bmatrix} 3 & 2 \\ -2 & 3 \end{bmatrix}\begin{bmatrix} x(t) \\ y(t) \end{bmatrix}. \quad (11.18)$$

For (11.17), we have $\alpha = -3 < 0$. Therefore, the equilibrium point $(x, y) = (0, 0)$ is stable. All trajectories spiral toward the origin; see Fig. 11.9a.

For (11.18), we have $\alpha = 3 > 0$. Therefore, the equilibrium point $(x, y) = (0, 0)$ is unstable. All trajectories spiral away from the origin; see Fig. 11.9b. □

Example 11.11 (Center (Case (3), $\alpha = 0$)) Consider

$$\frac{d}{dt}\begin{bmatrix} x(t) \\ y(t) \end{bmatrix} = \begin{bmatrix} 0 & -2 \\ 2 & 0 \end{bmatrix}\begin{bmatrix} x(t) \\ y(t) \end{bmatrix}.$$

Here we see that $\alpha = 0$ in (11.15). The phase portrait consists of circles or ellipses (which are periodic solutions) enclosing the equilibrium point $(x, y) = (0, 0)$ (Fig. 11.10). □

In summary, we have a total of nine types of phase portraits:

$$\left.\begin{array}{l} \text{parallel lines} \\ \text{all } \mathbb{R}^2 \end{array}\right\} \text{there three types have neutral stability}$$

$$\left.\begin{array}{l} \text{nodes} \\ \text{stars} \\ \text{saddle points} \\ \text{improper nodes} \\ \text{spirals} \\ \text{two semi-infinite lines} \end{array}\right\} \text{these five types are either stable or unstable}$$

We end this section by including the following example, which shows how to analyze and visualize local behaviors of more complicated 2×2 autonomous differential equations.

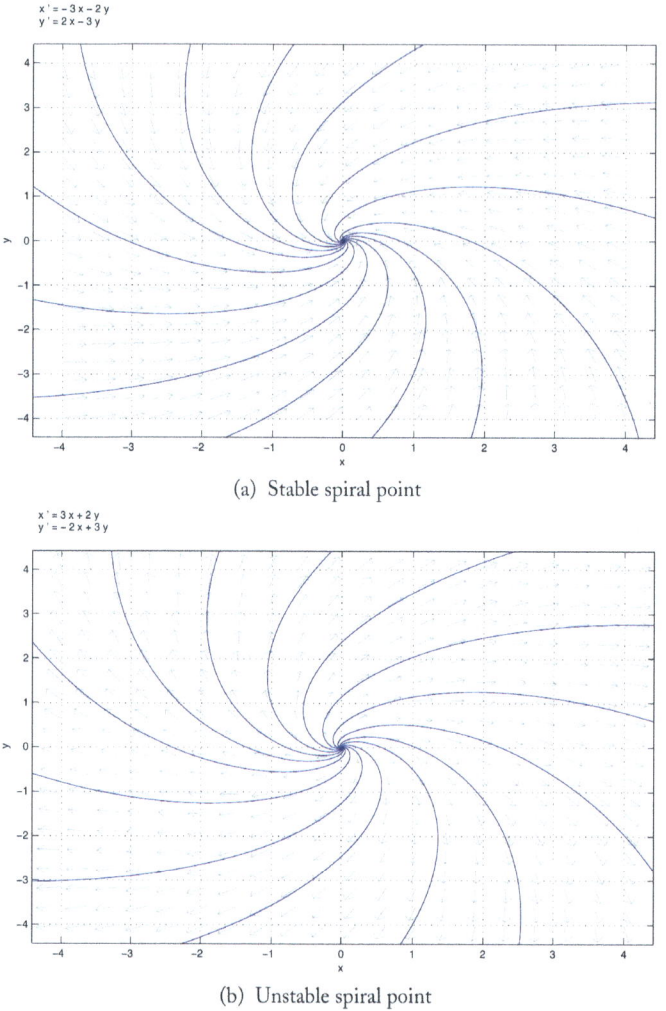

(a) Stable spiral point

(b) Unstable spiral point

Fig. 11.9 Phase portraits for Equations (11.17) and (11.18) in Example 11.10, as shown in, respectively, **a** and **b**. They correspond to Case (3), with $\alpha \neq 0$

Example 11.12 Consider

$$\frac{d}{dt}\begin{bmatrix} x(t) \\ y(t) \end{bmatrix} = \begin{bmatrix} x(t) + y(t) \\ -2x(t) - y(t) - x^2(t) \end{bmatrix}. \qquad (11.19)$$

Equilibrium points are determined by points (x, y) satisfying

$$x + y = 0, \quad -2x - y - x^2 = 0.$$

11.2 Index for Two-Dimensional Systems

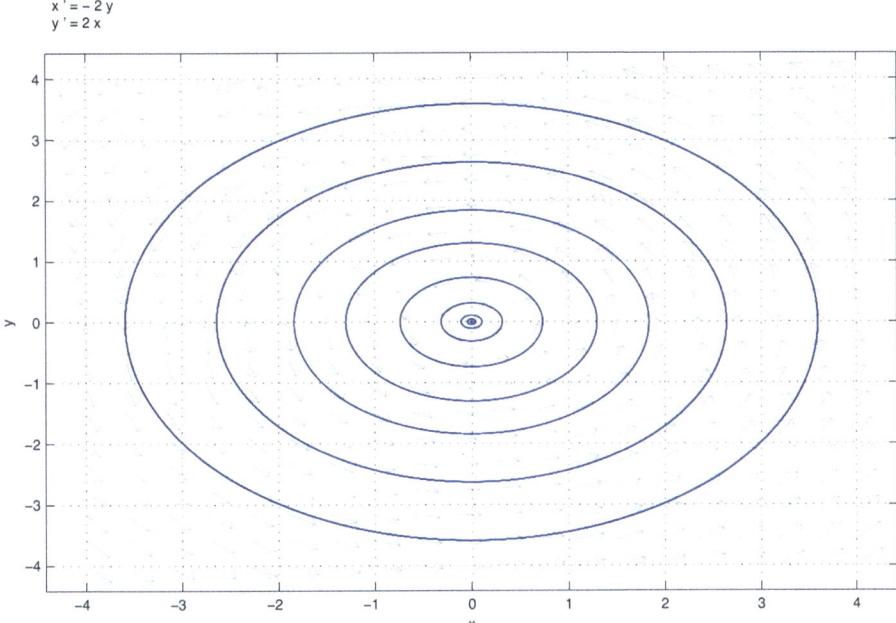

Fig. 11.10 The phase portrait for Example 11.11. This equilibrium point is a center (Case (3), $\alpha = 0$)

We obtain two equilibria:
$$(0, 0), (-1, 1).$$

The phase portrait of (11.19) is given in Fig. 11.11. □

Exercise 11.13 By linearization, determine the linearized system at $(0, 0)$ and $(-1, 1)$ of (11.19) and show that these equilibria are, respectively, center and saddle points. □

11.2 Index for Two-Dimensional Systems

A powerful method to study two-dimensional systems is the index theory. Consider

$$\begin{bmatrix} \dot{x} \\ \dot{y} \end{bmatrix} = \begin{bmatrix} f(x, y) \\ g(x, y) \end{bmatrix} \equiv V(x, y), \quad (x, y) \in \mathbb{R}^2, \tag{11.20}$$

where $V(x, y)$ denotes the vector field at (x, y). Then the angle formed by $V(x, y)$ with the x-axis is

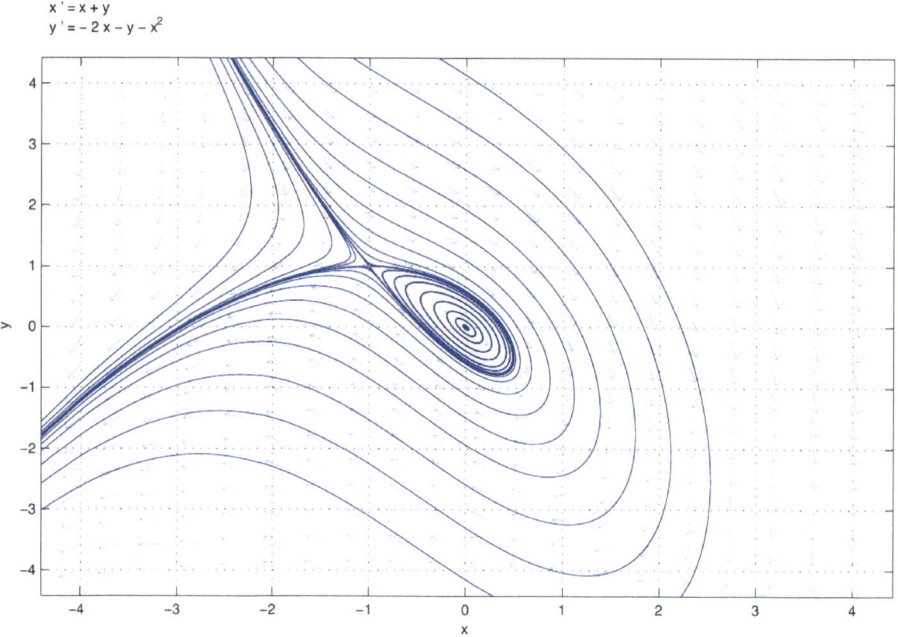

Fig. 11.11 The phase portrait for Example 11.12. Two equilibrium points are $(0, 0)$ and $(-1, 1)$. One can see that $(0, 0)$ is a center, while $(-1, 1)$ is a saddle point

$$\phi = \tan^{-1}\left(\frac{g(x, y)}{f(x, y)}\right), \quad -\pi < \phi < \pi,$$

where we can determine whether $-\pi < \phi < -\frac{\pi}{2}$, $-\frac{\pi}{2} < \phi < 0$, $0 < \phi < \frac{\pi}{2}$ or $\frac{\pi}{2} < \phi < \pi$ by checking which quadrant $V(x, y)$ points to. Thus,

$$\frac{d}{dt}\phi = \frac{\frac{d}{dt}(g/f)}{1 + (g/f)^2} = \frac{(fg' - gf')/f^2}{(g^2 + f^2)/f^2} = \frac{fg' - gf'}{f^2 + g^2}. \tag{11.21}$$

Let Γ be a simple closed curve in $\mathbb{R}^2$. We integrate $\frac{d\phi}{dt}$ along Γ a full circuit. The outcome will be an integral multiple of 2π, assuming that Γ does not contain any equilibrium point of (11.20) (which causes singularity in the denominator of (11.21)). Thus, we define

$$\text{ind}(\Gamma) \equiv \text{index of the curve } \Gamma \equiv \frac{1}{2\pi} \oint d\phi$$

$$= \frac{1}{2\pi} \oint \frac{fg' - gf'}{f^2 + g^2} dt = \frac{1}{2\pi} \oint \frac{fdg - gdf}{f^2 + g^2}.$$

In complex variable theory, the index of a curve is also called the *winding number*.

11.2 Index for Two-Dimensional Systems

Theorem 11.14 *Let Γ be a simple closed curve in $\mathbb{R}^2$ passing through no equilibrium points of (11.20).*

(i) *If Γ encloses (in its interior) a single node, star, improper node, center, or spiral point, then* $\mathrm{ind}(\Gamma) = 1$.
(ii) *If Γ encloses a single saddle point, then* $\mathrm{ind}\,(\Gamma) = -1$.
(iii) *If Γ itself is a closed orbit of (11.20) (and thus represents a periodic solution), then* $\mathrm{ind}\,(\Gamma) = 1$.
(iv) *If Γ does not enclose any equilibrium point of (A.20), then* $\mathrm{ind}\,(\Gamma) = 0$.
(v) *$\mathrm{ind}\,(\Gamma)$ is equal to the sum of the indices of all the equilibrium points enclosed within Γ.*

□

Instead of giving a proof of the above, we just visualize the properties stated in Theorem 11.14 through some illustrations.

Example 11.15 The index of a star is +1 regardless of its stability or instability. See Figs. 11.3 and 11.4. □

Example 11.16 The index of a saddle point is -1 . See Fig. 11.6. □

Example 11.17 Consider the Duffing oscillator

$$\ddot{x} - x + x^3 = 0. \qquad (11.22)$$

Its first-order form is

$$\begin{bmatrix} \dot{x} \\ \dot{y} \end{bmatrix} = \begin{bmatrix} y \\ x - x^3 \end{bmatrix} = V(x, y). \qquad (11.23)$$

The equation has three equilibria:

$$(0, 0), (1, 0), (-1, 0),$$

where $(0, 0)$ is a saddle point while $(1, 0)$ and $(-1, 0)$ are centers; see Fig. 11.12. □

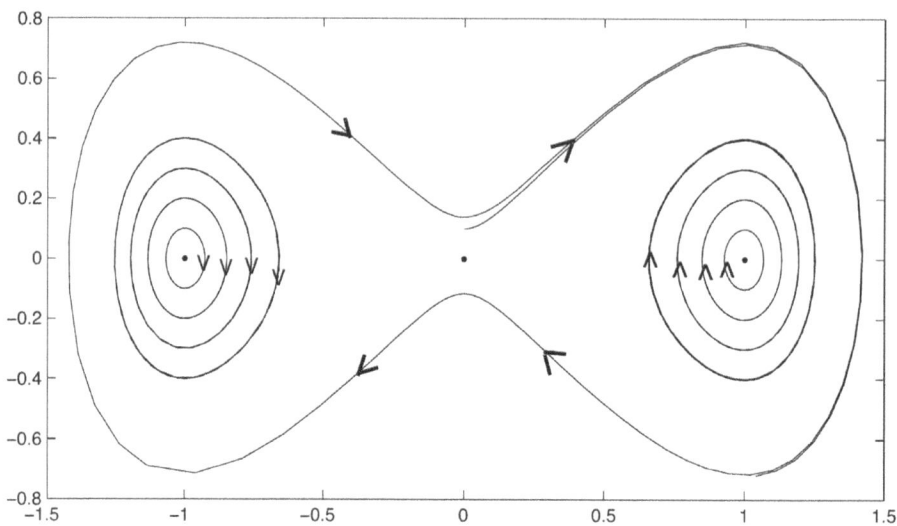

Fig. 11.12 Phase portrait for the Duffing oscillator in Example 11.17, where $(0, 0)$, $(1, 0)$, and $(-1, 0)$ are the three equilibrium points

11.3 The Poincaré Map for a Periodic Orbit in $\mathbb{R}^N$

We now consider a general N-dimensional autonomous system

$$\begin{cases} \dot{x}(t) = f(x(t)), \\ x(0) = x_0 \in \mathbb{R}^N \end{cases} \quad (11.24)$$

The local property of (11.24) near a nonequilibrium point $r \in \mathbb{R}^N$ (i.e., $f(r) \neq 0$) can be nicely described by the following.

Theorem 11.18 (The Flow Box Theorem [1]) *Let $\mathcal{O} \subseteq \mathbb{R}^N$ be an open neighborhood of a point x^0 such that $f(x^0) \neq 0$. Then there exists a local coordinate transformation $y = \psi(x^0)$ near x^0 such that (11.24), with respect to the new coordinates y, is transformed to*

$$\dot{y}(t) = \begin{bmatrix} 1 \\ 0 \\ \vdots \\ 0 \end{bmatrix}, \quad t > 0, \ y(0) = x^0. \quad (11.25)$$

Proof Since x^0 is not an equilibrium point for the differential equation $\dot{x} = f(x)$,

11.3 The Poincaré Map for a Periodic Orbit in $\mathbb{R}^N$

$$f(x^0) = \begin{bmatrix} a_1 \\ a_2 \\ \vdots \\ a_N \end{bmatrix} \neq \begin{bmatrix} 0 \\ 0 \\ \vdots \\ 0 \end{bmatrix}.$$

We can easily change the variables such that $a_1 \neq 0$. Let $\phi(x, t)$ be the solution flow of the differential equation (11.24). Define the coordinate transformation

$$x \leftrightarrow y, \quad x = \phi(0, y_2, \ldots, y_N, y_1) \quad \text{where} \quad y = \begin{bmatrix} y_1 \\ y_2 \\ \vdots \\ y_N \end{bmatrix}. \tag{11.26}$$

Note that y_1 above corresponds to the t variable: $y_1 = t$. Thus,

$$\dot{y}_1 = 1. \tag{11.27}$$

We now show that near $x = x^0$, the transformation (11.26) is $1-1$. We know that

$$\phi(y, 0) = y$$

and, thus,

$$D_y \phi(y, 0)\big|_{y=x^0} = D_y y = I_N = \begin{bmatrix} \frac{\partial \phi_1}{\partial y_1} & \frac{\partial \phi_1}{\partial y_2} & \cdots & \frac{\partial \phi_1}{\partial y_N} \\ \vdots & \vdots & & \vdots \\ \vdots & \vdots & & \vdots \\ \frac{\partial \phi_N}{\partial y_1} & \frac{\partial \phi_N}{\partial y_2} & \cdots & \frac{\partial \phi_N}{\partial y_N} \end{bmatrix}$$

$$= \begin{bmatrix} \frac{\partial \phi_1}{\partial y_1} & \frac{\partial \phi_1}{\partial y_2} & \cdots & \frac{\partial \phi_1}{\partial y_N} \\ \vdots & & & \\ \vdots & & I_{N-1} & \\ \frac{\partial \phi_N}{\partial y_1} & & & \end{bmatrix}, \tag{11.28}$$

where I_{n-1} is the $(N-1) \times (N-1)$ identity matrix. Now

$$D_y\phi(0, y_2, \ldots, y_N, y_1) = \begin{bmatrix} \frac{\partial \phi_1}{\partial t} & \frac{\partial \phi_1}{\partial y_2} & \cdots & \frac{\partial \phi_1}{\partial y_N} \\ \frac{\partial \phi_2}{\partial t} & \frac{\partial \phi_2}{\partial y_2} & & \frac{\partial \phi_1}{\partial y_N} \\ \vdots & \vdots & & \vdots \\ \frac{\partial \phi_N}{\partial t} & \frac{\partial \phi_N}{\partial y_2} & \cdots & \frac{\partial \phi_N}{\partial y_N} \end{bmatrix} \quad \text{(using } y_1 = t\text{)}$$

$$= \begin{bmatrix} a_1 & * & \cdots & * \\ a_2 & & & \\ \vdots & & I_{N-1} & \\ a_N & & & \end{bmatrix}, \text{ at } y = x^0, (\text{ by }((11.28))).$$

Thus,

$$\det D_y\phi(0, y_2, \ldots, y_N, y_1)\big|_{y=x^0} = a_1 \neq 0$$

and the map (11.26) is invertible in a small open set around $y = x^0$. So (11.26) is a well-defined local coordinate transformation.

Since the variables $y_2, y_3, \cdots, y_N$ in $\phi(0, y_2, \ldots, y_N, y_1)$ are just initial conditions (such that $x_2(0) = x_2^0 = y_2, x_3(0) = x_3^0 = y_3, \ldots, x_N(0) = x_N^0 = y_N$), we have

$$\frac{dy_2}{dt} = 0, \frac{dy_3}{dt} = 0, \ldots, \frac{dy_N}{dt} = 0. \quad (11.29)$$

Combining (11.27) with (11.29), we have completed the proof. □

Corollary 11.19 *Let x^0 be a nonequilibrium point of the differential equation $\dot{x} = f(x)$. Then in a neighborhood of x^0, there exist $N - 1$ independent integrals of motion, i.e., $N - 1$ functions $F_1(x), F_2(x), \ldots, F_{n-1}(x)$ such that*

$$\frac{d}{dt}F_j(x(t)) = 0, \quad j = 1, 2, \ldots, N - 1$$

and

$$\sum_{j=1}^{N-1} \lambda_j D_x F_j(x) \neq 0, \text{ for } x \text{ sufficiently close to } x^0,$$

for any $\lambda_1, \lambda_2, \ldots, \lambda_{N-1} \in \mathbb{R}$.

Proof The coordinate functions $y_2, y_3, \ldots, y_N$ in the flow box theorem are constant in a neighborhood of x^0. Thus, they are integrals of motion. □

Corollary 11.20 *Let x^0 be a nonequilibrium point of the differential equation $\dot{x} = f(x)$. Assume that $F(x)$ is an integral of motion such that F is nondegenerate near x^0, i.e., $D_x F(x)|_{x=x^0} \neq 0$. Then there exists a local coordinate system such that the differential equation $\dot{x} = f(x)$ is transformed to*

11.3 The Poincaré Map for a Periodic Orbit in $\mathbb{R}^N$

$$\dot{y}_1 = 1; \; y_2 = F(x) \text{ and } \dot{y}_2 = 0; \; \dot{y}_3 = \dot{y}_4 = \cdots = \dot{y}_N = 0. \tag{11.30}$$

Proof We know that y_1 is the same as the variable t, but $F(x)$ does not change along a trajectory. So $F(x)$ is independent of the y_1 variable. We have

$$D_y F(y) = \left[\frac{\partial F}{\partial y_1}, \frac{\partial F}{\partial y_2}, \ldots, \frac{\partial F}{\partial y_N}\right] \neq [0, 0, \ldots, 0]$$

by the nondegeneracy of F. Since $\frac{\partial F}{\partial y_1} = 0$, one of the $\frac{\partial F}{\partial y_2}, \ldots, \frac{\partial F}{\partial y_N}$ must be nonzero. Transform the coordinate system such that $\frac{\partial F}{\partial y_2} \neq 0$. Therefore, $y_2 = F(x)$ is uniquely solvable locally. This local coordinate system y is called the flow box coordinates.

We now consider periodic solutions of an autonomous differential equation. The solution $x(t)$ of

$$\begin{cases} \dot{x}(t) = f(x(t)), & t > 0, \\ x(0) = \xi \in \mathbb{R}^N \end{cases} \tag{11.31}$$

is denoted as $\phi(t, \xi)$. If this solution is periodic, then there exists a $T > 0$ such that

$$\phi(t + T, \xi) = \phi(t, \xi), \text{ for all } t > 0. \tag{11.32}$$

There is a smallest $T > 0$ satisfying (11.32) if $\phi(t, \xi)$ is not a constant state. We call the smallest such $T > 0$ the period of the solution $\phi(t, \xi)$. □

Lemma 11.21 ([1, p. 130]) *The solution $\phi(t, \xi)$ is a periodic solution with period T if and only if*

$$\phi(T, \xi) = \xi. \tag{11.33}$$

Proof of Sufficiency. If (11.33) is satisfied, then

$$\phi(t + T, \xi) = \phi(t, \phi(T, \xi)) \quad \text{(semi-group property)}$$
$$= \phi(t, \xi), \text{ for any } t > 0,$$

so (11.32) is satisfied and the solution is periodic.

(Necessity) If $\phi(t, \xi)$ satisfies (11.32), then set $t = 0$ therein, we obtain

$$\phi(T, \xi) = \phi(0, \xi) = \xi$$

so (11.33) is satisfied. □

Definition 11.22 Let $\phi(t, \xi)$ be a periodic solution of (11.31) with period T. The matrix $D_x \phi(t, x)|_{x=\xi}$ is called the monodromy matrix at ξ, and its eigenvalues are called the characteristic multipliers of the periodic solution $\phi(t, \xi)$.

Lemma 11.23 *([1, p. 130]) The monodromy matrix $D_x\phi(t, \xi)$ has 1 as its characteristic multiplier with eigenvector $f(\xi)$.*

Proof Differentiating the semi-group relation

$$\frac{d}{dt}\phi(\tau, \phi(t, x)) = \frac{d}{dt}\phi(t + \tau, x)$$

yields

$$D_y\phi(\tau, \phi(\tau, x)) = \frac{d}{dt}\phi(t, x) = \frac{d}{dt}\phi(t + \tau, x). \tag{11.34}$$

Set $t = 0$, $\tau = T$, and $x = \xi$ in (11.34). We obtain

$$D_y\phi(T, \xi)f(\xi) = f(\xi). \tag{11.35}$$

Since the periodic solution is not an equilibrium point, $f(\xi) \neq 0$. So $f(\xi)$ is an eigenvector corresponding to the eigenvalue 1. □

If $\xi_0 \in \mathbb{R}^n$ is an initial condition for a periodic solution $\phi(t, \xi_0)$, one is tempted to think that it might be possible to use the implicit function theorem (Theorem 1.7) to show that this T-periodic solution is unique in a neighborhood of ξ_0. Lemma 11.23 shows that this is not true as $D_x\phi(T, \xi_0) - I_N$ is not invertible. In fact, (initial conditions for) periodic solutions are never isolated. The periodic solution itself constitutes a one-dimensional manifold in a neighborhood of ξ_0 such that all points in the manifold correspond to an initial condition for a T-periodic solution.

In view of Lemma 11.23 and the above, one can introduce instead a cross section to the periodic solution according to the following.

Definition 11.24 Let $\phi(t, \xi_0)$ be a periodic solution of (11.31) with initial condition $\xi_0 \in \mathbb{R}^N$. Let $a \in \mathbb{R}^n$ be such that $a \cdot f(\xi_0) \neq 0$. Define the hyperplane passing ξ_0 with normal a as

$$\Sigma = \{x \in \mathbb{R}^n \mid a \cdot (x - \xi_0) = 0\}.$$

For any $\xi \in U \subseteq \Sigma$, for a small neighborhood U of ξ_0, let $\phi(T(\xi), \xi) \in \Sigma$ be the point on $\phi(t, \xi)$ with the smallest $T(\xi) > 0$. The map

$$P : \xi \in U \longrightarrow \phi(T(\xi), \xi) \in \Sigma$$

is called the *Poincaré map* and $T(\xi)$ is called the *first return time*. The hyperplane Σ is called the Poincaré section. □

One can see that Definition 11.24 is based on very geometric ideas.

11.3 The Poincaré Map for a Periodic Orbit in $\mathbb{R}^N$

One can show that the Poincaré map is *smooth* ([1, Lemma V.2.4, pp. 133-132]). We present a few more properties of the Poincaré map in the following.

Lemma 11.25 *([1, Lemma V.2.5, p. 132]) Let the characteristic multipliers of the monodromy matrix $D_x \phi(t, \xi_0)$ be $1, \lambda_2, \lambda_3, \ldots, \lambda_N$ for a periodic solution $\phi(t, \xi_0)$. Then the eigenvalues of the linearized Poincaré map $D_y P$ at ξ_0 are $\lambda_2, \lambda_3, \ldots, \lambda_N$.*

Proof Make a coordinate transformation such that $\xi_0 = \mathbf{0}$ and $f(\xi_0) = (1, 0, 0, \ldots, 0)^T$. Thus, $\sum$ corresponds to the hyperplane $x_1 = 0$. By Lemma 11.26, the monodormy matrix $M \equiv D_x \phi(T, \xi_0)$ has an eigenvalue 1 with eigenvector $f(\xi_0) = (1, 0, 0, \ldots, 0)$. Thus, M must take the form:

$$M = \begin{bmatrix} 1 & * & * & \cdots & * \\ \hline 0 & & & & \\ 0 & & & & \\ \vdots & & M' & & \\ 0 & & & & \end{bmatrix}$$

in order to satisfy

$$M \begin{bmatrix} 1 \\ 0 \\ \vdots \\ 0 \end{bmatrix} = \begin{bmatrix} 1 \\ 0 \\ \vdots \\ 0 \end{bmatrix}.$$

Since M has eigenvalues $1, \lambda_2, \lambda_3, \ldots, \lambda_N$, the $(N-1) \times (N-1)$ matrix M' must have eigenvalues $\lambda_2, \lambda_3, \ldots, \lambda_N$, belonging to the linearized Poincaré map at (the projection on $\sum$) point ξ_0. □

Lemma 11.26 *([1, Lemma V.4.7., p.134]) Let $F(\xi)$ be an integral as warranted by Corollary 11.20. If F satisfies $D_x F(x)|_{x=\phi(t,\xi_0)} \neq \mathbf{0}$ for the periodic solution $\phi(t, \xi_0)$, then the monodromy matrix $M \equiv D_x \phi(T, \xi_0)$ has a left eigenvector $D_x F(\xi_0)$ with eigenvalue 1. Consequently, M has eigenvalue 1 with multiplicity 2.*

Proof Since F remains constant on a trajectory,

$$F(\phi(t, \xi)) = F(\xi),$$

and so

$$F(\phi(T, \xi)) = F(\xi),$$
$$D_x F(\xi_0) D_x \phi(T, \xi_0) = D_x F(\xi_0);$$

this verifies that $D_x F(\xi_0)$ is a left eigenvector of M. We now choose a coordinate system such that $f(\xi_0)$ is the column vector $(1, 0, \ldots, 0)^T$ (T : transpose) as in the proof of

Lemma 11.25 and $D_x F(\xi_0)$ is the row vector $(0, 1, 0, \ldots, 0)$. (We note that the two vectors $f(\xi_0)$ and $D_x F(\xi_0)$ are independent if both are viewed as column vectors.) Thus, M takes the form:

$$M = \begin{bmatrix} 1 & * & * & * & \cdots & * \\ 0 & 1 & 0 & 0 & \cdots & 0 \\ 0 & * & * & * & \cdots & * \\ 0 & * & & & & \vdots \\ \vdots & \vdots & & & & \vdots \\ 0 & * & * & * & \cdots & * \end{bmatrix}.$$

The characteristic polynomial must satisfy

$$\det(M - \lambda I) = (\lambda - 1)^2 \cdot p(\lambda)$$

for some polynomial $p(\lambda)$ of degree $N - 2$.

Therefore, $\lambda = 1$ is an eigenvalue of multiplicity 2. □

We may now analyze the Poincaré map a little further using the geometric theory established above for a periodic solution with period T : $\phi(t + T, \xi_0) = \phi(t, \xi_0)$ with an integral $F(x)$, the monodromy matrix $D_x \phi(T, \xi_0)$ has eigenvalues $1, 1, \lambda_2, \lambda_3, \ldots, \lambda_N$. Let us now choose flow box coordinates $y_1, y_2, \ldots, y_N$ satisfying (11.30).

If $\sum$ is a Poincaré section of $\phi(t, \xi_0)$ passing ξ_0, then the trajectory $\phi(t, \xi_0)$ lies on the integral surface $F(x) = F(\xi_0)$, as Fig. 11.13 shows.

Now, using the flow box coordinates $y_1, y_2, \cdots, y_N$ satisfying (11.30), i.e., $\dot{y}_1 = 1$, $y_2 = F(y)$, $\dot{y}_3 = \dot{y}_4 = \cdots = \dot{y}_N = 0$; cf. Corollary 11.20. We can choose

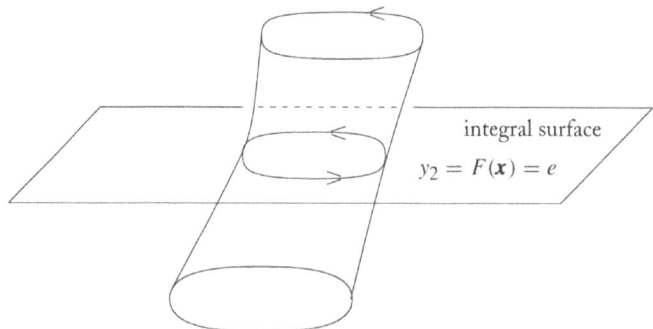

Fig. 11.13 A periodic orbit lying on an integral surface. This also illustrates the cylinder theorem (see Theorem 11.27 to follow), where a family of periodic orbits exist. This figure is adapted from [1, Fig. V.E.4, p. 136]

11.3 The Poincaré Map for a Periodic Orbit in $\mathbb{R}^N$

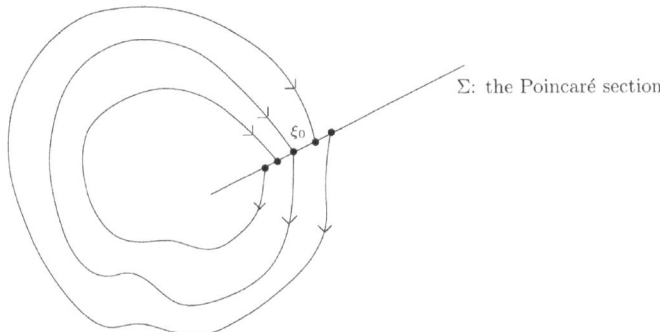

Fig. 11.14 Illustration of the Poincaré section

$$\sum : y_1 = 0, \text{ to be the Poincará section.}$$

The geometry now look like what is displayed in Fig. 11.14. Then $F(x) = F(\xi_0) = F(\phi(t, \xi_0)) = y_2$. So y_2 is constant on the periodic trajectory $\phi(t, \xi_0)$. Call this constant e. Define

$$\sum_e \equiv \text{ the intersection of } \sum \text{ with } y_2 = F(\xi_0).$$

On $\sum_e$, define coordinates $\widehat{y} = (y_3, y_4, \ldots, y_N)$ such that

$$\widehat{y}_1 = y_3, \quad \widehat{y}_2 = y_4, \ldots, \widehat{y}_{N-2} = y_N.$$

The Poincaré map thus becomes

$$P(e, \widehat{y}) = (e, Q(e, \widehat{y})), \quad (e = F(\xi_0) \text{ is fixed })$$

for some map $Q(e, \cdot)$ from a neighborhood N_e of the origin in $\sum_e$ to $\sum_e$, because $y_2 = F(\xi_0) = e$.

Therefore, $Q(e, \widehat{y})$ is a further reduction of the Poincaré map from $\sum$ to $\sum_e$. Q is called the Poincaré map on the integral surface. Q satisfies

$$Q(e, 0) = 0$$

because $(e, 0)$ corresponds to ξ_0 and is thus a fixed point of P.

When an integral of motion exists, then we often have a one-parameter family of periodic orbits instead of a single periodic orbit, under some additional assumptions. This is the following famous theorem.

Theorem 11.27 (The Cylinder Theorem) *Let $\phi\left(t, \xi^{0}\right)$ denote a periodic orbit with period T of the differential equation $\dot{x} = f(x)$. Assume that there is a nondegenerate integral of motion $F(x)$. Assume further that the eigenvalues satisfy*

$$\lambda_j \neq 1, \quad j = 3, 4, \ldots, N$$

for the monodromy matrix. Then there is a neighborhood of the periodic orbit $\phi\left(t, \xi^{0}\right)$, and we have a one-parameter family of periodic orbits parameterized by $e = F(x)$.

Proof Any periodic orbit corresponds to a fixed point of the Poincaré map

$$P(e, \hat{y}) = (e, Q(e, \hat{y})).$$

The given periodic orbit satisfies

$$P(e_1, 0) = (e_1, 0).$$

We want to solve

$$P(e, \hat{y}) = (e, Q(e, \hat{y})) = (e, \hat{y}), \text{ i.e., } Q(e, \hat{y}) = \hat{y}.$$

Applying the implicit function theorem to

$$g(e, z) = Q(e, z) - z, \text{ where } z \in \mathbb{R}^{N-2},$$

we have, near $e = e_1$,

$$D_z g(z)|_{z=0} = D_z Q(e, z) - D_z z = D_z Q(e, z) - I_{N-2}. \tag{11.36}$$

When $e = e_1$, $D_z Q(e_1)$, z has eigenvalues $\lambda_2, \lambda_3, \ldots, \lambda_N$, none of which is equal to 1. So, the $(N-2) \times (N-2)$ matrix on the right-hand side of (11.36) is invertible when $e = e_1$. Therefore, $g(e, z) = 0$ has a unique solution z for each given e near $e = e_1$. Such a z is a fixed point of $Q(e, \cdot)$ and, thus, corresponds to a periodic orbit. □

See an illustration in Fig. 11.13.

The dynamic properties of the continuous-time autonomous system can be studied in terms of the discrete Poincaré map P. For nonautonomous system 11.1 with periodic forcing and other cases, it is also possible to define certain discrete Poincaré maps. The chaotic behaviors of such continuous-time systems are then analyzed through P using Melnikov's method [2], Smale's horseshoe, and other ideas. We refer the readers to Guckenheimer and Holmes [3], Meyer and Hall [1] and Wiggins [4] for some further study.

References

1. K.R. Meyer and G.R. Hall, *Introduction to Hamiltonian Dynamical Systems and the N-Body Problem*, Springer, New York, 1992.
2. V.K. Melnikov, On the stability of the center for time periodic perturbations, *Trans. Moscow Math. Soc.* **12** (1963), 1–57.
3. J. Guckenheimer and P. Holmes, *Nonlinear Oscillations, Dynamical Systems and Bifurcations of Vector Fields*, Springer, New York, 1983.
4. S. Wiggins, *Introduction to Applied Nonlinear Dynamical Systems and Chaos*, 2nd ed., Springer, New York, 2003.

Chaotic Vibration of the Wave Equation Due to Energy Pumping and van der Pol Boundary Conditions

12

Beginning from this chapter, we will address chaotic vibration phenomena in some partial differential equations (PDEs). The onset of chaotic phenomena in systems governed by PDEs has fascinated scientists and mathematicians for many centuries. The most famous case in point is the Navier–Stokes equations and related models in fluid dynamics, where the occurrence of turbulence in fluids is well accepted as a chaotic phenomenon. Yet despite the diligence of numerous, most brilliant minds of mankind, and the huge amount of new knowledge gained through the vastly improved computational and experimental methods and facilities, at present, we still have not been able to rigorously prove that turbulence is indeed chaotic in a certain universal mathematical sense.

Nevertheless, for certain special partial differential equations, it is possible to rigorously prove the occurrence of chaos under certain given conditions. Here we mention the model of a vibrating string with nonlinear boundary conditions as studied by G. Chen, S.-B. Hsu, Y. Huang, J. Zhou, etc. in [1–10].

12.1 The Mathematical Model and Motivations

A linear wave equation

$$\frac{1}{c^2}\frac{\partial^2}{\partial t^2}w(x,t) - \frac{\partial^2}{\partial x^2}w(x,t) = 0, \quad 0 < x < L, t > 0 \tag{12.1}$$

describes the propagation of waves on an interval of length L. For convenience, set the wave speed $c = 1$ and the length $L = 1$ in (12.1) as such values have no essential effect as far as the mathematical analysis of chaotic vibration here is concerned. We thus have

$$w_{tt}(x,t) - w_{xx}(x,t) = 0, \qquad 0 < x < 1, \quad t > 0. \tag{12.2}$$

The two initial conditions are

$$w(x,0) = w_0(x), \quad w_t(x,0) = w_1(x), \qquad 0 < x < 1. \tag{12.3}$$

At the right end $x = 1$, assume a nonlinear boundary condition

$$w_x(1,t) = \alpha w_t(1,t) - \beta w_t^3(1,t); \qquad t > 0, \quad \alpha, \beta > 0. \tag{12.4}$$

At the left end $x = 0$, we choose the boundary condition to be

$$w_t(0,t) = -\eta w_x(0,t), \qquad t > 0; \quad \eta > 0, \quad \eta \neq 1. \tag{12.5}$$

Remark 12.1 Equation (12.5) says that negative force is fedback to the velocity at $x = 0$. An alternate choice would be

$$w_x(0,t) = -\eta w_t(0,t), \qquad t > 0; \quad \eta > 0, \quad \eta \neq 1,$$

which says negative velocity is fedback to force. □

Remark 12.2 An ordinary differential equation, called the van der Pol oscillator, is important in the design of classical servomechanisms:

$$\ddot{x} - (\alpha - \beta \dot{x}^2)\dot{x} + kx = 0; \qquad \alpha, \beta > 0, \tag{12.6}$$

where $x = x(t)$ is proportional to the electric current at time t on a circuit equipped with a van der Pol device. Then the energy at time t is $E(t) = \frac{1}{2}(\dot{x}^2 + kx^2)$ and

$$\frac{d}{dt}E(t) = \dot{x}(\ddot{x} + kx) = \dot{x}^2(\alpha - \beta \dot{x}^2),$$

so we have

$$E'(t) \begin{cases} \geq 0 \text{ if } |\dot{x}| \leq (\alpha/\beta)^{1/2}, \\ < 0 \text{ if } |\dot{x}| > (\alpha/\beta)^{1/2} \end{cases} \tag{12.7}$$

which is a desired *self-regulation effect*, i.e., energy will increase when $|\dot{x}|$ is small (which is unfit for operations), and energy will decrease when $|\dot{x}|$ is large in order to prevent electric current surge which may destroy the circuit. (This self-regulating effect is also called *self-excitation*.) A second version of the van der Pol equation is

$$\ddot{x} - (\alpha - 3\beta x^2)\dot{x} + kx = 0, \tag{12.8}$$

which may be regarded as a differentiated version of (12.6), satisfying a regulation effect similar to (12.7). Neither (12.6) nor (12.8) has any chaotic behavior as *the solutions tend to limit cycles according to the Poincaré–Bendixon theorem.* However, when a *forcing term*

12.1 The Mathematical Model and Motivations

$A\cos(\omega t)$ is added to the right-hand side of (12.6) or (12.8), solutions display chaotic behavior when the parameters A and ω enter a certain regime [11, 12]. □

With (12.4) and (12.5), we have, by (12.2), (12.4), and (12.5),

$$\begin{aligned}\frac{d}{dt}E(t) &= \frac{d}{dt}\int_0^1 \left[\frac{1}{2}w_x^2(x,t) + w_t^2(x,t)\right]dt \\ &= \int_0^1 [w_x(x,t)w_{xt}(x,t) + w_t(x,t)w_{tt}(x,t)]dx \\ &= \int_0^1 [w_x(x,t)w_{xt}(x,t) + w_t(x,t)w_{xx}(x,t)]dx \\ &(\Rightarrow \text{ integration by parts}) \\ &= w_t(x,t)w_x(x,t)|_{x=0}^{x=1} \\ &= \eta w_x^2(0,t) + w_t^2(1,t)[\alpha - \beta w_t^2(1,t)]. \end{aligned} \quad (12.9)$$

The contribution $\eta w_x^2(0,t)$ above, due to (12.5), is always nonnegative. Thus we see that the effect of (12.5) is to cause energy to increase. For this reason, the boundary condition (12.5) is said to be *energy-injecting* or *energy-pumping*. On the other hand, we have

$$w_t^2(1,t)[\alpha - \beta w_t^2(1,t)]\begin{cases} \geq 0 \text{ if } |w_t(1,t)| \leq (\alpha/\beta)^{1/2}, \\ < 0 \text{ if } |w_t(1,t)| > (\alpha/\beta)^{1/2}, \end{cases} \quad (12.10)$$

so the contribution of the boundary condition (12.4) to (12.9) is *self-regulating* because (12.10) works in exactly the same way as (12.7). Thus, we call (12.4) a *van der Pol, self-regulating, or self-exciting* boundary condition. Intuitively speaking, with the boundary condition (12.5) alone (and with the right-end boundary condition (12.4)) replaced by a conservative boundary condition such as $w(1,t) = 0$ or $w_x(1,t) = 0$ for all $t > 0$), it causes the well-known *classical linear instability*, namely, the energy grows with an exponential rate:

$$E(t) = \mathcal{O}(e^{kt}), \quad k = \frac{1}{2}\ln\left(\left|\frac{1+\eta}{1-\eta}\right|\right) > 0. \quad (12.11)$$

However, the self-regulating boundary condition (12.4) can hold the instability (12.11), *partly* in check by its regulation effect, for a large class of *bounded initial states* with bounds depending on the parameters α, β, and η. When α, β, and η match in a certain regime, chaos happens, which could be viewed as a *reconciliation between linear instability and nonlinear self-regulation*. Overall, there is a richness of nonlinear phenomena, including the following: the existence of *asymptotically periodic solutions, hysteresis, instability* of the type of unbounded growth, and *fractal invariant sets*.

A basic approach for the problems under consideration in this section is the *method of characteristics*. Let u and v be the *Riemann invariants* of (12.2) defined by

$$u(x,t) = \frac{1}{2}[w_x(x,t) + w_t(x,t)],$$
$$v(x,t) = \frac{1}{2}[w_x(x,t) - w_t(x,t)]. \qquad (12.12)$$

Then u and v satisfy a diagonalized first-order linear hyperbolic system

$$\frac{\partial}{\partial t}\begin{bmatrix} u(x,t) \\ v(x,t) \end{bmatrix} = \begin{bmatrix} 1 & 0 \\ 0 & -1 \end{bmatrix} \frac{\partial}{\partial x}\begin{bmatrix} u(x,t) \\ v(x,t) \end{bmatrix}, \qquad 0 < x < 1, \quad t > 0, \qquad (12.13)$$

with initial conditions

$$\left.\begin{array}{l} u(x,0) = u_0(x) \equiv \dfrac{1}{2}[w_0'(x) + w_1(x)], \\[6pt] v(x,0) = v_0(x) \equiv \dfrac{1}{2}[w_0'(x) - w_1(x)], \end{array}\right\} \qquad 0 < x < 1. \qquad (12.14)$$

The boundary condition (12.4), after converting to u and v and simplifying, becomes

$$u(1,t) = F_{\alpha,\beta}(v(1,t)), \qquad t > 0, \qquad (12.15)$$

where the relation $u = F_{\alpha,\beta}(v)$ is defined implicitly by

$$\beta(u-v)^3 + (1-\alpha)(u-v) + 2v = 0; \qquad \alpha, \beta > 0. \qquad (12.16)$$

Remark 12.3 For (12.16), we know that

(i) when $0 < \alpha \leq 1$, for each $v \in \mathbb{R}$, there exists a unique $u \in \mathbb{R}$;
(ii) when $\alpha > 1$, for each $v \in \mathbb{R}$, in general, there may exist two or three distinct $u \in \mathbb{R}$ satisfying (12.16). Thus $u = F_{\alpha,\beta}(v)$ is not a function relation.

Only case (i) will be treated in Sect. 12.2 while for case (ii), which contains hysteresis, the interested reader may refer to [5]. □

The boundary conditions (12.2), by (12.12), become

$$v(0,t) = G_\eta(u(0,t)) \equiv \frac{1+\eta}{1-\eta} u(0,t), \qquad t > 0. \qquad (12.17)$$

Equations (12.16) and (12.17) are, respectively, the *wave-reflection relations* at the right end $x = 1$ and the left end $x = 0$. The reflection of characteristics is depicted in Fig. 12.1.

Assume that $F_{\alpha,\beta}$ is well defined. Then a solution (u,v) of the system (12.13), (12.14), (12.15), and (12.17) can be expressed as follows:

For $0 \leq x \leq 1$ and $t = 2k + \tau$, with $k = 0, 1, 2, \ldots$, and $0 \leq \tau < 2$,

12.2 Chaotic Vibration of the Wave Equation

Fig. 12.1 Reflection of characteristics

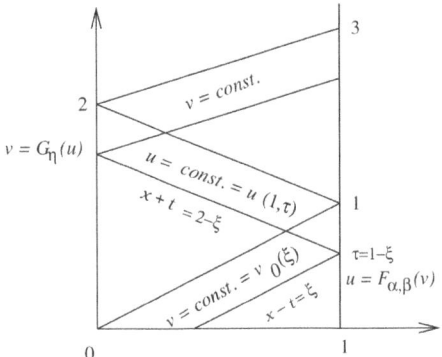

$$u(x,t) = \begin{cases} (F \circ G)^k(u_0(x+\tau)), & \tau \leq 1-x, \\ G^{-1} \circ (G \circ F)^{k+1}(v_0(2-x-\tau)), & 1-x < \tau \leq 2-x, \\ (F \circ G)^{k+1}(u_0(\tau+x-2)), & 2-x < \tau \leq 2; \end{cases}$$

(12.18)

$$v(x,t) = \begin{cases} (G \circ F)^k(v_0(x-\tau)), & \tau \leq x, \\ G \circ (F \circ G)^k(u_0(\tau-x)), & x < \tau \leq 1+x, \\ (G \circ F)^{k+1}(v_0(2+x-\tau)), & 1+x < \tau \leq 2, \end{cases}$$

where in the above, $F = F_{\alpha,\beta}$ and $G = G_\eta$, and $(G \circ F)^k$ represents the k-th iterate of the map $G \circ F$. From now on, we often abbreviate $F_{\alpha,\beta}$ and G_η, respectively, as F and G, in case no ambiguities will occur. We call the map $G_\eta \circ F_{\alpha,\beta}$, naturally, the *composite reflection relation*. This map $G_\eta \circ F_{\alpha,\beta}$ can be regarded as the *Poincaré section* of the PDE system because we can essentially construct the solution from $G_\eta \circ F_{\alpha,\beta}$ using (12.18).

From (12.18), it becomes quite apparent that the solutions $(u(x,t), v(x,t))$ will manifest chaotic behavior when the map $G \circ F$ is chaotic, in the sense of Devaney cf. Definition 6.27 in Chap. 6 and [13, p. 50], for example. We proceed with the discussion in the following section. The main source of reference is [4].

12.2 Chaotic Vibration of the Wave Equation

As mentioned in Remark 12.1, when $0 < \alpha \leq 1$, for each $v \in \mathbb{R}$ there exists a unique $u \in \mathbb{R}$ such that $u = F_{\alpha,\beta}(v)$. Therefore, the solution (u, v) to (12.13), (12.14), (12.15), and (12.17) is unique. When the initial condition (u_0, v_0) is sufficiently smooth satisfying compatibility conditions with the boundary conditions, then (u, v) will also be C^1-smooth on the spatiotemporal domain.

Let α and β be fixed, and let $\eta > 0$ be the only parameter that varies. To aid understanding, we include a sample graph of the map $G_\eta \circ F_{\alpha,\beta}$, with $\alpha = 1/2$, $\beta = 1$, and $\eta = 0.552$, in Fig. 12.2. We only need to establish that $G_\eta \circ F_{\alpha,\beta}$ is chaotic, because $F_{\alpha,\beta} \circ G_\eta$ is

Fig. 12.2 The graph of $u = G_\eta \circ F_{\alpha,\beta}(v)$. (Here $\alpha = 0.5, \beta = 1, \eta = 0.552$ are used.) Note that $\pm I_1 = v$-axis nonzero intercepts, $I_1 = [(1 + \alpha/\beta)]^{1/2}; \pm v_c =$ (local) critical points, $v_c = |(2 - \alpha)/3|[(1 + \alpha)/(3\beta)]^{1/2};$ $\pm M =$ local extremum values, $M = \frac{1+\eta}{1-\eta}\frac{1+\alpha}{3}[(1 + \alpha/3\beta)]^{1/2};$ $\pm I_2 = v$-values where the curve intersects with the line $u - v = 0$, and $[-I_2, I_2] \times [-I_2, I_2]$ is an invariant square when $M \leq I_2$

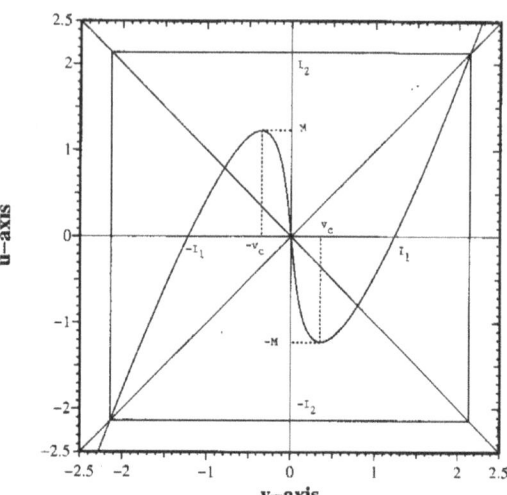

topologically conjugate to $G_\eta \circ F_{\alpha,\beta}$ through

$$F_{\alpha,\beta} \circ G_\eta = G_\eta^{-1} \circ (G_\eta \circ F_{\alpha,\beta}) \circ G_\eta$$

and, thus, the iterates $(F \circ G)^k$ or $(F \circ G)^{k+1}$ appearing in (12.18) do not need to be treated separately.

We note the following bifurcations: For fixed α: $0 < \alpha \leq 1$ and $\beta > 0$, let $\eta \in (0, 1)$ be varying.

(1) *Period-doubling bifurcation* (Theorem 4.1). Define

$$h(v, \eta) = -G_\eta \circ F(v)$$

and let

$$v_0(\eta) \equiv \eta[(1 + \eta)/2][(\alpha + \eta)/\beta]^{1/2}$$

which, for each η, represents a fixed point of h, i.e.:

$$h(v_0(\eta), \eta) = v_0(\eta).$$

Then the algebraic equation

$$\frac{1}{2}\left(\frac{1+\alpha\eta}{3\beta\eta}\right)^{1/2}\left[\frac{1+(3-2\alpha)\eta}{3\eta}\right] = \frac{1+\eta}{2}\left(\frac{\alpha+\eta}{\beta}\right)^{1/2} \quad (12.19)$$

has a unique solution $\eta = \eta_0$: $0 < \eta_0 \leq \boldsymbol{\eta_H}$, where

12.2 Chaotic Vibration of the Wave Equation

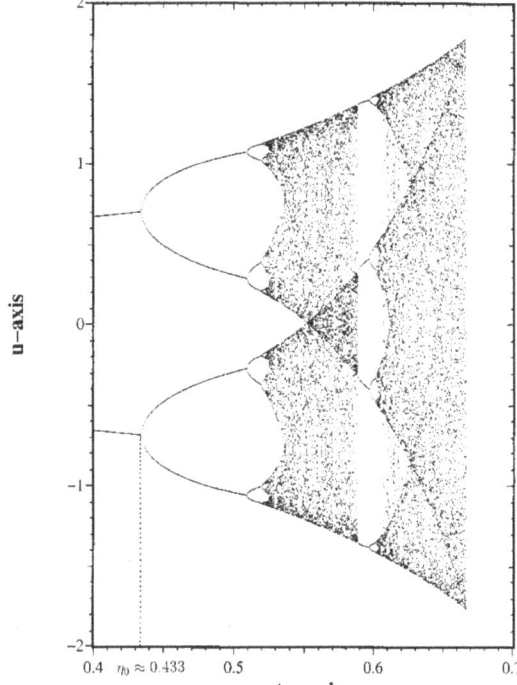

Fig. 12.3 The orbit diagram of $G_\eta \circ F_{\alpha,\beta}$, where $\alpha = 0.5$, $\beta = 1$, and η varies in $[0.4, 2/3]$. Note that the first period-doubling occurs near $\eta_0 \approx 0.433$, agreeing with the computational result of the solution η_0 satisfying Eq. (12.21). (Reprinted from [4, p. 433, Fig. 3], courtesy of World Scientific, Singapore)

$$\eta_H \equiv \left(1 - \frac{1+\alpha}{3\sqrt{3}}\right) \Big/ \left(1 + \frac{1+\alpha}{3\sqrt{3}}\right) \tag{12.20}$$

satisfying

$$\frac{\partial}{\partial v} h_1(v, \eta) \bigg|_{\substack{v=v_0(\eta_0) \\ \eta=\eta_0}} = -1 \tag{12.21}$$

which is the primary necessary condition for period-doubling bifurcation to happen, at $v = v_0(\eta_0)$, $\eta = \eta_0$. Furthermore, the other "accessory" conditions are also satisfied, and the bifurcationed period-2 solutions are attracting.

Consequently, there is a period-doubling route to chaos, as illustrated in the orbit diagram in Fig. 12.3.

(2) *Homoclinic orbits* (cf. Chap. 4). Let η_H be given by (12.19). If

$$\eta_H \leq \eta < 1, \tag{12.22}$$

then $M \geq I_1$ (cf. Fig. 12.2) and, consequently, the repelling fixed point 0 of $G_\eta \circ F$ has homoclinic orbits. Furthermore, if $\eta = \eta_H$, then there are *degenerate homoclinic orbits* (and, thus, homoclinic bifurcations [13, p. 125]).

Fig. 12.4 The graph of $G_\eta \circ F_{\alpha,\beta}$ with $\alpha = 0.5$, $\beta = 1$, and $\eta = 0.8$. Note that here $M > I_2$ (cf. Fig. 12.2) and $[-I_2, I_2] \times [-I_2, I_2]$ is no longer an invariant square for $G_\eta \circ F_{\alpha,\beta}$. On $[-I_2, I_2]$, what $G_\eta \circ F_{\alpha,\beta}$ has is a Cantor-like fractal invariant set

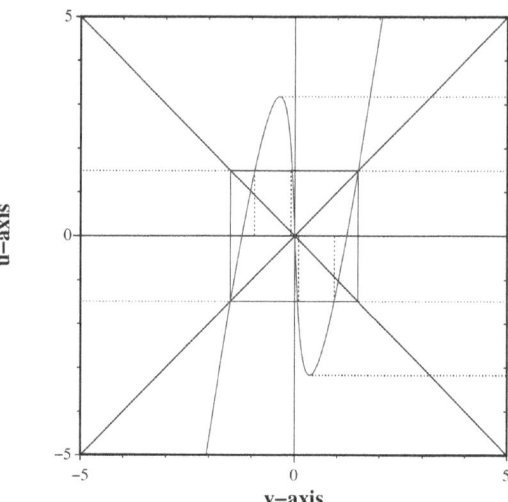

When $M > I_2$; cf. Fig. 12.2, then $[-I_2, I_2] \times [-I_2, I_2]$ is no longer an invariant square for the map $G \circ F$. What happens is exactly similar to the case of the quadratic map $f_\mu(x) = \mu x(1-x)$, for $0 \le x \le 1$, when $\mu > 4$ because part of the graph of f_μ will protrude above the unit square. See Fig. 12.4. It is easy to see that now the map $G \circ F$ has a Cantor-like fractal invariant set Λ on the interval $[-I_2, I_2]$, where $\Lambda = \bigcap_{j=1}^{\infty} (G \circ F)^k([-I_2, I_2])$. All the other points outside Λ are eventually mapped to $\pm\infty$ as the number of iterations increases.

We furnish a PDE example below.

Example 12.4 ([4, p. 435, Example 3.3]) Consider (12.13), (12.14), (12.15), and (12.17), where we choose

$$\alpha = 0.5, \beta = 1, \eta = 0.525 \approx \eta_H, \text{ satisfying (12.22)},$$

$$w_0(x) = 0.2 \sin\left(\frac{\pi}{2}x\right), \quad w_1(x) = 0.2 \sin(\pi x), \quad x \in [0, 1].$$

Two spatiotemporal profiles of u and v are plotted, respectively, in Figs. 12.5 and 12.6. Their rugged outlooks manifest chaotic vibration. □

Miscellaneous Remarks

(1) In this subsection, we have illustrated only the case $0 < \eta < 1$. When $\eta > 1$, the results are similar. See [4].
(2) With the nonlinear boundary condition (12.4), we can only establish that u and v are chaotic. From this, we can then show that w_x and w_t, i.e., the gradient of w, are also

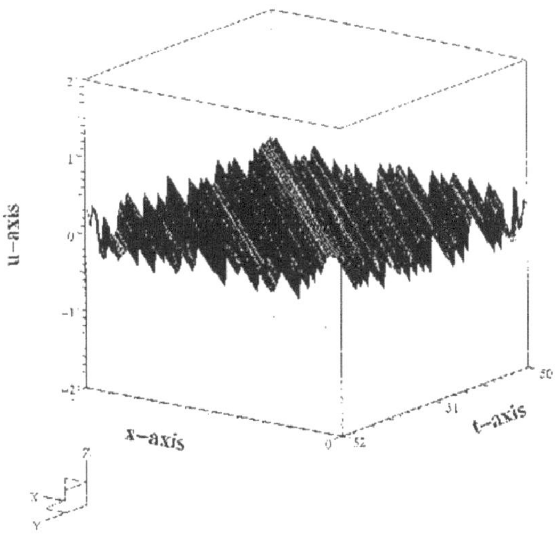

12.2 Chaotic Vibration of the Wave Equation

Fig. 12.5 The spatiotemporal profile of the u-component for Example 1.1, for $t \in [50, 52]$, $x \in [0, 1]$. (Reprinted from [4, p. 435, Fig. 7], courtesy of World Scientific, Singapore)

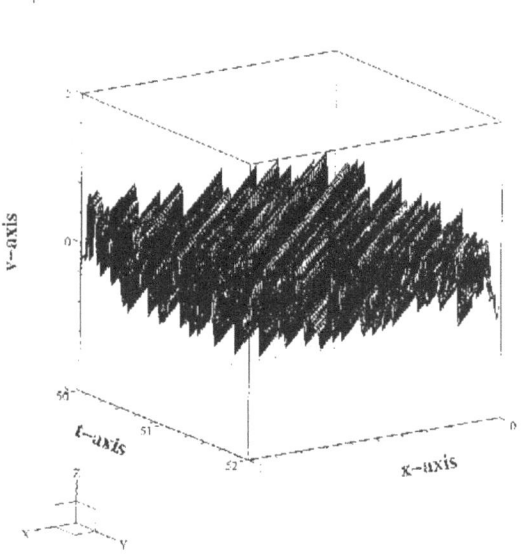

Fig. 12.6 The spatiotemporal profile of the v-component for Example 1.1, for $t \in [50, 52]$, $x \in [0, 1]$. (Reprinted from [4, p. 435, Fig. 8], courtesy of World Scientific, Singapore)

chaotic by a natural topological conjugacy, see [3, Sect. 5]. However, w itself is not chaotic because w is the time integral of w_t, which smooths out the oscillatory behavior of w_t.

In order to have chaotic vibration of w, one must use a differentiated boundary condition; see [4, Sect. 6]. This is actually an analog of (12.8).

(3) When the initial data (u_0, v_0) takes values outside the invariant square $[-I_2, I_2] \times [I_2, I_2]$, then part of u and v will diverge to $\pm\infty$ as $t \to \infty$. This behavior belongs to classical unbounded instability.

Further studies of the chaotic vibration of the wave equation for the case when there is boundary hysteresis or the case of nonisotropic spatiotemporal chaos and others may be found in [1–3, 6] and [7].

References

1. G. Chen, S.B. Hsu and T.W. Huang, Analyzing the displacement terms memory effect to prove the chaotic vibration of the wave equation, *Int. J. Bifur. Chaos* **12** (2002), 965–981. https://doi.org/10.1142/S0218127402005741
2. G. Chen, S.B. Hsu and J. Zhou, Linear superposition of chaotic and orderly vibrations on two serially connected strings with a van der Pol joint, *Int. J Bifur. Chaos* **6** (1996), 1509–1527.
3. G. Chen, S.B. Hsu and J. Zhou, Chaotic vibrations of the one-dimensional wave equation due to a self-excitation boundary condition. Part I, controlled hysteresis, *Trans. Amer. Math. Soc.* **350** (1998), 4265–4311. https://doi.org/10.1090/S0002-9947-98-02022-4
4. G. Chen, S.B. Hsu and J. Zhou, Chaotic vibrations of the one-dimensional wave equation due to a self-excitation boundary condition. Part II, energy injection, period doubling and homoclinic orbits, *Int. J. Bifur. Chaos* **8** (1998), 423–445. https://doi.org/10.1142/S0218127498001236
5. G. Chen, S.B. Hsu and J. Zhou, Chaotic vibrations of the one-dimensional wave equation due to a self-excitation boundary condition. Part III, natural hysteresis memory effects, *Int. J. Bifur. Chaos* **8** (1998), 447–470. https://doi.org/10.1142/S0218127498001236
6. G. Chen, S.B. Hsu and J. Zhou, Snapback repellers as a cause of chaotic vibration of the wave equation with a van der Pol boundary condition and energy injection at the middle of the span, *J. Math. Phys.* **39** (1998), 6459–6489. https://doi.org/10.1063/1.532670
7. G. Chen, S.B. Hsu and J. Zhou, Nonisotropic spatiotemporal chaotic vibration of the wave equation due to mixing energy transport and a van der Pol boundary condition, *Int. J. Bifur. Chaos* **12** (2002), 447–470. https://doi.org/10.1142/S0218127402005741
8. G. Chen, T. Huang, J. Juang, and D. Ma, Unbounded growth of total variations of snapshots of the 1D linear wave equation due to the chaotic behavior of iterates of composite nonlinear boundary reflection relations, G. Chen et al. (ed.), in *Control of Nonlinear Distributed Parameter Systems*, Marcel Dekker Lectures Notes on Pure Appl. Math., New York, 2001, 15–43. Cited on page(s) 205
9. Y. Huang, Boundary feedback anticontrol of spatiotemporal chaos for 1D hyperbolic dynamical systems, *Int. J. Bifur. Chaos* **14** (2004), 1705–1723. https://doi.org/10.1142/S021812740401031X
10. Y. Huang, J. Luo and Z.L. Zhou, Rapid fluctuations of snapshots of one-dimensional linear wave equation with a van der Pol nonlinear boundary condition, *Int. J. Bifur. Chaos* **15** (2005), 567–580. https://doi.org/10.1142/S0218127405012223
11. J. Guckenheimer and P. Holmes, *Nonlinear Oscillations, Dynamical Systems and Bifurcations of Vector Fields*, Springer, New York, 1983.
12. M. Levi, Qualitative analysis of the periodically forced relaxation oscillations, *Mem. Amer. Math. Soc.* **214** (1981), 1–147. https://doi.org/10.1007/BFb0086995
13. R.L. Devaney, *An Introduction to Chaotic Dynamical Systems*, 2nd ed., Addison-Wesley, New York, 1989. Cited on page(s) 19, 36, 75, 86, 209, 211

Necessary Conditions for Chaotic Vibrations

13

In this part, we mainly analyze some necessary conditions for the onset of chaotic vibrations in the systems governed by 1D wave equation $w_{tt} - c^2 w_{xx} = 0$, where c denotes wave propagation speed, associated with a generalized boundary condition. The PDE system, 1D wave equation with composite boundary conditions, has received considerable attention since it exhibits many interesting and complicated dynamical phenomena, such as limit cycles and chaotic behavior of (w_t, w_x) [1, 2]. Different from dynamics of a system of ODEs, this is a simple and useful infinite-dimensional model for the study of spatiotemporal behaviors as time evolves. For instance, the propagation of acoustic waves in a pipe satisfies the linear wave equation: $w_{tt} - w_{xx} = 0$. As we know, the solution of 1D wave equation describes a superposition of two traveling waves with arbitrary profiles, one propagating with unit speed to the left, while the other with unit speed to the right. The composite boundary conditions can create irregularly acoustical vibrations ([1–3]). This type of vibrations, for example, can be generated by noise signals radiated from underwater vehicles, and there are intensive research for the properties of acoustical vibrations in current literature (see, e.g., [4] and references therein). Hence, the study of this type of vibration is not only important but it also may lead to a better understanding of the dynamics of acoustic systems.

Here we mention the following models as studied by Li [5]. Recall the definition of chaos for this kind of system, which is first introduced in [2], as follows:

Definition 13.1 Consider an initial-boundary problem (S) governed by 1D wave equation $w_{tt} - c^2 w_{xx} = 0$ defined on a segment I, where $c > 0$ denotes the propagation speed of the wave. The system is said to be chaotic if there exists a large class of initial data (w_0, w_1) such that

(i) $|w_x| + |w_t|$ is uniformly bounded,
(ii) $V(t) \stackrel{def}{=} V_I(w_x(\cdot, t)) + V_I(w_t(\cdot, t)) < +\infty$ for all $t \geq 0$,
(iii) $\liminf\limits_{t \to +\infty} \frac{\ln V(t)}{t} > 0$.
□

Remark 13.2 When chaos occurs, the function $x \mapsto (|w_x| + |w_t|)(x, t)$ is uniformly bounded, whereas the length of the curve $\{(x, (|w_x| + |w_t|)(x, t)), \ x \in [0, 1]\}$ grows exponentially w.r.t. time t. Therefore, the system (S) must undergo extremely complex oscillations as time t increases.
□

Remark 13.3 There are lots of work about chaos studies, see, e.g., [6–8] and references therein. As we know, there is no a common mathematical definition for chaos, which is actually a challenge to give. However, Li–Yorke chaos is probably one of the most popular and acceptable notions of chaos. As we have shown in Chap. 12 that the solution (w_x, w_t) can be represented by two interval maps, say K_1 and K_2, respectively, then K_1 and K_2 have positive entropy when chaotic vibration occurs, which implies K_1 and K_2 are chaotic in the sense of Li–Yorke.
□

We here mainly consider the oscillation problems described by the following models:

$$\begin{cases} w_{tt}(x, t) - c^2 w_{xx}(x, t) = 0, \quad x \in (0, 1), \ t > 0, \\ w_x(0, t) = -\eta w_t(0, t), \ \eta \neq c^{-1}, \ t > 0, \\ w_x(1, t) = h(w_t(1, t)), \ t > 0, \\ w_0(x) = w(x, 0), \quad w_1(x) = w_t(x, 0), \quad 0 \leq x \leq 1, \end{cases} \quad (13.1)$$

$$\begin{cases} z_{tt}(x, t) - c^2 z_{xx}(x, t) = 0, \quad x \in (0, 1), \ t > 0, \\ z(0, t) = 0, \ t > 0, \\ z_x(1, t) = h(z_t(1, t)), \ t > 0, \\ z_0(x) = z(x, 0), \quad z_1(x) = z_t(x, 0), \quad 0 \leq x \leq 1, \end{cases} \quad (13.2)$$

where the function $h \in C^0(\mathbb{R})$ satisfies

(A1) $\phi : t \mapsto \frac{1}{2}\left(h(t) - \frac{1}{c}t\right)$ strictly monotonically decreases on $\mathbb{R}$,
(A2) $\phi(\mathbb{R}) = \mathbb{R}$.

It is clear that $\phi^{-1} : \mathbb{R} \to \mathbb{R}$ is well defined and strictly decreasing on $\mathbb{R}$. In fact, the model (13.1) is a generalized case of the van del Pol-type boundary condition, say $h(x) = \alpha x - \beta x^3$. Denote the total energy of system (13.1) as

$$E(t) = \frac{1}{2c^2} \int_0^1 \left[c^2 w_x^2(x, t) + w_t^2(x, t)\right] dx,$$

then the boundary conditions show

$$\frac{d}{dt}E(t) = \eta w_t^2(0, t) + w_t^2(1, t) \cdot \frac{h(w_t(1, t))}{w_t(1, t)}.$$

Hence $\eta > 0$ can cause the energy of system to increase. Moreover, the system (13.1) has a generalized self-excited mechanism if we assume that, roughly speaking, $\frac{h(y)}{y} > 0$ if $|y|$ is small and $\frac{h(y)}{y} < 0$ if $|y|$ is large. Thus if $\eta > 0$, the system (13.1) has a self-excited mechanism that supplies energy to the system itself, which could induce irregular oscillations. An interesting part is that the system (13.2) may have chaos even though there is no energy supplied at the boundary.

We now treat the initial-boundary problems (13.1) and (13.2) by using the *wave propagation method*. Let's take (13.1) as a start. It is well known that w, a solution of 1D wave equation, has the following form:

$$w(x, t) = L(x + ct) + R(x - ct), \tag{13.3}$$

where L and R are two C^1 functions. Then the gradient of w can be represented as follows:

$$w_t(x, t) = cL'(x + ct) - cR'(x - ct), \quad w_x(x, t) = L'(x + ct) + R'(x - ct), \tag{13.4}$$

for $x \in [0, 1]$, $t \geq 0$. Introduce two new variables

$$u(x, t) = L'(x + ct), \quad v(x, t) = R'(x - ct), \tag{13.5}$$

which are the *Riemann invariants*. It is evident that u and v remain constants along the lines $x + ct = const.$ and $x - ct = const.$, respectively, which are referred to as *characteristics*.

When $t > 0$ and $x = 0$, from the boundary condition at the left end $x = 0$,

$$v(0, t) = \frac{cw_x(0, t) - w_t(0, t)}{2c}$$
$$= \frac{-c\eta w_t(0, t) - w_t(0, t)}{2c} = -\frac{1 + c\eta}{2c} w_t(0, t)$$
$$= -\frac{1 + c\eta}{2}(u(0, t) - v(0, t)),$$

that is,

$$v(0, t) = \frac{c\eta + 1}{c\eta - 1} u(0, t) \stackrel{def}{=} \gamma(\eta) \cdot u(0, t). \tag{13.6}$$

Multiplication by the factor $\gamma(\eta)$ may be viewed as a linear map, also simply denoted as $\gamma(\eta)$.

When $t > 0$ and $x = 1$, from the boundary condition at the right end $x = 1$, it follows:

$$h(c(u(1, t) - v(1, t))) - \frac{1}{c} \cdot c(u(1, t) - v(1, t)) - 2v(1, t) = 0,$$

that is,
$$v(1,t) = \phi(c(u(1,t) - v(1,t))),$$
which determines a reflection relationship between $u(1,t)$ and $v(1,t)$ as follows:
$$u(1,t) = \frac{1}{c} \cdot \phi^{-1}(v(1,t)) + v(1,t) \stackrel{def}{=} \varphi(v(1,t)), \tag{13.7}$$
where
$$\varphi(x) = \frac{1}{c} \cdot \phi^{-1}(x) + x. \tag{13.8}$$
When $t = 0$, from the initial conditions,
$$u_0(x) \stackrel{def}{=} u(x,0) = \frac{cw_0'(x) + w_1(x)}{2c}, \quad v_0(x) \stackrel{def}{=} v(x,0) = \frac{cw_0'(x) - w_1(x)}{2c}, \tag{13.9}$$
which are referred to as the initial data of (u,v).

For $x \in [0,1]$ and $\tau \geq 0$, it follows from the boundary reflections (13.6) and (13.7) that
$$u\left(x, \tau + \frac{2}{c}\right) = L'(x + c\tau + 2) = u\left(1, \frac{1}{c}x + \tau + \frac{1}{c}\right) = \varphi\left(v\left(1, \frac{1}{c}x + \tau + \frac{1}{c}\right)\right)$$
$$= \varphi\left(R'(-x - c\tau)\right) = \varphi\left(v\left(0, \frac{1}{c}x + \tau\right)\right)$$
$$= \varphi\left(\gamma(\eta) \cdot u(x,\tau)\right),$$
and
$$v\left(x, \tau + \frac{2}{c}\right) = \gamma(\eta) \circ \varphi(v(x,\tau));$$
inductively,
$$u\left(x, \tau + \frac{2n}{c}\right) = (\varphi \circ \gamma(\eta))^n (u(x,\tau)), \quad v\left(x, \tau + \frac{2n}{c}\right) = (\gamma(\eta) \circ \varphi)^n (v(x,\tau)), \tag{13.10}$$
where the superscript $n \in \mathbb{N}$ denotes the nth iteration of a function. Analogously, for $x \in [0,1]$ and $\tau \in [0, \frac{2}{c})$,
$$u(x,\tau) = \begin{cases} u_0(x + c\tau), & 0 \leq c\tau \leq 1 - x, \\ \varphi(v_0(2 - x - c\tau)), & 1 - x < c\tau \leq 2 - x, \\ \varphi \circ \gamma(\eta)(u_0(x + c\tau - 2)), & 2 - x < c\tau < 2, \end{cases} \tag{13.11}$$
and
$$v(x,\tau) = \begin{cases} v_0(x - c\tau), & 0 \leq c\tau \leq x, \\ \gamma(\eta)(u_0(c\tau - x)), & x < c\tau \leq x + 1, \\ \gamma(\eta) \circ \varphi(v_0(2 + x - c\tau)), & x + 1 < c\tau < 2. \end{cases} \tag{13.12}$$

Equations (13.10)–(13.12) show that the system (13.1) is solvable and the dynamics of the solution (w, w_x, w_t) to Eq. (13.1) can be uniquely determined by the initial data and the following two functions:

$$\psi_\eta = \gamma(\eta) \circ \varphi, \quad g = \varphi \circ \gamma(\eta). \tag{13.13}$$

Note that $g = \gamma^{-1}(\eta) \circ \psi_\eta \circ \gamma(\eta)$, that is to say, there is topological conjugacy between ψ_η and g. Therefore, one only needs to consider one of them, say ψ_η.

13.1 Necessary Conditions for the Onset of Chaos

In this section, we first give a necessary condition for the onset of chaos in the following system:

$$\begin{cases} w_{tt}(x,t) - c^2 w_{xx}(x,t) = 0, & x \in (0,1), \ t > 0, \\ w_x(0,t) = -\eta w_t(0,t), \ \eta \neq c^{-1}, & t > 0, \\ w_x(1,t) = h(w_t(1,t)), & t > 0, \\ w_0(x) = w(x,0), \ w_1(x) = w_t(x,0), & 0 \leq x \leq 1, \end{cases} \tag{13.14}$$

where h satisfies hypotheses (A1) and (A2). In addition, assume that the function $t \mapsto \left(h(t) + \frac{1}{c}t\right)$ is piecewise monotone.

Theorem 13.4 *Suppose the system (13.14) is chaotic in the sense of Definition 13.1. Then h is not even and $\eta \neq 0$.*

Proof It is equivalent to prove that there is no chaos in the system (13.14) if $\eta = 0$ or h is an even function.

First, assume that h is an even function. Let $\eta \in \mathbb{R}$ and $\eta \neq c^{-1}$. Recall the function ψ_η given by (13.13):

$$\psi_\eta = \gamma(\eta) \circ \varphi, \quad \varphi(x) = \frac{1}{c} \cdot \phi^{-1}(x) + x.$$

Introduce a new map Q from $\mathbb{R}$ to $\mathbb{R}$ as follows:

$$Q(y) = \frac{1}{2}\left(h(y) + \frac{1}{c}y\right). \tag{13.15}$$

For $x \in \mathbb{R}$, let $y = \phi^{-1}(x)$; then

$$\psi_\eta(x) = \gamma(\eta)\left(\frac{1}{c}y + \phi(y)\right) = \frac{1}{2}\gamma(\eta)\left(\frac{1}{c}y + h(y)\right)$$
$$= \gamma(\eta) \cdot Q \circ \phi^{-1}(x).$$

Since ϕ^{-1} is strictly decreasing and $\gamma(\eta) \in \mathbb{R}^*$, ψ is monotonic if and only if Q is monotonic. Let $y_1, y_2 \in \mathbb{R}$ with $y_1 < y_2$. From the hypothesis of h being an even function, it follows

$$Q(y_2) - Q(y_1) = \frac{1}{2}\left(h(-y_2) - \frac{1}{c}(-y_2)\right) - \frac{1}{2}\left(h(-y_1) - \frac{1}{c}(-y_1)\right)$$
$$= \phi(-y_2) - \phi(-y_1)$$
$$> 0,$$

which implies Q strictly monotonically increases. Therefore, ψ is strictly monotonically increasing (decreasing) if $\gamma(\eta) < 0 (\gamma(\eta) > 0)$. It is easily seen that

$$w_t(x,t) = c\left[\gamma^{-1}(\eta) \circ \psi^n\left(\gamma(\eta)u(x,\tau)\right) - \psi^n\left(v(x,\tau)\right)\right],$$
$$w_x(x,t) = \gamma^{-1}(\eta) \circ \psi^n\left(\gamma(\eta)u(x,\tau)\right) + \psi^n\left(v(x,\tau)\right).$$

Hence the dynamics of (w_x, w_t) does not have chaos for the periodic system.

Next, assume $\eta = 0$. In other words, the system is free at the left end. In this case,

$$\psi_0(x) = -x - \phi^{-1}(x).$$

We will show that there is no periodic point of ψ_0 with period 2. Let $x_0 \in \mathbb{R}$ satisfy $x_0 < \psi_0(x_0)$. Define two functions as follows:

$$q : x \mapsto -x + \psi_0(x_0) + x_0, \quad k = \psi_0 - q.$$

Since $\phi^{-1}(\cdot)$ is strictly monotonically decreasing while

$$k(x) = -x - \phi^{-1}(x) - q(x) = \phi^{-1}(x_0) - \phi^{-1}(x),$$

and k is strictly monotonically increasing in $[x_0, \psi_0(x_0)]$, hence

$$k(\psi_0(x_0)) = \psi_0^2(x_0) - x_0 > k(x_0) = 0.$$

This implies that there are no periodic points of ψ_0 with period 2. By *Sharkovsky's theorem*, there are no periodic points of ψ_0 with period lager than 2. By virtue of the Main Theorem 6 in [9], there is no chaos in the system when $\eta = 0$. □

Next, we consider the following system governed by a 1D wave equation with a fixed left end:
$$\begin{cases} z_{tt}(x,t) - c^2 z_{xx}(x,t) = 0, & x \in (0,1),\ t > 0, \\ z(0,t) = 0,\ t > 0, \\ z_x(1,t) = h(z_t(1,t)),\ t > 0, \\ z_0(x) = z(x,0),\ z_1(x) = z_t(x,0),\ 0 \le x \le 1, \end{cases} \quad (13.16)$$

where h satisfies hypotheses $(A1)$ and $(A2)$.

Theorem 13.5 *Suppose the system* (13.16) *is chaotic in the sense of Definition 13.1. Then h is neither an even function nor an odd function.*

13.2 Sufficient Conditions for the Onset of Chaos

Proof Set
$$\psi_\infty(x) = x + \phi^{-1}(x).$$

By similar analysis as in Sect. 13.1, it is clear that (z_x, z_t) can be represented by iterations of ψ_∞ and initial data. Hence, the dynamics of (z_x, z_t) is completely determined by ψ_∞.

If h is even, ψ_∞ is monotonically monotone. If h is odd, $-\psi_\infty$ has no periodic points of period larger than 2. Therefore, chaos does not occur in the system 13.16 if h is either an even function or an odd function. □

Remark 13.6 Chaos can definitely happen in the system (13.16) if h is properly chosen. One can find more details later.

In this section, we have shown some necessary conditions for the onset of chaos. However, if chaos does not occur for that general boundary condition, the analysis for the necessary conditions would be useless. Therefore, we are now ready to study some sufficient conditions for the occurrence of chaotic vibrations of the 1D wave equation with a general nonlinear boundary condition.

13.2 Sufficient Conditions for the Onset of Chaos

First, consider the following system:

$$\begin{cases} w_{tt}(x,t) - c^2 w_{xx}(x,t) = 0, & x \in (0,1), \ t > 0, \\ w_x(0,t) = -\eta w_t(0,t), \ \eta \neq c^{-1}, & t > 0, \\ w_x(1,t) = h(w_t(1,t)), \ t > 0, \\ w_0(x) = w(x,0), \ w_1(x) = w_t(x,0), & 0 \leq x \leq 1, \end{cases} \quad (13.17)$$

where $h \in C^0(\mathbb{R})$ satisfies $h(0) = 0$, hypotheses (A1) and (A2). When $w_t(1,t) \equiv 0$, there should be no signals feedback to $w_x(1,t)$, that's to say the right end should be free. Therefore it is reasonable to let $h(0) = 0$.

According to the analysis in Sect. 13.1, we have known that the function ψ_η given by (13.13) plays a vital role in studying the dynamics of system (13.17). We first give two lemmas that are useful in analyzing dynamics of ψ_η, as follows:

Lemma 13.7 *Let I be a nondegenerate closed interval, $J \subseteq \mathbb{R}$ and $F : I \to J, G : J \to \mathbb{R}$. Assume that $V_J G < +\infty$ and F is piecewise monotone. Then*

$$V_{F(I)} G \leq V_I G \circ F < +\infty.$$

Proof Let $I_1 = [a, b]$ be a monotone interval of F. Without loss of generality, let F be monotonically increasing in I_1. Let

be a partition of I_1. Then

$$V_{I_1}[G \circ F, P] = V_{F(I_1)}[G, P'] \leq V_{F(I_1)}G,$$

where

$$P' : F(a) = F(t_0) \leq F(t_1) \leq \cdots \leq F(t_n) = F(b)$$

is a partition of $F(I_1)$. Consequently, $V_{I_1}G \circ F \leq V_{F(I_1)}G$. Conversely, if

$$P' : F(a) = q_0 \leq q_1 \leq \cdots \leq q_n = F(b)$$

is a partition of $F(I_1)$, then

$$P : a = \tilde{F}^{-1}(q_0) \leq \tilde{F}^{-1}(q_1) \leq \cdots \leq \tilde{F}^{-1}(q_n) \leq q_{n+1} = b,$$

where $\tilde{F}^{-1}(q) = \min\{x \in I_1, F(x) = q\}$ is a partition of I_1. That implies

$$V_{F(I_1)}[G, P'] = V_{I_1}[G \circ F, P] \leq V_{I_1}G \circ F.$$

Therefore, $V_{I_1}G \circ F = V_{F(I_1)}G$.

Let $(I_j)_{1 \leq j \leq n}$ be a finite sequence consisting of monotone intervals of F. Assume $\sqcup_{1 \leq j \leq n} I_j = I$ and $Card(I_i \cap I_j) \leq 1$ provided $i \neq j$. Then

$$V_I G \circ F = \sum_{j=1}^{n} V_{I_j} G \circ F = \sum_{j=1}^{n} V_{F(I_j)} G$$

$$\geq V_{\sqcup_{1 \leq j \leq n} F(I_j)} G = V_{F(I)} G.$$

$\square$

Lemma 13.8 *Let I be a closed interval and $F \in \mathcal{C}^0(I, I)$ be piecewise monotone. If there exist nondegenerate subintervals $I_1, I_2 \subseteq A$ with $Card(I_1 \cap I_2) \leq 1$ such that $I_2 \subseteq F(I_1)$ and $I_1 \cup I_2 \subseteq F(I_2)$, then*

$$\liminf_{n \to +\infty} \frac{\ln V_{I_1 \cup I_2} F^n}{n} \geq \ln \frac{1 + \sqrt{5}}{2}. \tag{13.18}$$

Proof Let two subintervals $I_1, I_2 \subseteq A$ satisfy the hypothesis. Let $n \in \mathbb{N}$. Put

$$x_n = V_{I_2} F^n, \quad y_n = V_{I_1 \cup I_2} F^n,$$

where $F^0 = Id_{\mathbb{R}}$ if $n = 0$. Note that $F(\cdot)$ is continuous and piecewise monotone, so it follows from Lemma 13.7 that

13.2 Sufficient Conditions for the Onset of Chaos

$$x_{n+1} = V_{I_2} F^n \circ F \geq V_{F(I_2)} F^n \geq V_{I_1 \cup I_2} F^n = y_n,$$

and

$$y_{n+2} = V_{I_1} F^{n+1} \circ F + V_{I_2} F^{n+1} \circ F$$
$$\geq V_{F(I_1)} F^{n+1} + V_{F(I_2)} F^{n+1}$$
$$\geq V_{I_2} F^{n+1} + V_{I_1 \cup I_2} F^{n+1} = x_{n+1} + y_{n+1}$$
$$\geq y_n + y_{n+1}.$$

Let $(z_n)_{n \in \mathbb{N}}$ be a *Fibonacci sequence*, i.e., $z_{n+2} = z_{n+1} + z_n$, with the initial data $z_0 = 0$, $z_1 = 1$. It is well known that

$$z_n = \frac{1}{\sqrt{5}} \left[\left(\frac{1+\sqrt{5}}{2} \right)^n - \left(\frac{1-\sqrt{5}}{2} \right)^n \right], \quad n \in \mathbb{N}.$$

It is clear that $y_1 > 0$ and $\forall n \in \mathbb{N}$, $y_n \geq y_1 z_n$. Therefore,

$$\liminf_{n \to +\infty} \frac{\ln V_{I_1 \cup I_2} F^n}{n} = \liminf_{n \to +\infty} \frac{\ln y_n}{n} \geq \lim_{n \to +\infty} \frac{\ln y_1 z_n}{n} = \ln \frac{1+\sqrt{5}}{2}.$$

This competes the proof. □

Proposition 13.9 *Let $h \in C^0(\mathbb{R})$ satisfy hypotheses (A1) – (A2) and ψ_η be given by (13.13). In addition, assume $h(0) = 0$ and*

(i) *the non-constant function Q defined by $Q : t \mapsto \frac{1}{2} \left(h(t) + \frac{1}{c} t \right)$ is piecewise monotone,*
(ii) *there is at least one solution to the equation $h(y) + c^{-1} y = 0$ in $\mathbb{R}^*$, where $\mathbb{R}^* \triangleq \mathbb{R} \setminus \{0\}$.*

Then there exist $A > 0$ and a nondegenerate interval $I \subseteq (0, +\infty) \setminus \{c^{-1}\}$ such that for any $\eta \in I$, $[0, A]$ is an invariant set of ψ_η and

$$\liminf_{n \to +\infty} \frac{\ln V_{[0,A]} \psi_\eta^n}{n} \geq \ln \frac{1+\sqrt{5}}{2}. \tag{13.19}$$

Proof We need to finish the following two steps:

(1) Determine an invariant interval $[0, A]$ of ψ_η.
(2) To make sure that the hypotheses of Lemma 13.8 holds.

It has been known that
$$\psi_\eta = \gamma(\eta) \cdot \varphi = \gamma(\eta) \cdot Q \circ \phi^{-1}.$$

Denote
$$S^- = \{ y < 0 \mid Q(y) = 0 \text{ and } \exists t \in (y, 0), Q(t) \neq 0 \}$$

and
$$S^+ = \{y > 0 \mid Q(y) = 0 \text{ and } \exists\, t \in (0, y),\, Q(t) \neq 0\}.$$

By hypotheses (i)–(ii), $S^- \neq \emptyset$ or $S^+ \neq \emptyset$. Without loss of generality, assume $S^- \neq \emptyset$. We take
$$\bar{y} = \max S^-, \quad A = -c^{-1}\bar{y}. \tag{13.20}$$

Then
$$\bar{y} < 0, \ A = \phi(\bar{y}) > 0, \ \psi_\eta(A) = 0. \tag{13.21}$$

Moreover, let
$$M = \max\{Q(t) \mid t \in [\bar{y}, 0]\}, \quad m = \min\{Q(t) \mid t \in [\bar{y}, 0]\}.$$

By the definition of S^-, it is evident that $M = 0$ or $m = 0$.

Case 1: Assume $M = 0$. This implies $m < 0$. Let
$$y_0 = \max\{t \in [\bar{y}, 0] \mid Q(t) = m\}, \quad x_0 = \phi(y_0).$$

For any $\eta \in [0, c^{-1})$, it is easily seen that
$$\psi_\eta(x_0) = \max\{\psi_\eta(x) \mid x \in [0, A]\} = m \cdot \gamma(\eta)$$

and $\psi_\eta(x) < \psi_\eta(x_0)$ provided that $x \in [0, x_0)$. We need to prove that $m \cdot \gamma(0) < A$. Proceed the proof by contradiction. Assume $m \cdot \gamma(0) \geq A$, which implies
$$\psi_0([0, x_0]) \cap \psi_0([x_0, A]) \supseteq [0, \psi_0(x_0)] \supseteq [0, A] = [0, x_0] \cup [x_0, A].$$

Thus, ψ_0 is turbulent. According to Lemma 3 in [10], ψ_0 has periodic points of all periods. But from the proof of Theorem 13.4, ψ_0 has no periodic points whose period is larger than 2. We have thus reached a contradiction. Since that $m \cdot \gamma(\cdot)$ is strictly increasing in $[0, c^{-1})$ and $m \cdot \gamma(\eta) \to +\infty$ as $\eta \to (c^{-1})^-$, it is reasonable to find
$$\bar{\eta} \triangleq \gamma^{-1}\left(\frac{m}{A}\right) \in (0, c^{-1}). \tag{13.22}$$

Then for $\eta \in [0, \bar{\eta}]$, we have
$$\forall x \in [0, A], \ 0 \leq \psi_\eta(x) \leq A,$$

that is to say, $[0, A]$ is an invariant set of ψ_η provided $\eta \in [0, \bar{\eta}]$. Let
$$\underline{\eta} = \begin{cases} \gamma^{-1}\left(\frac{m}{x_0}\right), & \text{if } \frac{m}{x_0} < -1, \\ 0, & \text{else,} \end{cases} \tag{13.23}$$

then $\underline{\eta} < \bar{\eta}$ and $\psi(x_0) > x_0$ if $\eta > \underline{\eta}$. For $\eta \in [\underline{\eta}, c^{-1})$, consider the following set:

13.2 Sufficient Conditions for the Onset of Chaos

$$S(\eta, x_0) = \{x \in (0, x_0) \mid \gamma(\eta) \cdot \varphi(x) = x_0\}.$$

It is clear that $S(\eta, x_0)$ is closed. Since x_0 is uniquely determined by h and h is independent of η, the following function is well defined:

$$\alpha(\eta) \triangleq \max S(\eta, x_0), \quad \eta \in [\underline{\eta}, c^{-1}). \tag{13.24}$$

We first prove that $\alpha(\cdot)$ is strictly monotonically decreasing in $[\underline{\eta}, c^{-1})$. Let $t_1, t_2 \in (\underline{\eta}, c^{-1})$ with $t_1 < t_2$. From the definition of $\alpha(\cdot)$, it follows

$$\begin{aligned} \gamma(t_1) \cdot \varphi(\alpha(t_1)) &= x_0, \quad \gamma(t_1) \cdot \varphi(x) > x_0, \quad x \in (\alpha(t_1), x_0], \\ \gamma(t_2) \cdot \varphi(\alpha(t_2)) &= x_0, \quad \gamma(t_2) \cdot \varphi(x) > x_0, \quad x \in (\alpha(t_2), x_0]. \end{aligned} \tag{13.25}$$

Since

$$\gamma(t_1) \cdot \varphi(\alpha(t_2)) < \gamma(t_2) \cdot \varphi(\alpha(t_2)) = x_0, \quad \gamma(t_1) \cdot \varphi(x_0) > x_0,$$

by applying the continuity of $\gamma(t_1) \cdot \varphi(\cdot)$ there exists $x \in (\alpha(t_2), x_0)$ such that $\gamma(t_1) \cdot \varphi(x) = x_0$. According to the definition of $\alpha(t_1)$, we have

$$\alpha(t_2) < x \leq \alpha(t_1).$$

Therefore, $\alpha(\cdot)$ is strictly monotonically decreasing on $[\underline{\eta}, c^{-1})$.

Next, we prove that $\alpha(\cdot)$ is continuous from the left. Let $t_0 \in (\underline{\eta}, x_0)$ be fixed. Since α is monotone, we have

$$\alpha(t_0) \leq L \stackrel{def}{=} \lim_{t \to t_0^-} \alpha(t) < +\infty. \tag{13.26}$$

By the continuity of $\eta \mapsto \gamma(\eta)$ and $x \mapsto \varphi(x)$, we obtain

$$\lim_{\varepsilon \to 0^+} \gamma(t_0 - \varepsilon) = \gamma(t_0), \quad \lim_{\varepsilon \to 0^+} \varphi(\alpha(t_0 - \varepsilon)) = \varphi(L), \tag{13.27}$$

which gives $\gamma(t_0) \cdot \varphi(L) = \gamma(t_0) \cdot \varphi(\alpha(t_0)) = x_0$. From the definition of $\alpha(\cdot)$ and (13.26), it follows that

$$L \leq \alpha(t_0) \leq L = \lim_{t \to t_0^-} \alpha(t).$$

Therefore, $\alpha(\cdot)$ is continuous from the left.

In particular, $\psi_{\bar{\eta}}(\alpha(\underline{\eta})) = \gamma(\underline{\eta}) \cdot \varphi(\alpha(\bar{\eta})) = x_0$ and $\psi_{\bar{\eta}}(x_0) = \gamma(\underline{\eta}) \cdot \varphi(x_0) = A$. Define a function as follows:

$$K(s) = \alpha(\underline{\eta} - s) - \psi^2_{\bar{\eta}-s}(x_0). \tag{13.28}$$

It is easily seen that $K(\cdot)$ is continuous from the right at $s = 0$. Note that

$$K(0) = \alpha(\underline{\eta}) - \psi_{\bar{\eta}}(\psi_{\bar{\eta}}(x_0)) = \alpha(\underline{\eta}) - \psi_{\bar{\eta}}(A) = \alpha(\underline{\eta}) > 0,$$

hence there exists $\rho_0 \in (0, \bar{\eta} - \underline{\eta})$ such that

$$\forall s \in [0, \rho_0], \ K(s) = \alpha(\bar{\eta} - s) - \psi^2_{\bar{\eta}-s}(x_0) > 0. \tag{13.29}$$

Denote

$$I = [\bar{\eta} - \rho_0, \bar{\eta}] \tag{13.30}$$

and

$$J_1 = [\alpha(\eta), x_0], \ J_2 = [x_0, \psi_\eta(x_0)], \ \eta \in I. \tag{13.31}$$

Let $\eta \in I$ be fixed. By (13.29),

$$\psi_\eta(J_1) \supseteq [\psi_\eta(\alpha(\eta)), \psi_\eta(x_0)] = J_2,$$

$$\psi_\eta(J_2) \supseteq [\psi^2_\eta(x_0), \psi_\eta(x_0)] \supseteq [\alpha(\eta), \psi_\eta(x_0)] \supseteq J_1 \cup J_2,$$

which is to say hypotheses of Lemma 13.8 hold. Therefore, for any $\eta \in I$, by applying Lemma 13.8 we have

$$\liminf_{n \to +\infty} \frac{\ln V_{[0,A]} \psi^n_\eta}{n} \geq \liminf_{n \to +\infty} \frac{\ln V_{J_1 \cup J_2} \psi^n_\eta}{n} \geq \ln \frac{1 + \sqrt{5}}{2}. \tag{13.32}$$

For the case $m = 0$, one just needs to consider $\eta > c^{-1}$. The proof is similar, so we omit it. □

Theorem 13.10 *Consider the system* (13.17) *with* (A1) − (A2). *Assume* $h(0) = 0$ *and*

(i) *the non-constant function* $Q : t \mapsto \frac{1}{2}\left(h(t) + \frac{1}{c}t\right)$ *is piecewise monotone;*
(ii) *there is at least one solution to the equation* $h(y) + c^{-1}y = 0$ *in* $\mathbb{R}^*$.

Then there exists a nondegenerate interval I *such that for all* $\eta \in I$ *the system* (13.17) *is chaotic in the sense of Definition 13.1.*

Proof Without loss of generality, assume the equation $h(y) + c^{-1}y = 0$ has at least one solution in $(-\infty, 0)$. We continue to use the symbols given in the proof of Proposition 13.9, such as A, x_0, $\alpha(\eta)$ and so on. Let I and J_1, J_2 be given by (13.30) and (13.31), respectively. Take $\eta \in I$.

Let (w_0, w_1) be the initial data of system (13.17) and (w, w_x, w_t) be a solution of system (13.17). Recall the *Riemann invariants* as follows:

$$u(x, t) = \frac{cw_x(x, t) + w_t(x, t)}{2c}, \ v(x, t) = \frac{cw_x(x, t) - w_t(x, t)}{2c}, \ x \in [0, 1], \ t \geq 0,$$

and $u_0(\cdot) = u(\cdot, 0)$ and $v_0(\cdot) = v(\cdot, 0)$ are the initial data. Let $t > 2c^{-1}$, $n = [\frac{ct}{2}]$, and $\tau = t - 2nc^{-1}$ be fixed. It is clear that $0 \leq c\tau < 2$. By using (13.10), (13.11), and (13.12), we obtain

13.2 Sufficient Conditions for the Onset of Chaos

$$u(x,t) = \begin{cases} \gamma^{-1}(\eta) \circ \psi_\eta^n \left(\gamma(\eta) \cdot u_0(x + c\tau)\right), & 0 \leq c\tau \leq 1 - x, \\ \gamma^{-1}(\eta) \circ \psi_\eta^{n+1} \left(v_0(2 - x - c\tau)\right), & 1 - x < c\tau \leq 2 - x, \\ \gamma^{-1}(\eta) \circ \psi_\eta^n \left(\gamma(\eta) \cdot u_0(x + c\tau - 2)\right), & 2 - x < c\tau < 2, \end{cases} \quad (13.33)$$

and

$$v(x,t) = \begin{cases} \psi_\eta^n \left(v_0(x - c\tau)\right), & 0 \leq c\tau \leq x, \\ \psi_\eta^n \left(\gamma(\eta) \cdot u_0(c\tau - x)\right), & x < c\tau \leq x + 1, \\ \psi_\eta^{n+1} \left(v_0(2 + x - c\tau)\right), & x + 1 < c\tau < 2. \end{cases} \quad (13.34)$$

Since $[0, A]$ is an invariant set of ψ_η, $|u| + |v|$ is uniformly bounded if

$$Range(\gamma(\eta) \cdot u_0) \cup Range(v_0) \subseteq [0, A]. \quad (13.35)$$

In addition, assume that u_0 and v_0 are piecewise monotone and

$$[0, \alpha(\eta)] \subseteq Range(\gamma(\eta) \cdot u_0) \cap Range(v_0). \quad (13.36)$$

Consider the total variations of $u(\cdot, t)$ and $v(\cdot, t)$ on $[0, 1]$. When $0 \leq c\tau < 1$, from (13.10)–(13.12) and Lemma 13.7, it follows that

$$V_{[0,1]}u(\cdot, t) = V_{[0,1-c\tau]}u(\cdot, t) + V_{[1-c\tau,1]}u(\cdot, t)$$
$$= V_{[c\tau,1]}\gamma^{-1}(\eta) \circ \psi_\eta^n \circ \gamma(\eta) \circ u_0 + V_{[1-c\tau,1]}\gamma^{-1}(\eta) \circ \psi_\eta^{n+1} \circ v_0$$
$$= \left|\gamma^{-1}(\eta)\right| \left(V_{[c\tau,1]}\psi_\eta^n \circ \gamma(\eta) \circ u_0 + V_{[1-c\tau,1]}\psi_\eta^{n+1} \circ v_0\right),$$

$$V_{[0,1]}v(\cdot, t) = V_{[0,c\tau]}v(\cdot, t) + V_{[c\tau,1]}v(\cdot, t)$$
$$= V_{[0,c\tau]}\psi_\eta^n \circ \gamma(\eta) \circ u_0 + V_{[0,1-c\tau]}\psi_\eta^n \circ v_0.$$

Note that $0 < \left|\gamma^{-1}(\eta)\right| < 1$, by Lemma 13.7 and (13.36), we obtain

$$V_{[0,1]}u(\cdot, t) + V_{[0,1]}v(\cdot, t) \geq \left|\gamma^{-1}(\eta)\right| V_{\gamma(\eta) \cdot u_0([0,1])}\psi_\eta^n$$
$$\geq \left|\gamma^{-1}(\eta)\right| V_{[0,\alpha(\eta)]}\psi_\eta^n \geq \left|\gamma^{-1}(\eta)\right| V_{\psi_\eta^2[0,\alpha(\eta)]}\psi_\eta^{n-2}$$
$$\geq \left|\gamma^{-1}(\eta)\right| V_{J_1 \cup J_2}\psi_\eta^{n-2}.$$

Then Propositions 13.9 and (13.32) show that

$$\frac{\ln\left(V_{[0,1]}u(\cdot, t) + V_{[0,1]}v(\cdot, t)\right)}{t} \geq \frac{\ln\left|\gamma^{-1}(\eta)\right|}{t} + \frac{\ln\left(V_{J_1 \cup J_2}\psi_\eta^{n-2}\right)}{t}$$
$$\geq \frac{c \ln\left(V_{J_1 \cup J_2}\psi_\eta^{n-2}\right)}{2n + 2} + \frac{\ln\left|\gamma^{-1}(\eta)\right|}{t} \quad (13.37)$$
$$\geq \frac{c}{2} \ln \frac{1 + \sqrt{5}}{2}, \quad \text{as } t \to +\infty.$$

When $1 \leq c\tau < 2$, in the same way, we can obtain

$$\frac{\ln\left(V_{[0,1]}u(\cdot,t) + V_{[0,1]}v(\cdot,t)\right)}{t} \geq \frac{c\ln\left(V_{J_1\cup J_2}\psi_\eta^{n-1}\right)}{2n+2} + \frac{\ln|\gamma^{-1}(\eta)|}{t} \quad (13.38)$$

$$\geq \frac{c}{2}\ln\frac{1+\sqrt{5}}{2}, \text{ as } t \to +\infty.$$

It is evident that

$$V_{[0,1]}u(\cdot,t) + V_{[0,1]}v(\cdot,t) \leq \max\{1, c^{-1}\}\left(V_{[0,1]}w_x(\cdot,t) + V_{[0,1]}w_t(\cdot,t)\right). \quad (13.39)$$

Hence

$$\liminf_{t \to +\infty} \frac{\ln\left(V_{[0,1]}w_x(\cdot,t) + V_{[0,1]}w_t(\cdot,t)\right)}{t} \geq \frac{c}{2}\ln\frac{1+\sqrt{5}}{2} > 0.$$

Therefore, the system (13.17) is chaotic in the sense of Definition 13.1. □

We next consider the oscillation problem governed by a 1D wave equation with a fixed end. We will show that only the effect of self-regulation effect at one of the two ends can cause the onset of chaos. But if the fixed end is replaced by a free end, the system never has chaos. As we know, both the fixed end and the free end are energy-conserving boundary condition. Even though either a fixed end or a free end has the same effect to the energy of the system, the systems may show completely different dynamics: one with chaos, the other without chaos. Thus the relationship between chaos and energy of the system is much more complicated.

Theorem 13.11 *Consider an initial-boundary problem governed by the following 1D wave equation with a fixed end:*

$$\begin{cases} z_{tt}(x,t) - c^2 z_{xx}(x,t) = 0, & x \in (0,1), \ t > 0, \\ z(0,t) = 0, \ t > 0, \\ z_x(1,t) = h(z_t(1,t)), \ t > 0, \\ z_0(x) = z(x,0), \ z_1(x) = z_t(x,0), \ 0 \leq x \leq 1. \end{cases} \quad (13.40)$$

We can choose a piecewise affine map h such that the system is chaotic in the sense of Definition 13.1.

Proof Without loss of generality, we take $c = 1$. Define

$$h_\theta(x) = \begin{cases} -2x + 2(\theta^2 - 3), & x \in [2(\theta^2 - 3), -2), \\ (1 - \theta^2)x, & x \in [-2, 0], \\ (\theta - 1)x, & x \in (0, 1], \\ -2x + (\theta + 1), & x \in (1, \theta + 1], \\ -x, & \text{else,} \end{cases} \quad (13.41)$$

13.2 Sufficient Conditions for the Onset of Chaos

where $\theta \in (0, 1)$ is a fixed parameter. It is clear that $h_\theta \in C^0(\mathbb{R})$ and

$$\phi_\theta : x \mapsto \frac{1}{2}(h_\theta(x) - x)$$

is strictly monotonically decreasing and $\phi_\theta(\mathbb{R}) = \mathbb{R}$. Then ϕ_θ^{-1} exists. Let

$$Q_\theta(y) = \frac{1}{2}(h_\theta(y) + y).$$

Then

$$\psi_\theta(x) = x + \phi_\theta^{-1}(x) = Q_\theta \circ \phi_\theta^{-1}(x).$$

Let $u = \frac{z_x + z_t}{2}$ and $u = \frac{z_x - z_t}{2}$ be the *Riemann invariants* of system (13.40). For some initial date (u_0, v_0), suppose (u, v) is solved for $t \le 2$. Then for $t = 2n + \tau$ with $\tau \in [0, 2)$ and $n \in \mathbb{N}$, we have

$$u(x, t) = \psi_\theta^n(u(x, \tau)), \quad v(x, t) = \psi_\theta^n(v(x, \tau)), \quad x \in [0, 1]. \tag{13.42}$$

Set

$$I_1 = \left[-\theta - 1, \frac{1}{2}\theta - 1\right], \quad I_2 = \left[\frac{1}{2}\theta - 1, 0\right], \quad I_3 = \left[0, \theta^2\right].$$

Each of them is a monotone intervals of ψ_θ. A simple calculation shows that

$$\psi_\theta(-\theta - 1) = 0, \quad \psi_\theta(\frac{1}{2}\theta - 1) = \frac{1}{2}\theta, \quad \psi_\theta(0) = 0, \quad \psi_\theta(\theta^2) = \theta^2 - 2.$$

Let $\theta \in (0, 1)$ be sufficiently small such that $\frac{1}{2}\theta \ge \theta^2$ and $\theta^2 - 2 \le -\theta - 1$, which implies

$$I_3 \subseteq \psi_\theta(I_1) \cap \psi_\theta(I_2), \quad I_1 \cup I_2 \subseteq \psi_\theta(I_3).$$

Let $n \in \mathbb{N}$ and

$$x_n = V_{I_1}\psi_\theta^n, \quad y_n = V_{I_2}\psi_\theta^n, \quad z_n = V_{I_3}\psi_\theta^n.$$

Lemma 13.8 shows that

$$x_{n+1} \ge V_{\psi_\theta(I_1)}\psi_\theta^n \ge V_{I_3}\psi_\theta^n = z_n, \quad y_{n+1} \ge z_n, \quad z_{n+2} \ge x_{n+1} + y_{n+1},$$

which implies

$$z_{n+2} \ge 2z_n.$$

Inductively, we obtain

$$z_{2n+2} \ge 2^n z_2 > 0, \quad z_{2n+1} \ge 2^n z_1 > 0.$$

Therefore,

$$\liminf_{n \to +\infty} \frac{\ln z_n}{n} \ge \frac{1}{2}\ln 2 > 0. \tag{13.43}$$

A simple calculation shows that

$$\psi_\theta(x) = 0, \ x \in (-\infty, -\theta - 1] \cup [2(3 - \theta^2), +\infty].$$

By (13.42), $|u| + |v|$ is always uniformly bounded for any initial date. Analogous to the proof of Theorem 13.10, there exists a large class of initial data such that

$$\liminf_{t \to +\infty} \frac{\ln \left(V_{[0,1]} z_x(\cdot, t) + V_{[0,1]} z_t(\cdot, t) \right)}{t} \geq \frac{1}{4} \ln 2 > 0.$$

Therefore, the proof is complete. □

13.3 Applications

As we have stated in the previous sections, chaos never happen if the feedback control law is even. In particular, consider the system described by the following model:

$$\begin{cases} z_{tt}(x, t) - z_{xx}(x, t) = 0, \ x \in (0, 1), \ t > 0, \\ z_x(0, t) = -\eta z_t(0, t), \ \eta \neq 1, \text{ or } z(0, t) = 0, \ t > 0, \\ z_z(1, t) = \cos(z_t(1, t)) - 1, \ t > 0, \\ z_0(x) = z(x, 0), \ z_1(x) = z_t(x, 0), \ 0 \leq x \leq 1. \end{cases} \quad (13.44)$$

The feedback control law at the right end is periodic and even. Therefore, there is no chaos in the system (13.44). Moreover, this system is neither asymptotically stable, nor stable in the sense of Lyapunov. An interesting question is how the system dynamics behaves of the above system.

Our second example is a problem with perturbations at boundaries, as described by the following system:

$$\begin{cases} w_{tt}(x, t) - c^2 w_{xx}(x, t) = 0, \ x \in (0, 1), \ t > 0, \\ w_x(0, t) = -\eta w_t(0, t), \ \eta \neq c^{-1}, \ t > 0, \\ w_x(1, t) = -c^{-1} w_t(1, t) + \varepsilon \sin(w_t(1, t)), \ t > 0, \\ w_0(x) = w(x, 0), \ w_1(x) = w_t(x, 0), \ 0 \leq x \leq 1. \end{cases} \quad (13.45)$$

If $\varepsilon = 0$, for any initial data (w_0, w_1) and $\eta \neq c^{-1}$, by using the wave propagation method, we can obtain

$$\forall t > 2, x \in [0, 1], \ w_x(x, t) = w_t(x, t) = 0.$$

This shows that system (13.45) has global asymptotic stability.

Assume $\varepsilon \neq 0$ and $|\varepsilon|$ is sufficiently small. Define

$$h_\varepsilon(y) = -c^{-1} y + \varepsilon \sin(y).$$

13.4 Numerical Simulations

It is evident that (i) $y \mapsto h_\varepsilon(y) + c^{-1}y$ is piecewise monotone; (ii) $y \mapsto h_\varepsilon(y) - c^{-1}y$ is strictly monotonically decreasing; and (iii) there are infinite solutions to the equation $h_\varepsilon(y) + c^{-1}y = 0$ in $\mathbb{R}^*$. These show the hypotheses of Theorem 13.10 hold. Therefore, system (13.45) is chaotic in the sense of Definition 13.1.

The above analysis shows that this kind of system is no longer stable under small perturbations at one of the boundaries. That is to say, this kind of system does not have structural stability.

13.4 Numerical Simulations

In this section, we will give some numerical simulations to validate the theoretical results of this chapter. We first consider the following system:

$$\begin{cases} z_{tt}(x,t) - z_{xx}(x,t) = 0, & x \in (0,1), \ t > 0, \\ z(0,t) = 0, \ t > 0, \\ z_x(1,t) = h(z_t(1,t)), \ t > 0, \\ z_0(x) = z(x,0), \ z_1(x) = z_t(x,0), \ 0 \le x \le 1, \end{cases} \quad (13.46)$$

where

$$h(x) = \begin{cases} -2x - \frac{11}{2}, & x \in [-\frac{11}{2}, -2), \\ \frac{3}{4}x, & x \in [-2, 0], \\ -\frac{1}{2}x, & x \in (0, 1], \\ -2x + \frac{3}{2}, & x \in (1, \frac{3}{2}], \\ -x, & \text{else.} \end{cases} \quad (13.47)$$

Choose

$$w_0 = 0, \ w_1(x) = \frac{1}{4} \sin^4(2\pi x), \ x \in [0,1],$$

as the initial data of system (13.46). Set

$$\psi_\infty(x) = \begin{cases} \frac{1}{3}x - \frac{11}{6}, & x \in (\frac{1}{4}, \frac{11}{2}], \\ -7x, & x \in (0, \frac{1}{4}], \\ -\frac{1}{3}x, & x \in (-\frac{3}{4}, 0], \\ \frac{1}{3}x + \frac{1}{2}, & x \in [-\frac{3}{2}, -\frac{3}{4}], \\ 0, & \text{else.} \end{cases} \quad (13.48)$$

The solution (z, z_x, z_t) can be represented by the iterations of this ψ_∞ and initial data (z_0, z_1). Note that h and ψ_∞ are piecewise affine, thus the feedback control at the boundaries is easy to implement and the solutions of system are relatively simple to be solved and represented.

We present the graphics in some detail, for z_x, z_t for $98 \le t \le 100$. Figures 13.1 and 13.2 show that z_x, z_t are extremely oscillatory at all points in space and time.

Fig. 13.1 The profile of z_x for $t \in [98, 100]$. (Reprinted from [5, p. 2616, Fig. 1])

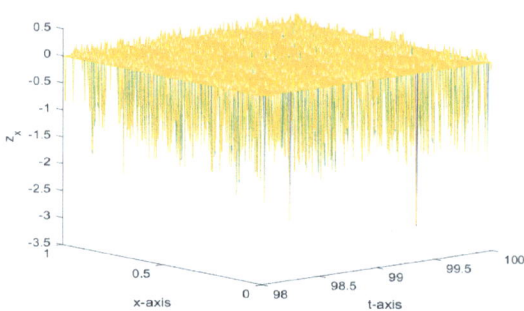

Fig. 13.2 The profile of z_t for $t \in [98, 100]$. (Reprinted from [5, p. 2616, Fig. 1])

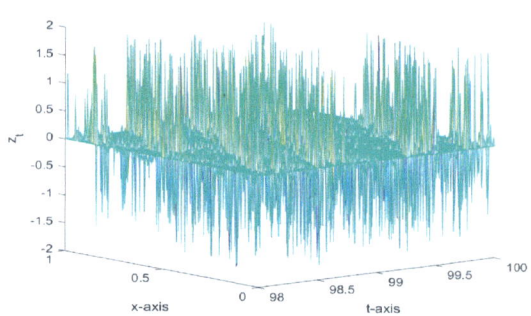

References

1. G. Chen, S.B. Hsu and J. Zhou, Chaotic vibrations of the one-dimensional wave equation due to a self-excitation boundary condition. Part II, energy injection, period doubling and homoclinic orbits, *Int. J. Bifur. Chaos* **8** (1998), 423–445. https://doi.org/10.1142/S0218127498001236
2. Y. Huang, Growth rates of total variations of snapshots of the 1D linear wave equation with composite nonlinear boundary reflection, *Int. J. Bifur. Chaos* **13** (2003), 1183–1195. https://doi.org/10.1142/S0218127403007138
3. Y. Huang, A new characterization of nonisotropic chaotic vibrations of the one-dimensional linear wave equation with a van der Pol boundary condition, *J. Math. Anal. Appl.* **288** (2003), no. 1, 78–96. https://doi.org/10.1016/S0022-247X(03)00562-6
4. P.C. Etter, Underwater Acoustic Modeling and Simulation, *Spon Press*, London, New Yoek, 2003.
5. L.L. Li, Chaotic oscillations of 1D wave equation with or without energy-injections[J]. Electronic Research Archive, 30(7): 2600–2617, 2022. https://doi.org/10.3934/era.2022133
6. T.Y. Li and J.A. Yorke, Period three implies chaos, *Amer. Math. Monthly* **82** (1975), 985–992. https://doi.org/10.2307/2318254
7. Mata Štefánková, Inheriting of chaos in uniformly convergent nonautonomous dynamical systems on the interval, *Discrete & Continuous Dynamical Systems*, **36(6)**, pp. 3435–3443, 2016
8. Jose S. Cánocas, Tönu Puu, Manuel Ruiz Marín, Detecting chaos in a duopoly model via symbolic dynamics, *Discrete & Continuous Dynamical Systems-B*, **13(2)**, pp. 269–278, 2010.

9. G. Chen, T. Huang and Y. Huang, Chaotic behavior of interval maps and total variations of iterates, *Int. J. Bifur. Chaos* **14** (2004), 2161–2186. https://doi.org/10.1142/S0218127404010242
10. L. Block, Homoclinic points of mappings of the interval, *Proc. Amer. Math. Soc.* **72** (1978), 576–580. https://doi.org/10.1090/S0002-9939-1978-0509258-X

Chaotic Vibrations of a Multi-Dimensional Wave Equation

Much of the existing, rigorous proofs for the occurrence of chaos in PDEs are in one space dimension. Very few results are known for problems in multi-dimensional settings. The study of chaotic vibrations for multi-dimensional PDEs due to nonlinear boundary conditions is challenging. Not only the equation itself but the geometric shape of the domain matters in analyzing the complex dynamics.

For hyperbolic PDEs, the ray tracing method is mostly ineffective in multi-dimensional space. Thus, boundary reflection relations become too complicated. In this part, we show several methods to deal with the oscillation problems of multi-dimensional PDEs, say 3D or 2D wave equations, namely,

$$\frac{1}{c^2}\frac{\partial^2 w(x,t)}{\partial t^2} - \nabla^2 w(x,t) = 0, \ x \in \Omega \subseteq \mathbb{R}^n, \ t > 0, \tag{14.1}$$

where $n = 2$ or $n = 3$.

Case I: We first think about the case $n = 3$ in (14.1) with certain spherical geometry. Let Ω be a 3D annular domain bounded by two concentric spherical surfaces

$$\Omega = \{X \in \mathbb{R}^3 \mid r_1 < |X| < r_2\}.$$

Thanks to the independence of the polar and azimuth angles θ and ϕ, the *Laplacian* in spherical coordinates is reduced to

$$\nabla^2 = \frac{\partial^2}{\partial r^2} + \frac{2}{r}\frac{\partial}{\partial r} = \frac{1}{r^2}\frac{\partial}{\partial r}\left(r^2 \frac{\partial}{\partial r}\right). \tag{14.2}$$

Then,

$$\frac{1}{c^2}\frac{\partial^2}{\partial t^2} - \nabla^2 = \frac{1}{c^2}\frac{\partial^2}{\partial t^2} - \frac{1}{r^2}\frac{\partial}{\partial r}\left(r^2\frac{\partial}{\partial r}\right)$$
$$= \left[\frac{1}{c}\frac{\partial}{\partial t} - \left(\frac{\partial}{\partial r} + \frac{1}{r}\right)\right]\left[\frac{1}{c}\frac{\partial}{\partial t} + \left(\frac{\partial}{\partial r} + \frac{1}{r}\right)\right] \quad (14.3)$$
$$= \mathcal{L}_1 \circ \mathcal{L}_2,$$

i.e., the wave operator factors into a product analogously as the factoring of

$$\frac{1}{c^2}\frac{\partial^2}{\partial t^2} - \frac{\partial^2}{\partial x^2} = \left(\frac{1}{c}\frac{\partial}{\partial t} - \frac{\partial}{\partial x}\right)\left(\frac{1}{c}\frac{\partial}{\partial t} + \frac{\partial}{\partial x}\right) \quad (14.4)$$

for the 1D wave equation. For the next step, we need to introduce two new variables, so-called *Riemann invariants* as mentioned in Appendix B, as follows:

$$u(r,t) = \mathcal{L}_1(w) = \frac{1}{c}w_t(x,t) - (\frac{\partial}{\partial r}w(x,t) + \frac{1}{r}w(x,t))$$

and

$$v(r,t) = \mathcal{L}_2(w) = \frac{1}{c}w_t(x,t) + (\frac{\partial}{\partial r}w(x,t) + \frac{1}{r}w(x,t)).$$

Then we obtain two transport equations

$$\mathcal{L}_2(u) = 0, \quad \mathcal{L}_1(v) = 0. \quad (14.5)$$

To solve Eq. (14.5), we need another two new variables

$$\tilde{u} = ru, \quad \tilde{v} = rv.$$

Straightforward calculations yield

$$\frac{1}{c}\frac{\partial \tilde{u}}{\partial t} + \frac{\partial \tilde{u}}{\partial r} = r\mathcal{L}_2(u) = 0$$

and

$$\frac{1}{c}\frac{\partial \tilde{v}}{\partial t} - \frac{\partial \tilde{v}}{\partial r} = r\mathcal{L}_1(v) = 0.$$

They imply that $\tilde{u}$ and $\tilde{v}$ have the following form, respectively:

$$\tilde{u} = L(r+t), \quad \tilde{v} = R(r-t),$$

for some one-dimensional maps L and R, which are uniquely determined by the initial data and the boundary conditions at $r = r_1$ and $r = r_2$. In fact, if we set

$$\tilde{w} = rw, \quad (14.6)$$

then
$$\frac{1}{c^2}\frac{\partial^2 \tilde{w}}{\partial t^2} - \frac{\partial^2 \tilde{w}}{\partial r^2} = r\mathcal{L}_1 \circ \mathcal{L}_2(w) = 0. \tag{14.7}$$

Therefore, the 3D wave equation is reduced to a 1D problem on the interval (r_1, r_2). Consequently, the 1D methodology can be extended to cover the case of the 3D angular-independent wave equation

$$\frac{1}{c^2}\frac{\partial^2 w(x,t)}{\partial t^2} - \frac{1}{r^2}\frac{\partial}{\partial r}\left(r^2 \frac{\partial w(x,t)}{\partial r}\right) = 0, \ r_1 < r < r_2, \ t > 0,$$

on an annular domain. However, this method would fail when Ω is no longer an annular domain, say Ω is a rectangular domain.

Case II: When $n \geq 1$ and $n \neq 3$, the method introduced in Case I cannot be applied to n-dimensional wave equation. Take $n = 2$ as an example. In fact, under the assumption of independence of the angular, the *Laplacian* in polar coordinates is reduced to

$$\nabla^2 = \frac{\partial^2}{\partial r^2} + \frac{1}{r}\frac{\partial}{\partial r}.$$

However, the wave operator

$$\frac{1}{c^2}\frac{\partial^2}{\partial t^2} - \left(\frac{\partial^2}{\partial r^2} + \frac{1}{r}\frac{\partial}{\partial r}\right)$$

does not admit a factoring as (14.3) such as the 3D case. For a general case of $n \geq 2$, assume Ω is a generalized annular domain:

$$\Omega = \{X \in \mathbb{R}^n \mid r_1 < |X| < r_2\}.$$

Due to the independence of angular variable, the *Laplacian* in n-dimensional spherical coordinates is reduced to

$$\nabla^2 = \frac{\partial^2}{\partial r^2} + \frac{n-1}{r}\frac{\partial}{\partial r}. \tag{14.8}$$

It is easy to see that the wave operator

$$\frac{1}{c^2}\frac{\partial^2}{\partial t^2} - \left(\frac{\partial^2}{\partial r^2} + \frac{n-1}{r}\frac{\partial}{\partial r}\right)$$

admits a factoring as (14.3) if, and only if, $n = 3$. In other words, this method of reduction of dimensionality is only valid for the 3D wave equations.

Through reviewing the discussions in the preceding paragraph, several key elements have come to our attention:

(i) The operator-factoring method and the resulting Riemann invariants offer a useful idea for the possibility of treating multi-dimensional problems.

(ii) The geometry/shape of multi-dimensional domains poses a new challenge. For example, if the boundary of the annular domain Ω was formed by circles that are *not concentric*, then the method will not work.

(iii) The dimension n of the domain also matters.

Motivated by the above understanding, especially item **(i)**, we here begin to study a *second-order factorizable* partial differential equation in 2D in the form:

$$\left(\frac{\partial}{\partial t} + a\frac{\partial}{\partial x} + b\frac{\partial}{\partial y}\right)\left(\frac{\partial}{\partial t} + c\frac{\partial}{\partial x} + d\frac{\partial}{\partial y}\right) w(x, y, t) = 0. \tag{14.9}$$

Nevertheless, on a bounded domain we again face challenging and complicated ray tracing and reflection situations as the coefficients a, b, c, and d are somewhat arbitrary (as long as the two 3D vectors $(a, b, 1)$ and $(c, d, 1)$ are linearly independent). Equation (14.9) has two directions for its rays:

$$\ell_1 = (a, b, 1), \quad \ell_2 = (c, d, 1),$$

i.e., straight lines in the (x, y, t)-space satisfying, respectively,

$$\frac{x - x_1}{a} = \frac{y - y_1}{b} = \frac{t - t_1}{1} = \sigma, \quad \frac{x - x_2}{c} = \frac{y - y_2}{d} = \frac{t - t_2}{1} = \tau, \quad (\sigma, \tau) \in \mathbb{R}^2, \tag{14.10}$$

for any given (x_1, y_1, t_1), (x_2, y_2, t_2). For a general 2D bounded domain Ω with boundary Γ, a ray can travel an *arbitrarily short time* before it hits the boundary $\Gamma \times \mathbb{R}_+$, while other rays may not have hit any boundary point on $\Gamma \times \mathbb{R}_+$ at all. Also, the formation of *foci* (where many rays converge) and also possibly that of *caustics* causing the development of *singularities* is a real concern of the involved technical complexity.

In view of the above, we concede to consider only a special case of (14.9), namely,

$$\left(\frac{\partial}{\partial t} - \frac{\partial}{\partial x} - \frac{\partial}{\partial y}\right)\left(\frac{\partial}{\partial t} + \frac{\partial}{\partial x} + \frac{\partial}{\partial y}\right) w(x, y, t) = 0, \quad (x, y) \in \Omega, \ t > 0, \tag{14.11}$$

as the governing equation. The above leads to

$$w_{tt} - \nabla^2 w - 2w_{xy} = 0, \quad \left(\nabla^2 = \partial_x^2 + \partial_y^2\right). \tag{14.12}$$

It has the advantage that it somehow *resembles* the wave equation.

As noted in item **(ii)** previously, in order to make our problem tractable, the choice of Ω is important. Here we choose

$$\Omega = \{(x, y) \in \mathbb{R}^2 \mid -1 < x + y < 1, \ -1 < x - y < 1\}. \tag{14.13}$$

Its boundary Γ, where $\Gamma = \partial\Omega$, consists of four parts:

$$\Gamma = \Gamma_1 \cup \Gamma_2 \cup \Gamma_3 \cup \Gamma_4,$$

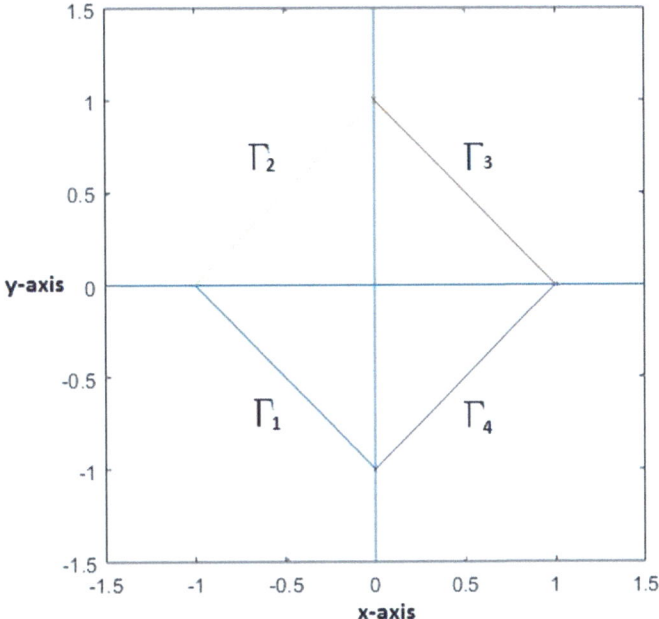

Fig. 14.1 The domain Ω

where
$$\Gamma_1 = \{(x, y) \in \Omega \mid x + y = -1\}, \quad \Gamma_2 = \{(x, y) \in \Omega \mid x - y = -1\},$$
$$\Gamma_3 = \{(x, y) \in \Omega \mid x + y = 1\}, \quad \Gamma_4 = \{(x, y) \in \Omega \mid x - y = 1\},$$

see Fig. 14.1.

Remark 14.1 In order to make the *method of characteristics* and *ray tracing* work for a generally given bounded convex domain Ω, what is needed is the following assumption. For any given ray satisfying (14.10), let $P_1 = (a_1, b_1, t^{(1)})$, $P_2 = (a_2, b_2, t^{(2)})$ be two points on the same ray, such that $p_1 = (a_1, b_1)$, $p_2 = (a_2, b_2)$ are their respective projections on the (x, y)-plane. Consider

$$S = \{|\overrightarrow{p_1 p_2}| \mid p_1, p_2 \in \Gamma; \overrightarrow{p_1 p_2} \text{ is the } (x, y)\text{-plane projection of a ray } \overrightarrow{P_1 P_2}\}.$$

If
$$\inf S = \sup S > 0, \tag{14.14}$$

then every ray starting on a boundary point on $\Gamma \times \mathbb{R}_+$ will hit another boundary point on $\Gamma \times \mathbb{R}_+$ in exactly the same duration of time.

One can easily check that for the domain Ω in (14.13), condition (14.14) is satisfied.

Remark 14.2 Equation of the type (14.12) is called a *non-strictly hyperbolic equation*, as its (negative) elliptic part

$$\nabla^2 + 2\frac{\partial^2}{\partial x \partial y}$$

satisfies the condition

$$\sum_{i=1}^{2} \xi_i^2 + 2\xi_1\xi_2 = (\xi_1 + \xi_2)^2 \geq 0, \ \forall \xi = (\xi_1, \xi_2) \in \mathbb{R}^2,$$

but not

$$\sum_{i=1}^{2} \xi_i^2 + 2\xi_1\xi_2 > 0, \ \forall \xi \in \mathbb{R}^2.$$

The reader may find some literature about relevant non-strictly hyperbolic equations in [1–3].

Now, we describe the boundary conditions. On the boundary Γ_1, we have a linear boundary condition

$$w_t(x, y, t) = -\eta \left(w_x(x, y, t) + w_y(x, y, t) \right), \ (x, y) \in \Gamma_1, \ t > 0, \ 0 < \eta \neq 1. \quad (14.15)$$

When $(x, y) \in \Gamma_3$, there is a nonlinear boundary condition

$$w_t(x, y, t) = \alpha \left(w_x(x, y, t) + w_y(x, y, t) \right) - \beta \left(w_x(x, y, t) + w_y(x, y, t) \right)^3, \ t > 0, \quad (14.16)$$

where

$$0 < \alpha < 1, \quad \beta > 0. \quad (14.17)$$

On Γ_2 and Γ_4, we have the Dirichlet boundary conditions:

$$w(t, x, y) = 0, \ (x, y) \in \Gamma_2 \cup \Gamma_4, \ t > 0.$$

Thus, the overall system is

$$\begin{cases} w_{tt} = \Delta w + 2w_{xy}, \ (x, y) \in \Omega, \ t > 0, \\ w_t = -\eta(w_x + w_y), \ (x, y) \in \Gamma_1, \ t > 0, \\ w_t = \alpha(w_x + w_y) - \beta(w_x + w_y)^3, \ (x, y) \in \Gamma_3, \ t > 0, \\ w(x, y, t) = 0, \ (x, y) \in \Gamma_2 \cup \Gamma_4, \ t > 0, \\ w(x, y, 0) = w_0(x, y), \ w_t(x, y, 0) = w_1(x, y), \ (x, y) \in \overline{\Omega}, \end{cases} \quad (14.18)$$

where $0 < \eta \neq 1, 0 < \alpha < 1, \beta > 0$ and the initial data w_0 and w_1 satisfy

$$w_0 \in C^2(\overline{\Omega}), \quad w_1 \in C^1(\overline{\Omega}); \quad (14.19)$$

14 Chaotic Vibrations of a Multi-Dimensional Wave Equation

and
$$w_0(x, y) = w_1(x, y) = 0, \quad (x, y) \in \Gamma_2 \cup \Gamma_4;$$
$$w_1(x, y) = -\eta \left(\frac{\partial w_0}{\partial x} + \frac{\partial w_0}{\partial y} \right), \quad (x, y) \in \Gamma_1;$$
$$w_t = \alpha \left(\frac{\partial w_0}{\partial x} + \frac{\partial w_0}{\partial y} \right) - \beta \left(\frac{\partial w_0}{\partial x} + \frac{\partial w_0}{\partial y} \right)^3, \quad (x, y) \in \Gamma_3.$$

Remark 14.3 When $\eta = 1$, the system is not well-posed [4]. □

Remark 14.4 For a linear second-order PDE of the form
$$a_{00}(x_1, x_2, t) w_{tt} - \sum_{i,j=1}^{2} a_{ij}(x_1, x_2, t) w_{x_i x_j} = 0, \quad ((x_1, x_2) = (x, y))$$

let
$$S(x_1, x_2, t) = c$$
denote its *characteristic surface* [5]. Then S satisfies
$$a_{00}(x_1, x_2, t) \left(\frac{\partial S}{\partial t} \right)^2 - \sum_{i,j=1}^{2} a_{ij}(x_1, x_2, t) \frac{\partial S}{\partial x_i} \frac{\partial S}{\partial x_j} = 0. \quad (14.20)$$

It is well known that as a Cauchy problem, PDEs with initial data defined on characteristic surfaces may lack the existence and uniqueness of solutions, and solutions may have discontinuities across characteristic surfaces.

Here, for our PDE problem (14.18), the boundary condition (14.18)$_4$, namely,
$$w(x, y, t) = 0, \text{ on } (\Gamma_2 \times \{t > 0\}) \cup (\Gamma_4 \times \{t > 0\}),$$

the surfaces
$$S_1 \equiv \{(x, y, t) \mid x - y = -1, t > 0, (x, y) \in \overline{\Omega}\} = \Gamma_2 \times \{t > 0\}$$

and
$$S_2 \equiv \{(x, y, t) \mid x - y = 1, t > 0, (x, y) \in \overline{\Omega}\} = \Gamma_4 \times \{t > 0\}$$

satisfy
$$S_j(x, y, t) = x - y = 2j - 3 = \text{constant, for } j = 1, 2.$$

Thus,
$$\frac{\partial S_j}{\partial x} = 1, \quad \frac{\partial S_j}{\partial y} = -1, \quad \frac{\partial S_j}{\partial t} = 0, \quad j = 1, 2.$$

We see from (14.20) that for the non-strictly hyperbolic equation (14.18)$_1$ and $k = 1, 2,$

$$a_{00}(x_1, x_2, t)\left(S_{k,t}\right)^2 - \sum_{i,j=1}^{2} a_{ij}(x_1, x_2, t)\left(S_{k,x_i}\right)\left(S_{k,x_j}\right)$$
$$= 0 - [1 \cdot 1 - 2 \cdot 1 \cdot 1 + 1 \cdot 1] = 0.$$

Therefore, S_1 and S_2 are actually characteristic surfaces and, in general, (14.18) has no solutions. This is indeed true. In what follows, the reader can clearly see that *compatibility conditions* need to be imposed on the functions at $(\Gamma_2 \cup \Gamma_4) \cap (\Gamma_1 \cup \Gamma_3)$. Then the initial and boundary data for (14.18) become consistent. Then under sufficient smoothness of the data w_0 and w_1 in (14.18)$_5$, *existence and uniqueness of solutions become self-evident* by Theorem 14.11 and Remark 14.12. □

At time t, the energy of the system (14.18) is

$$E(t) = \frac{1}{2}\int_\Gamma w_t^2 + (w_x + w_y)^2 dS. \tag{14.21}$$

By applying Green's formula, we see the rate of change of energy is

$$E'(t) = \sqrt{2}\eta \int_{\Gamma_1} (w_x + w_y)^2 d\sigma + \sqrt{2}\int_{\Gamma_3} (w_x + w_y)^2 \left(\alpha - \beta(w_x + w_y)^2\right) d\sigma. \tag{14.22}$$

More details about the derivation of (14.22) can be found later. We can find that if $\eta > 0$, energy is injected to the system from Γ_1. For this reason, we refer to (14.15) as an *energy-injecting (or pumping) boundary condition*. Note that the nonlinearities are distributed on the entire Γ_3, and the sign of the second term on the RHS of (14.22) is dependent of the integral, we may call (14.16) a *distributed self-regulating* boundary condition.

We now give the proof of (14.22).

Theorem 14.5 *Consider the system* (14.18) *and the energy function*

$$\forall t > 0, \quad E(t) = \frac{1}{2}\int_\Omega w_t^2 + (w_x + w_y)^2 dS.$$

Then for sufficiently smooth data the derivative of the energy functional is given by the following:

$$\forall t > 0, \quad E'(t) = \sqrt{2}\eta \int_{\Gamma_1} (w_x + w_y)^2 d\sigma + \sqrt{2}\int_{\Gamma_3} (w_x + w_y)^2 \left(\alpha - \beta(w_x + w_y)^2\right) d\sigma.$$

Proof Let w be at least C^2-smooth. Consider the vector field

$$\mathbb{H} = (w_x + w_y, w_x + w_y).$$

Then
$$div\,(\mathbb{H}) = \nabla^2 w + 2w_{xy}.$$

For $t > 0$, we have
$$E'(t) = \int_\Omega w_t div\,(\mathbb{H}) + \mathbb{H} \cdot \nabla w_t dS.$$

Since
$$div\,(w_t \mathbb{H}) = w_t div\,(\mathbb{H}) + \mathbb{H} \cdot \nabla w_t.$$

Applying Green's formula, we have
$$E'(t) = \int_\Omega w_t div\,(\mathbb{H}) + \mathbb{H} \cdot \nabla w_t dS$$
$$= \int_\Omega div\,(w_t \mathbb{H})\,dS$$
$$= \int_\Gamma (w_t \mathbb{H} \cdot \mathbf{n})\,d\sigma.$$

Note that $\mathbb{H} \cdot \mathbf{n} = 0$ on the boundaries $\Gamma \setminus (\Gamma_1 \cup \Gamma_3)$. By applying the boundary conditions on Γ_1 and Γ_3, respectively, we have

$$\forall t > 0, \ E'(t) = \sqrt{2}\eta \int_{\Gamma_1} (w_x + w_y)^2 d\sigma + \sqrt{2} \int_{\Gamma_3} (w_x + w_y)^2 \left(\alpha - \beta(w_x + w_y)^2\right) d\sigma.$$

$\square$

14.1 Preliminary Analysis

Recall that we have the following system:

$$\begin{cases} w_{tt} = \Delta w + 2w_{xy}, & (x, y) \in \Omega, \ t > 0, \\ w_t = -\eta (w_x + w_y), & (x, y) \in \Gamma_1, \ t > 0, \\ w_t = \alpha(w_x + w_y) - \beta(w_x + w_y)^3, & (x, y) \in \Gamma_3, \ t > 0, \\ w(x, y, t) = 0, & (x, y) \in \Gamma_2 \cup \Gamma_4, \ t > 0, \\ w(x, y, 0) = w_0(x, y) \in C^2(\overline{\Omega}), \ w_t(x, y, 0) = w_1(x, y) \in C^1(\overline{\Omega}), \end{cases} \quad (14.23)$$

where
$$0 < \eta \neq 1, \ 0 < \alpha < 1, \ \beta > 0. \tag{14.24}$$

Define two linear operators:
$$\mathcal{L}_1 = \frac{\partial}{\partial t} + \frac{\partial}{\partial x} + \frac{\partial}{\partial y}, \ \mathcal{L}_2 = \frac{\partial}{\partial t} - \frac{\partial}{\partial x} - \frac{\partial}{\partial y}. \tag{14.25}$$

If w is a C^2 function, we have

$$\mathcal{L}_2(w) = w_t - w_x - w_y, \quad \mathcal{L}_1\mathcal{L}_2(w) = w_{tt} - w_{xx} - w_{yy} - 2w_{xy} = 0.$$

Similarly, we have $\mathcal{L}_2\mathcal{L}_1(w) = 0$. Therefore, we can rewrite the first equation of system (14.23) as

$$\mathcal{L}_1\mathcal{L}_2(w) = \mathcal{L}_2\mathcal{L}_1(w) = 0. \tag{14.26}$$

Let u and v be the Riemann invariants of (14.23) defined by

$$u = \frac{1}{2}\mathcal{L}_1(w) = \frac{w_t + w_x + w_y}{2}, \quad v = \frac{1}{2}\mathcal{L}_2(w) = \frac{w_t - w_x - w_y}{2}. \tag{14.27}$$

Consequently,

$$w_x + w_y = u - v; \quad w_t = u + v.$$

Therefore, the first equation of system (14.23) can be written as

$$\mathcal{L}_2 u = 0 \quad \text{or} \quad \mathcal{L}_1 v = 0.$$

For $t > 0$, the boundary condition on Γ_1 can be represented as a reflection relation between u and v:

$$v(x, y, t) = \frac{\eta + 1}{\eta - 1} u(x, y, t) := G_\eta(u(x, y, t)), \quad (x, y) \in \Gamma_1. \tag{14.28}$$

Moreover, the nonlinear condition on Γ_3 is equivalent to the relation of u and v:

$$\beta(u(x, y, t) - v(x, y, t))^3 + (1 - \alpha)(u(x, y, t) - v(x, y, t)) + 2v(x, y, t) = 0, \quad (x, y) \in \Gamma_3.$$

Since $\alpha < 1$ and $\beta > 0$, let $f = u(x, y, t) - v(x, y, t)$, then $f = p(v)$ satisfies the cubic equation:

$$\beta f^3 + (1 - \alpha) f + 2v = 0. \tag{14.29}$$

Remark 14.6 Since we have $\beta > 0$ and $0 < \alpha < 1$, the real solution f is uniquely defined by *Cardano's formula*

$$f = \left(-\frac{v}{\beta} + \sqrt{D}\right)^{1/3} + \left(-\frac{v}{\beta} - \sqrt{D}\right)^{1/3}, \quad D = \frac{(1-\alpha)^3}{27\beta^3} + \frac{v^2}{\beta^2} > 0.$$

$\square$

Therefore, the reflection relation between u and v on Γ_3 takes the form:

$$u(x, y, t) = v(x, y, t) + p(v(x, y, t)) := F_{\alpha,\beta}(v(x, y, t)), \quad (x, y) \in \Gamma_3. \tag{14.30}$$

14.1 Preliminary Analysis

On $\Gamma_2 \cup \Gamma_4$, we have $w_t = 0$, $w_x + w_y = 0$, implying

$$u(x, y, t) = v(x, y, t) = 0, \quad (x, y) \in \Gamma_2 \cup \Gamma_4, \quad t > 0.$$

Consequently, for given smooth initial data $w_0 \in C^2(\overline{\Omega})$ and $w_1 \in C^1(\overline{\Omega})$, the system (14.23) is equivalent to a system of two coupled first-order equations as follows:

$$\begin{cases} \mathcal{L}_1(v) = \mathcal{L}_2(u) = 0, \quad (x, y) \in \Omega, \ t > 0, \\ v(x, y, t) = G_\eta(u(x, y, t)), \quad (x, y) \in \Gamma_1, \ t > 0, \\ u(x, y, t) = F_{\alpha,\beta}(v(x, y, t)), \quad (x, y) \in \Gamma_3, \ t > 0, \\ u(x, y, t) = v(x, y, t) = 0, \quad (x, y) \in \Gamma_2 \cup \Gamma_4, \ t > 0, \\ u(x, y, 0) = u_0(x, y) \in C^1(\overline{\Omega}), \ v(x, y, 0) = v_0(x, y) \in C^1(\overline{\Omega}), \end{cases} \quad (14.31)$$

where the initial data u_0 and v_0 can now be derived in the form:

$$u_0 = \frac{w_1 + \frac{\partial w_0}{\partial x} + \frac{\partial w_0}{\partial y}}{2}, \quad v_0 = \frac{w_1 - \frac{\partial w_0}{\partial x} - \frac{\partial w_0}{\partial y}}{2}. \quad (14.32)$$

In order to ensure that u and v are C^1 functions, we need u_0 and v_0 to be in C^1, and also satisfy some *compatible conditions*:

(1)

$$\forall (\tilde{x}, \tilde{y}) \in \Gamma_1, \ v_0(\tilde{x}, \tilde{y}) = G_\eta(u_0(\tilde{x}, \tilde{y})), \quad \forall (\tilde{x}, \tilde{y}) \in \Gamma_3, \ u_0(\tilde{x}, \tilde{y}) = F_{\alpha,\beta}(v_0(\tilde{x}, \tilde{y})),$$

$$\forall (\tilde{x}, \tilde{y}) \in \Gamma_2 \cup \Gamma_4, \ u_0(\tilde{x}, \tilde{y}) = v_0(\tilde{x}, \tilde{y}) = 0;$$

(2)

$$\forall (\tilde{x}, \tilde{y}) \in \Gamma_1, \forall \vec{v} \in \mathbb{R}^2, \ D_{\vec{v}} v_0(x, y)|_{(\tilde{x}, \tilde{y})} = D_{\vec{v}} G_\eta(u_0(x, y))|_{(\tilde{x}, \tilde{y})},$$

and

$$\forall (\tilde{x}, \tilde{y}) \in \Gamma_3, \forall \vec{u} \in \mathbb{R}^2, \ D_{\vec{u}} u_0(x, y)|_{(\tilde{x}, \tilde{y})} = D_{\vec{u}} F_{\alpha,\beta}(v_0(x, y))|_{(\tilde{x}, \tilde{y})},$$

where for a vector $\vec{\alpha} = (\alpha_1, \alpha_2)$ ($\vec{\alpha} = (\alpha_1, \alpha_2, \alpha_3)$), the directional derivative along the direction $\vec{\alpha}$ is defined by

$$D_{\vec{\alpha}} = \alpha_1 \frac{\partial}{\partial x} + \alpha_2 \frac{\partial}{\partial y} \left(D_{\vec{\alpha}} = \alpha_1 \frac{\partial}{\partial x} + \alpha_2 \frac{\partial}{\partial y} + \alpha_3 \frac{\partial}{\partial t} \right).$$

Remark 14.7 Note that the two boundary conditions in (14.31) are "reflection" boundary conditions that result from wave reflections on the boundaries. □

Therefore, proving the well-posedness of the main system (14.23) is equivalent to proving the well-posedness of system (14.31). From now on, we will focus on system (14.31).

Lemma 14.8 *Let u and v be given by (14.27). Then, u is constant along the direction $\vec{l_1}$ and v is constant along the direction $\vec{l_2}$, where*

$$\vec{l_1} = (-1, -1, 1), \quad \vec{l_2} = (1, 1, 1). \tag{14.33}$$

Proof It follows from the fact

$$D_{\vec{l_1}} u = \left(\frac{\partial}{\partial t} - \frac{\partial}{\partial x} - \frac{\partial}{\partial y}\right) u = \mathcal{L}_2 u = 0. \tag{14.34}$$

Similarly, we have $D_{\vec{l_2}} v = 0$. □

Next, we show the existence of solutions of system (14.31) on the spatiotemporal domain $\overline{\Omega} \times [0, 2]$.

Lemma 14.9 *Let $t \in [0, 2]$ and $(x, y) \in \overline{\Omega}$, i.e., $-1 \leq x + y \leq 1$ and $-1 \leq x - y \leq 1$. Then, $u(x, y, t)$ and $v(x, y, t)$ can be uniquely solved.*

Proof First, recall that for $t > 0$,

$$\forall (x, y) \in \Gamma_1, \ v(x, y, t) = G_\eta(u(x, y, t)),$$
$$\forall (x, y) \in \Gamma_3, \ u(x, y, t) = F_{\alpha,\beta}(v(x, y, t)).$$

Note that the points (x, y, t) and $(x + t, y + t, 0)$ are on the same characteristics along $\vec{l_1}$. By applying Lemma 14.8, we have

$$u(x, y, t) = u(x + t, y + t, 0) = u_0(x + t, y + t), \quad t \leq \frac{1 - x - y}{2}. \tag{14.35}$$

When $\frac{1-x-y}{2} < t \leq \frac{1-x-y}{2} + 1$, we have

$$\left(x + \frac{1 - x - y}{2}, y + \frac{1 - x - y}{2}\right) \in \Gamma_3. \tag{14.36}$$

Also, (x, y, t) and $\left(x + \frac{1-x-y}{2}, y + \frac{1-x-y}{2}, t - \frac{1-x-y}{2}\right)$ are on the same characteristics along $\vec{l_1}$. $(x + \frac{1-x-y}{2}; y + \frac{1-x-y}{2}, t - \frac{1-x-y}{2})$ and $(1 - y - t, 1 - x - t, 0)$ are on the same characteristics along $\vec{l_2}$. By applying the reflection relation (14.30) and Lemma 14.8, we have

14.1 Preliminary Analysis

$$u(x, y, t) = u\left(x + \frac{1-x-y}{2}, y + \frac{1-x-y}{2}, t - \frac{1-x-y}{2}\right)$$

$$= F_{\alpha,\beta}\left(v\left(x + \frac{1-x-y}{2}, y + \frac{1-x-y}{2}, t - \frac{1-x-y}{2}\right)\right) \quad (14.37)$$

$$= F_{\alpha,\beta}\left(v\left(1 - y - t, 1 - x - t, 0\right)\right)$$

$$= F_{\alpha,\beta}\left(v_0\left(1 - y - t, 1 - x - t\right)\right).$$

When $\frac{1-x-y}{2} + 1 < t \leq 2$, note that

$$\left(x + \frac{1-x-y}{2} - 1, y + \frac{1-x-y}{2} - 1\right) \in \Gamma_1,$$

$$\left(x + \frac{1-x-y}{2}, y + \frac{1-x-y}{2}\right) \in \Gamma_3, \quad (14.38)$$

from the reflection relations (14.28), (14.30) and Lemma 14.8, we have

$$u(x, y, t) = u\left(x + \frac{1-x-y}{2}, y + \frac{1-x-y}{2}, t - \frac{1-x-y}{2}\right)$$

$$= F_{\alpha,\beta}\left(v\left(x + \frac{1-x-y}{2}, y + \frac{1-x-y}{2}, t - \frac{1-x-y}{2}\right)\right)$$

$$= F_{\alpha,\beta}\left(v\left(x + \frac{1-x-y}{2} - 1, y + \frac{1-x-y}{2} - 1, t - \frac{1-x-y}{2} - 1\right)\right) \quad (14.39)$$

$$= F_{\alpha,\beta} \circ G_\eta\left(u\left(x + \frac{1-x-y}{2} - 1, y + \frac{1-x-y}{2} - 1, t - \frac{1-x-y}{2} - 1\right)\right)$$

$$= F_{\alpha,\beta} \circ G_\eta\left(u\left(x + t - 2, y + t - 2, 0\right)\right)$$

$$= F_{\alpha,\beta} \circ G_\eta\left(u_0\left(x + t - 2, y + t - 2\right)\right).$$

So u can be solved as

$$u(t, x, y) = \begin{cases} u_0(x+t, y+t), & 0 \leq t \leq \frac{1-x-y}{2}, \\ F_{\alpha,\beta}\left(v_0(1-y-t, 1-x-t)\right), & \frac{1-x-y}{2} < t \leq \frac{1-x-y}{2} + 1, \\ F_{\alpha,\beta} \circ G_\eta\left(u_0(x+t-2, y+t-2)\right), & \frac{1-x-y}{2} + 1 < t \leq 2. \end{cases}$$

Similarly, v can be solved as

$$v(t, x, y) = \begin{cases} v_0(x-t, y-t), & 0 \leq t \leq \frac{x+y+1}{2}, \\ G_\eta\left(u_0(t-y-1, t-x-1)\right), & \frac{x+y+1}{2} < t \leq \frac{x+y+1}{2} + 1, \\ G_\eta \circ F_{\alpha,\beta}\left(v_0(x+2-t, y+2-t)\right), & \frac{x+y+1}{2} + 1 < t \leq 2. \end{cases}$$

For uniqueness, suppose there is another pair of solution (u', v'), we set $(r, s) = (u' - u, v' - v)$. Then (r, s) will satisfy (14.31) with zero initial data. From the explicit solution formulas we obtained above, we have $(r, s) = (0, 0)$. So the solution is unique. $\square$

Lemma 14.10 *For* $t \geq 0$ *and* $(x, y) \in \overline{\Omega}$, *we have* $u(x, y, t+2) = F_{\alpha,\beta} \circ G_\eta (u(x, y, t))$, $v(x, y, t) = G_\eta \circ F_{\alpha,\beta} (v(x, y, t))$.

Proof

$$\begin{aligned}
u(x, y, t+2) &= u\left(x + \frac{1-x-y}{2}, y + \frac{1-x-y}{2}, t+2 - \frac{1-x-y}{2}\right) \\
&= F_{\alpha,\beta}\left(v\left(x + \frac{1-x-y}{2}, y + \frac{1-x-y}{2}, t+2 - \frac{1-x-y}{2}\right)\right) \\
&= F\left(v\left(x + \frac{1-x-y}{2} - 1, y + \frac{1-x-y}{2} - 1, t+2 - \frac{1-x-y}{2} - 1\right)\right) \\
&= F \circ G_\eta \left(u\left(x + \frac{1-x-y}{2} - 1, y + \frac{1-x-y}{2} - 1, t+1 - \frac{1-x-y}{2}\right)\right) \\
&= F \circ G (u(x, y, t)).
\end{aligned}$$
(14.40)

Similarly, we have $v(x, y, t+2) = G_\eta \circ F_{\alpha,\beta} (v(x, y, t))$. $\square$

Theorem 14.11 *The system* (14.31) *is uniquely solvable on* $\overline{\Omega} \times [0, +\infty)$. *Moreover, for any* $t \geq 0$, *we can write* $t = 2n + \tau$ *where* $n \in \mathbb{N}$ *and* $\tau \in [0, 2)$. *Then the solution of* (14.31) *is given by*

$$u(x, y, t) = \begin{cases} (F_{\alpha,\beta} \circ G_\eta)^n (u_0(x + \tau, y + \tau)), & 0 \leq \tau \leq \frac{1-x-y}{2}, \\ (F_{\alpha,\beta} \circ G_\eta)^n (F_{\alpha,\beta} (v_0(1 - y - \tau, 1 - x - \tau))), & \frac{1-x-y}{2} < \tau \leq \frac{1-x-y}{2} + 1, \\ (F_{\alpha,\beta} \circ G_\eta)^n (F_{\alpha,\beta} \circ G_\eta (u_0(x + \tau - 2, y + \tau - 2))), & \frac{1-x-y}{2} + 1 < \tau \leq 2 \end{cases}$$
(14.41)

and

$$v(x, y, t) = \begin{cases} (G_\eta \circ F_{\alpha,\beta})^n (v_0(x - \tau, y - \tau)), & 0 \leq \tau \leq \frac{x+y+1}{2}, \\ (G_\eta \circ F_{\alpha,\beta})^n (G_\eta (u_0(\tau - y - 1, \tau - x - 1))), & \frac{x+y+1}{2} < \tau \leq \frac{x+y+1}{2} + 1, \\ (G_\eta \circ F_{\alpha,\beta})^n (G_\eta \circ F_{\alpha,\beta} (v_0(x + 2 - \tau, y + 2 - \tau))), & \frac{x+y+1}{2} + 1 < \tau \leq 2, \end{cases}$$
(14.42)

where $(F_{\alpha,\beta} \circ G_\eta)^n$ *represents the n-times iterative composition of* $F_{\alpha,\beta} \circ G_\eta$ *and* $(G_\eta \circ F_{\alpha,\beta})^n$ *represents the n-times iterative composition of* $G_\eta \circ F_{\alpha,\beta}$.

Proof Let $t \geq 0$, there exist unique $\tau \in [0, 2)$ and an integer $n \in \mathbb{N}$ such that $t = 2n + \tau$. For $(x, y) \in \Omega$, by applying Lemmas 14.8–14.10 and by induction, we have

$$u(x, y, t) = u(x, y, \tau + 2n) = (F_{\alpha,\beta} \circ G_\eta)^n (u(x, y, \tau))$$
(14.43)

and

$$v(x, y, t) = v(x, y, \tau + 2n) = (G_\eta \circ F_{\alpha,\beta})^n (v(x, y, \tau)).$$
(14.44)

Proof of the uniqueness is similar to that in the proof of Lemma 14.9. $\square$

14.1 Preliminary Analysis

Remark 14.12 (i) From (14.41) and (14.42), u and v are chaotic if $F \circ G$ or $G \circ F$ are chaotic.

(ii) After we have obtained the explicit formulas of (u, v), (w_x, w_y, w_t) can be computed by
$$w_x + w_y = u - v; \quad w_t = u + v.$$
Together with the initial data, we can solve for w by the formula:
$$w(x, y, t) = \int_0^t (u + v) dt + w_0(x, y).$$
From this, we then also obtain w_x and w_t. □

To summarize, we find that the solution (u, v) is fully determined by the maps $G \circ F(\cdot)$ and $F \circ G(\cdot)$. Before introducing the properties of the composite function $H_\eta(\cdot)$, we display the graphics of the composite functions $G_\eta \circ F_{\alpha,\beta}(\cdot)$ and $F_{\alpha,\beta} \circ G_\eta(\cdot)$ for certain values of η, α, and β. See Figs. 14.2 and 14.3. Since $F \circ G = G^{-1} \circ (G \circ F) \circ G$, these two maps are topologically conjugate. So we only need to study one of them. Let's focus on $G \circ F(\cdot)$. From now on, we fix α and β. So $G \circ F(\cdot)$ is a family of maps with a varying parameter η, denoted as
$$H_\eta(\cdot) \triangleq G_\eta \circ F_{\alpha,\beta}(\cdot). \tag{14.45}$$
Moreover, for the case $\eta > 1$, we can apply the transformation $H_{\frac{1}{\eta}}(\cdot) = -H_\eta(\cdot)$. For this reason, from now on we will only study the map $H_\eta(\cdot)$ for the case $\eta \in (0, 1)$.

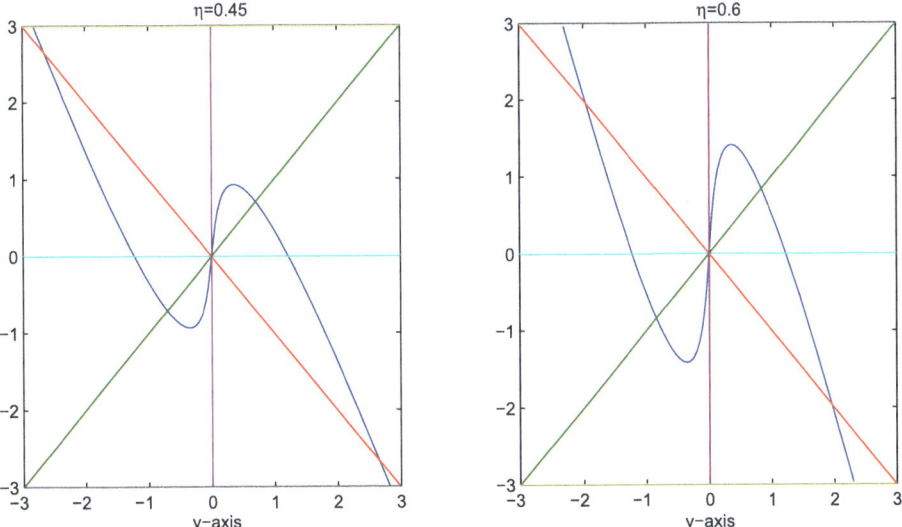

Fig. 14.2 Graphics of $G \circ F(v)$, when $\alpha = 0.5$, $\beta = 1$ and (left) $\eta = 0.45$, (right) $\eta = 0.6$

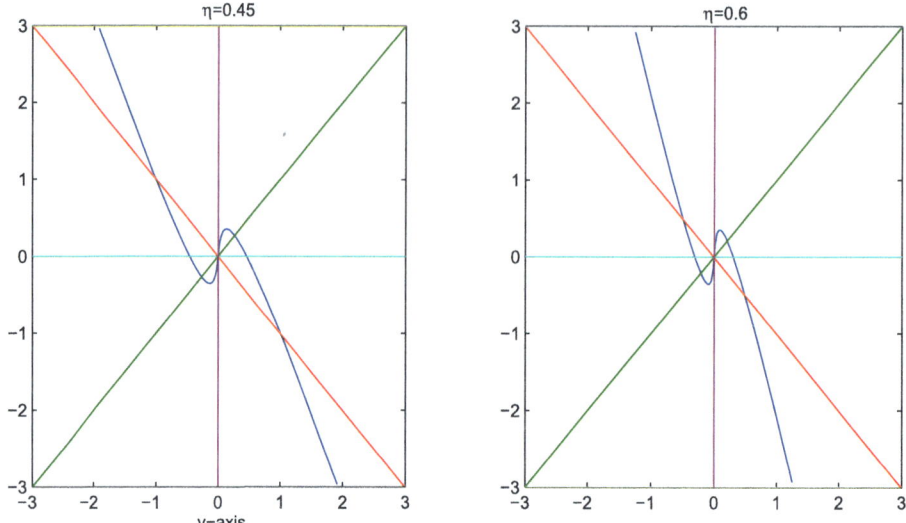

Fig. 14.3 Graphics of $G \circ F(v)$, when $\alpha = 0.5$, $\beta = 1$ and (left) $\eta = 0.45$, (right) $\eta = 0.6$

14.2 Chaotic Dynamics of the Composite Maps

Recall from (14.28),(14.29), (14.30), and (14.45), the definition of the composite reflection map H_η. The basic properties of H_η and the analysis of H_η are already available in [6]. Here, we just show some conclusions and parameters needed in the following section.

Let
$$\eta_0 = \frac{\sqrt{\alpha^2 + 3} - \alpha}{3}, \quad \eta_1 = \frac{3\sqrt{3} - (1 + \alpha)}{3\sqrt{3} + 1 + \alpha}, \tag{14.46}$$

and η_2 is the unique positive solution to the following equation:

$$(\frac{1}{\eta_2} - 1)\sqrt{\frac{1}{\eta_2} + \alpha} = 2(\frac{1 + \alpha}{3})^{\frac{3}{2}}. \tag{14.47}$$

Since

$$(\frac{1}{\eta_1} - 1)\sqrt{\frac{1}{\eta_1} + \alpha} = (\frac{1}{\eta_1} + 1)\frac{1 + \alpha}{3\sqrt{3}}\sqrt{\frac{1}{\eta_1} + \alpha} > 2\frac{1 + \alpha}{3\sqrt{3}}\sqrt{1 + \alpha} = 2(\frac{1 + \alpha}{3})^{\frac{3}{2}},$$

we have
$$0 < \eta_0 < \eta_1 < \eta_2 < 1. \tag{14.48}$$

14.3 Chaotic Vibration Phenomenon of the PDE System

Proposition 14.13 *Let $0 < \alpha < 1$, $\beta > 0$ be fixed, and $\eta \in (0, \eta_2]$ is a varying variable. Then the following holds for every integer n:*

(i) for every $\eta \in (0, \eta_0)$,

$$V_I(H_\eta^n) \leq C, \tag{14.49}$$

where C is a positive constant and $I = [0, \delta]$ or $[-\delta, 0]$, δ is a positive constant;

(ii) for every $\eta \in [\eta_0, \eta_1)$,

$$\lim_{n \to \infty} V_{I_\varepsilon}(H_\eta^n) = \infty, \tag{14.50}$$

where $I_\varepsilon = [0, \varepsilon]$ or $[-\varepsilon, 0]$, for any $\varepsilon > 0$. Here the rate of growth with respect to n is not exponential;

(iii) for every $\eta \in [\eta_1, \eta_2]$,

$$V_{I_\varepsilon}(H_\eta^n) \geq c_1(exp(c_2 n)), \text{ as } n \to \infty, \tag{14.51}$$

for some positive constants c_1 and c_2. Thus, the rate of growth is exponential.

$\square$

14.3 Chaotic Vibration Phenomenon of the PDE System

Recall the PDE system considered in this chapter:

$$\begin{cases} w_{tt} = \Delta w + 2w_{xy}, \ (x, y) \in \Omega, \ t > 0, \\ w_t = -\eta(w_x + w_y), \ (x, y) \in \Gamma_1, \ t > 0, \\ w_t = \alpha(w_x + w_y) - \beta(w_x + w_y)^3, \ (x, y) \in \Gamma_3, \ t > 0, \\ w(x, y, t) = 0, \ (x, y) \in \Gamma_2 \cup \Gamma_4, \ t > 0, \\ w(x, y, 0) = w_0(x, y), \ w_t(x, y, 0) = w_1(x, y), \ (x, y) \in \overline{\Omega}. \end{cases} \tag{14.52}$$

To our knowledge, there is no universally accepted definition of chaos for PDEs in 2D. Following [7], where those authors characterized the chaotic behavior by the growth rate of the total variation, we give a suitable definition of chaos for system (14.52) here.

First, recall that a *simple curve* $\mathcal{C}$ in a 2D domain Ω is defined through a continuous function g from a real number interval $I = [a, b]$ to Ω. The image $g(I)$ is called a *curve*. The adjective "simple" here means g is injective. More specifically, $\mathcal{C}$ is the set of all $g(s)$ when $s \in [a, b]$, where

$$g(s) = (x_\mathcal{C}(s), y_\mathcal{C}(s)) \in \Omega.$$

Definition 14.14 We say that a PDE system of w on the 2D domain Ω is chaotic or has chaotic vibration phenomenon, if there exists at least one direction $\vec{l}$ in $\mathbb{R}^3$, such that for any simple curve $\mathcal{C}$ with $g(a), g(b) \in \Gamma$ and $g(\xi) \in \Omega$, for any $\xi \in (a, b)$, the directional derivative $D_{\vec{l}} w$ satisfies

(i) $D_{\vec{l}} w (x_{\mathcal{C}}(s), y_{\mathcal{C}}(s), t)$ is uniformly bounded;
(ii) $V_{[a,b]}\left(D_{\vec{l}} w (x_{\mathcal{C}}(\cdot), y_{\mathcal{C}}(\cdot), t)\right)$ increases exponentially with time t. $\square$

With the above prerequisites ready, we are now in a position to state the final main theorem of this section.

Theorem 14.15 *Consider the system (14.23). Let $0 < \alpha < 1$, $\beta > 0$ be fixed for $\eta \in [\eta_1, \eta_2]$, where η_1 and η_2 are given by (14.46) and (14.47), respectively. Then, for a certain class of initial conditions the system (14.52) is chaotic.*

Proof Let $\eta \in [\eta_1, \eta_2]$. H_η has an invariant interval $[-M, M]$, where M is a local maximum of H_η given by

$$M = \frac{1+\alpha}{3} \cdot \frac{1+\eta}{1-\eta} \cdot \sqrt{\frac{1+\alpha}{3\beta}}. \tag{14.53}$$

Choose the initial data $w_0 = 0$ and $w_1 \in C^1(\overline{\Omega})$, satisfying

$$\forall (x, y) \in \Gamma, \quad w_1(x, y) = 0, \quad \forall (x, y) \in \Omega, \quad w_1(x, y) > 0. \tag{14.54}$$

Furthermore, assume that

$$Range(w_1) \cup Range(\frac{\eta+1}{\eta-1} \cdot w_1) \cup Range(\frac{\eta+1}{\eta-1}(F_{\alpha,\beta}(w_1))) \subset [-M, M]. \tag{14.55}$$

Consider the direction vector $\vec{l} = (-\frac{1}{2}, -\frac{1}{2}, \frac{1}{2})$, and let

$$v = D_{\vec{l}} w. \tag{14.56}$$

Consider any simple curve $\mathcal{C}$ in Ω with

$$g(a), g(b) \in \Gamma, \quad g(\xi) \in \Omega, \text{ for any } \xi \in (a, b). \tag{14.57}$$

Under assumption (14.55), we have

$$\forall t \geq 0, \forall s \in [a, b], \quad |v(x_{\mathcal{C}}(s), y_{\mathcal{C}}(s), t)| \leq M, \tag{14.58}$$

which is to say that $D_{\vec{l}} w (x_{\mathcal{C}}(s), y_{\mathcal{C}}(s), t)$ is uniformly bounded. Moreover, for given any $t \geq 0$, let $t = 2n + \tau$ where $\tau \in [0, 2)$ and $n \in \mathbb{N}$, we have

$$\forall \xi \in [a, b], \quad v(x_{\mathcal{C}}(\xi), y_{\mathcal{C}}(\xi), t) = \left(G_\eta \circ F_{\alpha,\beta}\right)^n (v(x_{\mathcal{C}}(\xi), y_{\mathcal{C}}(\xi), \tau)). \tag{14.59}$$

It follows from Proposition 14.13 (iii) that there exist constants $c_1 > 0$ and $c_2 > 0$ such that for any $\epsilon > 0$

$$V_{[0,\epsilon]} (H_\eta)^n \geq c_1 e^{c_2 n}, \quad n \in \mathbb{N}. \tag{14.60}$$

Under assumptions (14.54) and (14.57), we have an $\epsilon_0 > 0$ such that

$$[0, \epsilon_0] \subset Range\, (v\, (x_{\mathcal{C}}(\cdot), y_{\mathcal{C}}(\cdot), \tau)). \tag{14.61}$$

Take $\epsilon = \epsilon_0$ in (14.60). Consequently, we have

$$\begin{aligned} V_{[a,b]}\left(D_{\vec{l}}\, w\, (x_{\mathcal{C}}(\cdot), y_{\mathcal{C}}(\cdot), t)\right) &= V_{[a,b]}\, (v\, (x_{\mathcal{C}}(\cdot), y_{\mathcal{C}}(\cdot), t)) \\ &= V_{[a,b]}\left((G_\eta \circ F_{\alpha,\beta})^n\, (v\, (x_{\mathcal{C}}(\cdot), y_{\mathcal{C}}(\cdot), \tau))\right) \\ &\geq c_1 e^{c_2 n} \geq c_1 e^{c_2 \frac{t-\tau}{2}}. \end{aligned} \tag{14.62}$$

Thus, system (14.52) is chaotic. □

14.4 Numerical Simulations

In this section, we present numerical simulations for system (14.18) to illustrate the theoretical results:

Throughout, we fix $\alpha = 0.5, \beta = 1$ and let η be a varying parameter. The initial conditions are chosen as follows:

$$\forall (x, y) \in \overline{\Omega}, \ w_0(x, y) = 0, \ w_1(x, y) = \frac{1}{10} \cdot \left((x + y)^2 - 1\right)^3 \cdot \left((x - y)^2 - 1\right)^3, \tag{14.63}$$

which satisfy the conditions in the proof of Theorem 14.15.

We can obtain the three critical parameter values by following our established recipes:

$$\eta_0 \approx 0.434, \quad \eta_1 \approx 0.552, \quad \eta_2 \approx 0.667. \tag{14.64}$$

Theorem 14.15 shows that when $\eta \in [0.552, 0.667]$ the system (14.18) is chaotic. To verify this, in our numerical simulation, we compare two cases: $\eta = 0.45$ and $\eta = 0.6$.

Numerical simulations for w_t are provided in Figs. 14.4 and 14.5.

Since chaotic vibration is a *dynamics* process, we provide video animations for visualization, viewable at the following URLs:
https://www.dropbox.com/s/m9o7zffmj39ewd7/nonchaotic.mp4?dl=0, for $\eta = 0.45$ in time duration [0, 20];
https://www.dropbox.com/s/4exocjirixmre7g/chaotic2.mp4?dl=0, for $\eta = 0.6$ in time duration [0, 20].

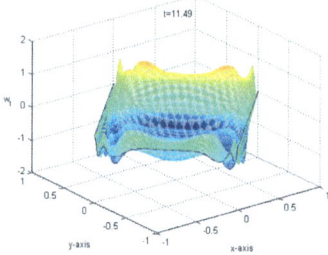 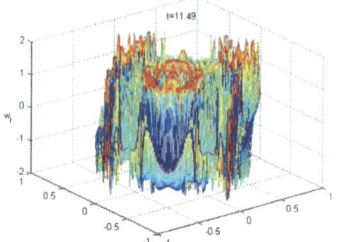

Fig. 14.4 The profiles of w_t at t=11.49 for $\eta = 0.45$(left) and $\eta = 0.6$(right)

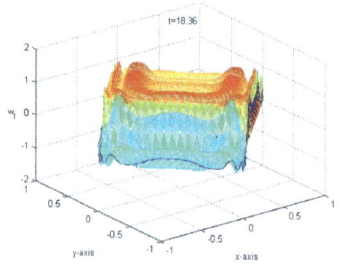 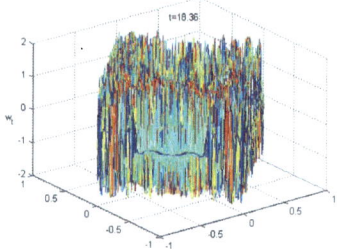

Fig. 14.5 The profiles of w_t at t=18.36 for $\eta = 0.45$(left) and $\eta = 0.6$(right)

These numerical simulations are also consistent with our Theorem 14.15.

Concluding Remarks

Chaos in multi-dimensional dynamic processes manifests complex behavior and phenomena. Here we get to see some of these through rigorously justified simulations and video animations. But there is too much we have not been able to rigorously treat or to even just simulate computationally. Nevertheless, the model we have rigorously studied here is definitely a two-dimensional problem which can't be reduced to 1D case. More novel ideas and methodologies need to be developed and constructed in order to understand better multi-dimensional chaotic vibration phenomena.

References

1. F. Colombini, E. De Giorgi and S. Spagnolo, Sur les équations hyperboliques avec des coefficients qui ne dépendent que du temps, Annali della Scuola Normale Superiore di Pisa-Classe di Scienze, Volume 6, number 3, pp. 511–559, 1979.
2. F. Colombini, S. Spagnolo, "An example of a weakly hyperbolic Cauchy problem not well posed in C^∞," Acta Mathematica, Acta Math. 148(none), pp. 243–253, 1982.

3. F. Colombini, E. Jannelli, and S. Spagnolo, "Well-posedness in the Gevrey classes of the Cauchy problem for a non-strictly hyperbolic equation with coefficients depending on time." Annali della Scuola Normale Superiore di Pisa - Classe di Scienze 10.2, pp. 291–312, 1983.
4. J. Liu, Y. Huang, H. Sun and M. Xiao, "Numerical methods for weak solution of wave equation with van der Pol type nonlinear boundary conditions", Numer. Methods Partial Differ. Equ., vol. 32, no. 2, pp. 373–398, 2016.
5. R. Courant and D. Hilbert, Methods of Mathematical Physics, Vol.II, Partial Differential Equations, Wiley-Interscience, New York, 1962.
6. L.L. Li, J. Tian, G. Chen, Chaotic Vibration of a Two-dimensional Non-strictly Hyperbolic Equation. Canadian Mathematical Bulletin. 61(4):768-786, 2018. https://doi.org/10.4153/CMB-2018-012-1
7. Y. Huang, Growth rates of total variations of snapshots of the 1D linear wave equation with composite nonlinear boundary reflection, *Int. J. Bifur. Chaos* **13** (2003), 1183–1195. https://doi.org/10.1142/S0218127403007138

Index

A
Accumulation point, 73
Aperiodic, 101
Autonomous system, 189

B
Backward orbit, 67
Bifurcation
 pitchfork, 46
 saddle-node, 59
 tangent, 43
Bi-Lipschitz map, 155

C
Cantor set, 72
Chaos
 higher dimensional, 154
 in the sense of P. Diamond, 161
 route to, 37
Chaotic
 in the sense of Devaney, 95
 in the sense of exponential growth of total
  variations of iterates, 29
 in the sense of Li-Yorke, 95
Chaotic maps
 rapid fluctuations of, 149
Chaotic vibrations, 227

Characteristic multipliers, 209
Characteristics, 229
Conley–Moser condition, 124, 133
Connection, 158
Continuous-time difference equation, 171
Contraction, 139
Covering diagram, 25
Crossing number, 158

D
Delay equations, 187
Dense, 80
Diameter, 136
Dimension of rapid fluctuations, 153
Duffing oscillator, 205

E
Equilibrium points, 190
 non-isolated, 196, 197

F
Feigenbaum constant, 12
First return time, 210
Fixed point, 5
 attracting, 5
 Fix(f), 5

repelling, 5
Flow box coordinates, 209
Forward orbit, 67
Fractal, 72, 135

H
Hausdorff
  dimension, 137
  distance, 140
  s-dimensional measure, 137
Henon map, 131
Heteroclinic, 63
Homeomorphism, 69, 101
Homoclinic, 63
Homoclinic bifurcations, 223
Homoclinic orbits, 63
  degenerate, 64
  nondegenerate, 64
Hopf bifurcation, 52
  normal form of, 57
Hyperbolic fixed point, 51
Hyperbolicity, 179
Hysteresis, 220

I
Index theory, 203
Infinite-Dimensional Discrete Dynamical
    System (I3DS), 171
Iterated Function Systems(IFS), 139
Itinerary, 78

J
Jordan canonical form, 190

K
Koch
  curve, 135
  island, 136
  snowflake, 136
  star, 136

L
Left-shift, 79
Lipschitz conjugacy, 154
L-map, 174

Local unstable set, 63
Logistic map, 2
Lotka-Volterra system, 162
Lyapunoff exponent, 67

M
May, Robert, 2
Menger sponge, 148
Method of characteristics, 219
Monodromy matrix, 209

N
Neimark–Sacker bifurcation, 52
Newton's algorithm, 3
Node
  stable, 192
  stable star, 192
  unstable, 192
  unstable star, 192
Non-strictly hyperbolic equation, 252

O
ω-limit point of x, 97
ω-limit set, 97
$\omega(x, f)$, 97, 181
Orbit diagram, 11
Orthogonal matrix, 140

P
Packing dimension, 146
Packing measure, 146
Perfect set, 73
Period-doubling cascade, 12
Periodic point, 5
  hyperbolic, 5
  Per(f), 5
  Per$_k(f)$, 5
Periodic solution, 209
Phase portrait, 192
Poincaré
  map, 210
  section, 210
Population model, 1
  Malthusian law, 1
  modified, 2
Preconnection, 158

Predator-prey model, 163
Prime period, 5

Q
Quadratic map, 2
Quasi-shift invariant set, 155

R
Riemann invariants, 219

S
Saddle point, 192, 196
Self-excitation, 218
Self-regulation effect, 218
Self-similar, 144
Semi-conjugacy, 102
Sensitive dependence on initial data, 83
Sensitivity constant, 83
Sharkovski ordering, 31
Shift invariant set, 106
Sierpinski
   carpet, 147
   sieve, 135
   triangle, 135
Sierpiński Wacław, 135
Similitude, 140
Smale horseshoe, 119
Snap-back repeller, 114
Star of David, 136
Stefan cycle, 33

T
Ternary representation, 72

Theorem
   Brouwer's fixed point, 5
   cylinder, 214
   flow box, 206
   implicitfunction, 8
   intermediate value, 7
   inverse function, 8
   mean value, 7
   period doubling bifurcation, 38
   Poincaré–Bendixon, 218
   Sharkovski, 31
   Tychonov, 87
Topological
   conjugacy, 70, 102
   horseshoes, 158
   semi-conjugacy, 105
Topologically
   mixing, 186
   transitive, 82
Total variation, 23
Transition matrix, 101

U
Unimodal map, 5
Upper semi-continuous, 185

V
Van der Pol device, 218

W
Wave equation, 217
Wave-reflection relations, 220
Weakly topological mixing, 91

The manufacturer's authorised representative in the EU is Springer Nature Customer Service Centre GmbH, Europaplatz 3, 69115 Heidelberg, Germany. If you have any concerns regarding our products, please contact ProductSafety@springernature.com

Printed and bound by CPI Group (UK) Ltd, Croydon, CR0 4YY

23/03/2026

02076446-0014